# The Moon – our sister planet

# The Moon –
# our sister planet

PETER H. CADOGAN

CAMBRIDGE UNIVERSITY PRESS

CAMBRIDGE

LONDON   NEW YORK   NEW ROCHELLE
MELBOURNE   SYDNEY

Published by the Press Syndicate of the University of Cambridge
The Pitt Building, Trumpington Street, Cambridge CB2 1RP
32 East 57th Street, New York, NY 10022, USA
296 Beaconsfield Parade, Middle Park, Melbourne 3206, Australia

First published 1981

Printed in Great Britain by
Fakenham Press Limited, Fakenham, Norfolk

*British Library cataloguing in publication data*

Cadogan, Peter H.
  The Moon – our sister planet
  1. Moon
  I. Title
  523.3  QB581  80–41564

  ISBN 0 521 23684 3 hard covers
  ISBN 0 521 28152 0 paperback

# Contents

*Preface*                                                    *page* vii

**1  The Moon from a distance**                                    1

  1.1   Why study the Moon?                             1
  1.2   Basic motion and eclipses                        2
  1.3   The Moon in Ancient Greece                       9
  1.4   The history of lunar cartography                14
  1.5   Flybys, hardlanders and orbiters                23
  1.6   Lunar geology                                   36
  1.7   Impact versus volcanism                         48
  1.8   Lunar craters                                   57
  1.9   Crater counting                                 69

**2  The Moon landers**                                           78

  2.1   Surveyor                                        78
  2.2   The early Apollos                               87
  2.3   Hadley–Apennine                                 97
  2.4   Cayley–Descartes                               107
  2.5   Taurus–Littrow                                 113
  2.6   Apollo's orbital tasks                         120
  2.7   Sample collection and documentation            123
  2.8   The Russian landers                            135

**3  Moon rocks and minerals**                                   141

  3.1   An introduction to lunar petrology             141
  3.2   The *mare* basalts                             151
  3.3   The characterisation of highland rock types    161
  3.4   Highland rock textures                         169
  3.5   Exotic lunar minerals                          176
  3.6   The origins of lunar rocks                     182

**4    The regolith**                                          194

   4.1    The fabric of the regolith                       194
   4.2    Solar wind chemistry                             202
   4.3    Cosmic ray chemistry                             214
   4.4    Radiation damage                                 222
   4.5    Cratering by micrometeorites                     231
   4.6    Meteorite remains                                239
   4.7    The search for lunar life                        246

**5    Global measurements**                                   251

   5.1    The shape of the Moon                            251
   5.2    Gravity, tides and mascons                       259
   5.3    Remote sensing                                   272
   5.4    The lunar atmosphere                             285
   5.5    Seismology and lunar structure                   293
   5.6    Magnetic studies                                 306
   5.7    The Moon's heat                                  317

**6    The Moon's history and future**                         328

   6.1    The age of the Moon                              328
   6.2    Dating individual rocks                          336
   6.3    The Moon's origin                                353
   6.4    A summary of lunar history                       360
   6.5    The future of lunar exploration                  372

   *Glossary*                                            381
   *Selected Bibliography*                               387
   *Index*                                               390

# Preface

It is hard to believe that more than ten years have now passed since men first set foot on the Moon. And yet the economic recession, the sudden need to find new energy sources and the explosive revolution in microelectronics have made those heady days of Apollo a distant memory.

With the resumption of the US manned space program in the form of the reusable Space Shuttle, it is appropriate to reflect on what we really managed to achieve, at least as far as science is concerned, by sending all those billions of dollars into space. Was the mammoth Apollo effort simply a pointless exercise in public relations, or did it cast a significant new light on our understanding of the Universe and of man's place within it?

My personal interest in lunar exploration was sparked off initially by the dramatic crash landing of Lunik 2, back in the summer of 1959. As far as my active participation in the Apollo program was concerned, however, this had to wait until the opportunity presented itself to work, with Geoffrey Eglinton, on the first samples from Apollo 11. It was a truly awesome moment when those few ounces of extraterrestrial dust first saw the light of day in the clean air facility of the University of Bristol. It certainly was hard to believe that our brightly shining Moon could be covered all over by such dark-coloured soil.

Working on lunar samples was an exhilarating experience, if only because they had never been analysed before and no-one knew quite what to expect. There was even a distinct possibility that my own research project, and hence my Ph.D., might fizzle out completely through lack of substantial results. Fortunately for me this was not to be and I was able to submit my thesis in a totally new scientific discipline: solar wind carbon chemistry.

At that time there was still considerable popular interest in Apollo. When the soil was put on view to the general public, for example, the queue stretched half way down Park Street. But it was not very long before people started to have a niggling feeling that those boffins in their ivory towers had discovered nothing at all of any significance, and that all that money had been totally wasted. The problem was, of course, that only rarely can scientists provide instant answers to simple questions. If a question is simple to ask, it is usually difficult to answer.

But this is not always the case, because some questions can be answered very quickly. There is no life or water on the Moon, and there are no precious minerals. Others, however, are more difficult. We still do not know, for example, where the Moon came from. And this is hardly surprising because we don't know where the Earth came from either, and we have been studying the Earth for centuries. The truth is that, as the

Apollo program progressed, the nature of the questions changed as new discoveries were made. And this is really what this book is all about. The intention is to answer not only those questions that were posed long before the space program began, but to show how much more we have learnt, from the experiments left behind and from the rocks brought back by the astronauts.

When I left Bristol, I spent a number of years with Grenville Turner at the University of Sheffield trying to establish some sort of temporal lunar perspective. By dating Moon rocks, we attempted to relate ancient events on the Moon to the earliest events of terrestrial history and to some more abstract ones elsewhere in the Solar System, as recorded in meteorites. Perhaps the greatest contribution made by Apollo, then, was that it allowed us to see our world from a distance. We have landed on another planet and it is now only a matter of time before we venture farther afield.

I would like to thank Professor Geoffrey Eglinton and Dr Grenville Turner for providing me with opportunities for participation in their lunar sample analysis programs, which enabled me to write this book in the first place.

Many people have assisted me by supplying diagrams and photographs, and these are acknowledged in the appropriate figure captions. I would like to thank the following in particular: The World Data Center A for Rockets and Satellites, Greenbelt, Maryland; P. Butler, Curator, Lunar Receiving Laboratory, JSC; C. P. Florensky, Vernadsky Institute of Geochemistry and Analytical Chemistry, Moscow; Pergamon Press; The British Library; Associated Press; G. J. Wasserburg; S. E. Haggerty; G. M. Brown; C. T. Pillinger and E. Whitaker.

Finally, I would like to thank my wife, Christine, who not only put up with the long gestation period for this book, but also typed the entire manuscript at least twice.

# 1. The Moon from a distance

## 1.1 Why study the Moon?

Anyone taking the trouble to look upwards on a clear moonlit night cannot fail to be moved by the sight of our sister planet hanging up there in the sky like a silver ball and casting ghostly shadows on the Earth below.

Few inanimate objects can have had such an influence over religion and the arts. And never is the Moon's visual impact more impressive than when it is full and just above the distant horizon. Under these conditions, the apparent proximity of the lunar disc to familiar terrestrial objects makes the Moon appear to be very much larger and closer than normal. But the fact that our unaided eyes still cannot discern more than just a few dark blotches on its surface (a beautiful lady to some, a rabbit to others, but a man's face to most) demonstrates that this effect must indeed be just an optical illusion. The light from the Moon cannot really become magnified as it passes at a grazing angle through our atmosphere.

But why should the Moon be an object worthy of scientific study at all? Why cannot we be content just to gaze up at it in awe rather than attempt to probe its innermost secrets? Some would say that the scientific study of the Moon, particularly during the last ten years or so, has destroyed any precious illusions about the Moon that we might have held and given us very little in return. But this would be a very shortsighted view. The Moon is our nearest astronomical neighbour and as such is an ideal testing ground for theories of planetary evolution. It would be a foolish man indeed who would ignore this golden opportunity for advancing our understanding of the Universe just for the sake of keeping the Moon mysterious.

The Apollo program has certainly taken the Moon away from the realms of science fiction and thereby dispelled early thoughts about selenites, lunar UFO bases and dried up river beds. But, in return, the Moon has become as tangible to us as the Earth beneath our feet. For, not only do we now know a great deal about the nature and history of another of the Solar System's planetary bodies, we have also had the opportunity to see our own world in its true cosmic perspective. Indeed, the awesome photographs of the Earth taken from a quarter of a million miles may well turn out to be the most long lasting of Apollo's legacies.

What, then, are the reasons for studying the Moon? What disciplines have been involved and of what possible benefit is the resulting knowledge to mankind? Is there any prospect of our being able to exploit the Moon's resources, for example, and will we ever be in a position to use the Moon as a stepping stone to the 'stars'?

1

*The scope of lunar science*

First of all there are the practical advantages which have been gained from studying the Moon. Before the days of precise timekeeping, for example, the motion of the Moon across the sky provided the easiest means for monitoring long time intervals. Similarly, the Moon's position in the sky was used by ocean navigators to calculate their longitude. And its tidal influence was, of course, also of crucial importance to seagoing nations. But it could be, in fact, that this influence may extend beyond the inanimate world of the oceans. For the lives of certain animals, particularly those living in a marine environment, are apparently regulated by biological rhythms linked to phases of the Moon. So here at least the 'stars' may be exerting some influence over life here on Earth.

But of all the sciences today, biology is surely the one which was least affected by the successes of Apollo. We may have learnt a few things about biological responses to weightlessness and the hardiness of terrestrial bacteria to the lunar environment, but we are still just as ignorant about extraterrestrial life (if indeed such a thing exists) as we ever were.

The study and exploration of the Moon has, however, pushed forward the frontiers in such widely separated disciplines as geochemistry and celestial mechanics, high energy physics and mineralogy, meteoritics and geomorphology, and astronomy and geophysics.

It is the purpose of this book, then, to present this broad perspective on lunar science (from ancient times to the post-Apollo era) with no particular emphasis being placed on any one scientific discipline. For it is this thoroughly interdisciplinary approach to lunar science which has proved to be so exciting.

Lunar geochemists frequently become involved in solar physics, physicists have to appreciate the geological implications of their measurements, and volcanologists are forced into becoming meteoriticists. Just as the success of the Apollo hardware demonstrated just what could be achieved by zero defect technology, so did this consortium approach to science greatly accelerate the pace of scientific discovery.

So much then for generalities, but where should we begin? Well the obvious place to start is with the Moon's basic properties. And what better way to appreciate these concepts than to introduce them in a historical context.

## 1.2   Basic motion and eclipses

The advent of spaceflight, and of the Apollo program in particular, has seen a veritable explosion in lunar science. Every year tens of thousands of pages are devoted to the subject of the Moon in the technical journals and a complete lunar bibliography would now fill a moderate sized library. But the scientific study of the Moon began not tens, or even hundreds, of years ago, but dates back to at least the second millenium B.C., when Northern Europe was still in the Stone Age and the Egyptian civilisation was at its peak.

Table 1.1. *Properties of the Moon's orbit*

| | |
|---|---|
| Synodic month | 29.53 days |
| Sidereal month | 27.32 days |
| Draconic (nodical) month | 27.21 days |
| Anomalistic month | 27.55 days |
| Metonic cycle (repeat phase/day) | 19 years |
| Mean distance from Earth | 238 857 miles |
| Distance at perigee | 226 426 miles |
| Distance at apogee | 252 731 miles |
| Mean eccentricity | 0.0549 |
| Period of regression of nodes | 18.61 years |
| Period of rotation of line of apsides | 8.85 years |
| Eclipse year | 346.6 days |
| Saros cycle (repeat eclipses) | 18 years 10/11 days |
| Mean inclination of orbit to ecliptic | 5° 9' |
| Mean inclination of lunar equator to ecliptic | 1° 32' |

*Sidereal and synodic months*

The most readily observable motion of the Moon, namely its daily rising in the east and setting in the west, has nothing whatever to do with the motion of the Moon itself, but is instead a consequence of the Earth's rotation on its axis. This rotation is similarly responsible for the daily westward motion of the Sun, the planets and the stars. Only Polaris, the star which is nowadays almost directly overhead at the North Pole, is completely exempt from it. But although all heavenly bodies appear to revolve around the earth approximately once a day, the Moon is gradually left behind. This is because the Moon is really moving in its orbit around the Earth in an easterly direction. The time taken for it to be completely lapped by the stars is known as the sidereal month, and this month is 27.32 days long. Every 27.32 days, then, the Moon returns to approximately the same position in the sky relative to the distant stars (Table 1.1).

But the Moon will not look the same after this period of time because it shines not by its own light, but by the light of the Sun. And the relative positions of the Sun, the Earth and the Moon are slightly different after only 27.32 days have elapsed, because of the year long revolution of the Earth around the Sun. For there to be a full moon, for example, the Sun, the Earth and the Moon must be in a straight line. But the direction of the Earth's revolution around the Sun is such that it will be slightly more than a sidereal month, 29.53 days in fact, before the Sun, the Earth and the Moon are once more in a straight line, and the next full moon can occur (Figure 1.1). This period of time is known as the synodic month and, as the phase of the Moon is easier to recognise than its position in the sky relative to the stars, the synodic month has always been of more use than the sidereal one as a timekeeper.

Unfortunately, the year does not contain an exact number of synodic months, nor sidereal months for that matter. In some years there are 13 full moons while in others there are only 12. So a major problem for early

Figure 1.1. *Synodic and sidereal months*. The Moon takes 29.53 days to revolve around the Earth and return to the same position relative to the Earth–Sun line. In other words this is the elapse time between successive new moons, a period known as the synodic month. But, because of the revolution of the Earth around the Sun, the Moon will return to the same position in the sky relative to the stars when it is still a waning crescent. The sidereal month is therefore somewhat shorter than the synodic month, and is in fact only 27.32 days long.

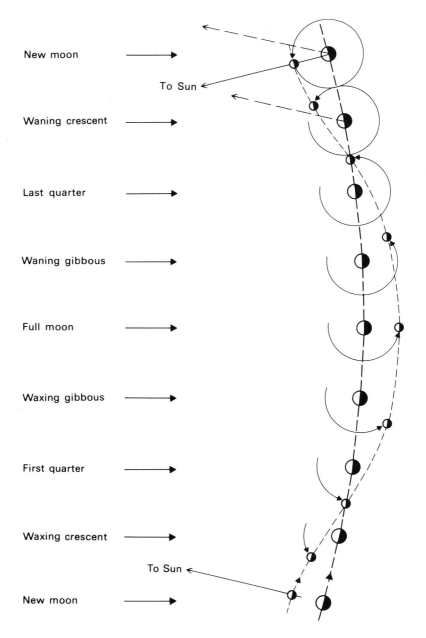

New moon

To Sun

Waning crescent

Last quarter

Waning gibbous

Full moon

Waxing gibbous

First quarter

Waxing crescent

To Sun

New moon

astronomers would have been to find a way to use the Moon's motion to subdivide the solar calendar. This would doubtless have been a useful aid to agricultural planning. But could they have achieved this? Well it so happens that 235 synodic months contain exactly the same number of days as 19 solar years. And therefore a particular phase of the Moon will recur on the same calendar date every 19 years. So this period (known since 433 B.C. as the Metonic cycle) may well have been the means by which early man succeeded in combining the two calendars. We avoid this problem altogether today by inventing calendar months which bear no relation whatsoever to the phase of the Moon or its position in the sky. And this explains why Easter Day may fall any time between the end of March and the end of April. Because it was decided by convention to make it the first Sunday following the first full moon after 21 March, the vernal equinox.

*Eclipses*

The apparent path of the Sun across the sky is to all intents and purposes fixed and is known as the ecliptic. But as the Sun does not revolve around the Earth the ecliptic is really just the plane of the Earth's orbit around the Sun. The Earth's equatorial plane (the plane passing through the Earth's equator) is inclined to this ecliptic plane by an angle of $23\frac{1}{2}°$ and it is this inclination which is responsible for our seasons. At midsummer in the Northern Hemisphere, for example, the Earth's North Pole is tilted towards the Sun and therefore the Sun is seen to be at its highest at this time of year (Figure 1.2).

Figure 1.2. *The inclination of the Moon's orbital plane.* The plane of the Earth's equator is inclined to the ecliptic by an angle of $23\frac{1}{2}°$. This means that during summer in the Northern Hemisphere the North Pole is tilted towards the Sun and consequently the days can be very long at high northern latitudes. The plane of the Moon's orbit, however, is inclined by about 5° to the ecliptic and the rotation of the pole of this orbit ensures that every 18.61 years the Moon is overhead at latitude $28\frac{1}{2}°$ N (i.e. $23\frac{1}{2}° + 5°$).

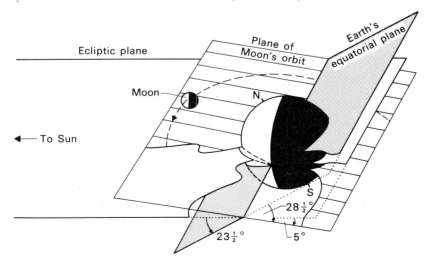

The Moon's orbit, unlike the Earth's, is not firmly fixed in space and its apparent path across the sky is therefore seen to be more tortuous than that of the Sun. But the plane of the Moon's orbit is always inclined to the ecliptic by about 5°.

And, although small, this inclination accounts for the fact that we do not see a solar eclipse every new moon, nor a lunar eclipse every full one. For eclipses can only occur when the Sun, the Earth and the Moon are in the same plane. In other words, the Moon must be in the process of crossing the ecliptic when a new or full moon is due. This restriction accounts of course for the origin of the name 'ecliptic' (Figure 1.7).

But, as the Moon's orbit is not fixed in space, the points in the sky where the Moon's path crosses the ecliptic (known as the lunar nodes) gradually move westwards around the ecliptic, making one complete revolution of the heavens in 18.61 years. So the inclination of the Moon's orbital plane to the Earth's equator oscillates between $18\frac{1}{2}°$ $(23\frac{1}{2}° - 5°)$ and $28\frac{1}{2}°$ $(23\frac{1}{2}° + 5°)$ and back again every 18.61 years (Figure 1.3). This regression of the lunar nodes complicates matters for anyone who wishes to predict eclipses. Not only does he have to find the positions of the nodes, he must also find a way to track their motion across the sky. But this may well have been achieved in prehistoric times by marking out the extreme rising and setting directions of the Sun and Moon on the distant horizon.

*Stone alignments*

As the motion of the Moon is that much more complex than that of the Sun, let us first consider the rising and setting directions of the Sun.

The $23\frac{1}{2}°$ inclination of the ecliptic to the equator means that the Sun

Figure 1.3. *The regression of the nodes.* The plane of the Moon's orbit around the Earth is inclined by about 5° to the ecliptic. But the pole of the Moon's orbit (in other words the direction perpendicular to its orbital plane) rotates slowly around the ecliptic pole. The direction of this rotation is such that a line passing through the points in the sky where the Moon's path crosses the ecliptic (known as the lunar nodes) moves slowly westwards, making one complete revolution of the Earth every 18.61 years. The time between successive passages of the Moon through the ascending node is called the draconic (or nodical) month and, at 27.21 days, is slightly shorter than the sidereal one. The time between successive passages of the Sun through the ascending node is called the eclipse year (346.6 days), because only during nodal passages can eclipses occur.

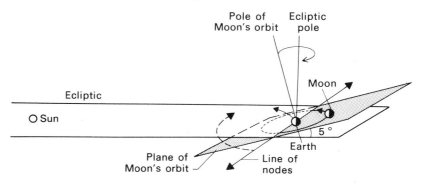

does not always rise due east. Only at the equinoxes, in fact, does the Sun rise due east and set due west, while in the Northern Hemisphere summer the Sun rises to the north of east and sets to the north of west. But the actual rising direction will also depend on the latitude of the observer. On the Arctic Circle, for example, the midsummer Sun just grazes the northern horizon.

Extreme solar rising and setting directions were recorded in prehistory as stone alignments, the very existence of which indicate that an accurate solar calendar was in use in these times. Studies of the Moon may have been carried out in a very similar manner but would have required considerably greater understanding and dedication. The inclination of the Moon's orbital plane to the Earth's equator oscillates between $18\frac{1}{2}°$ and $28\frac{1}{2}°$ and therefore 'most northerly risings of the Moon do not occur at the same place on the horizon every month. The most extreme northerly rising is in fact only observed when the Moon is $28\frac{1}{2}°$ north of the Earth's equator. And this is only true once every 18.61 years. The existence of such lunar alignments at, for example, Stonehenge, is more controversial but does suggest that prehistoric man was already interested in eclipse predictions. For the timing of this most northerly rising of the Moon allows us to locate the lunar nodes and track them around the ecliptic. Really accurate lunar observations would have revealed that the inclination of the Moon's orbital plane to the ecliptic is not quite constant. Instead, it varies by about $9'$ on either side of the mean value of $5°9'$. This variation is due to the gravitational pull of the Sun on the Earth and the Moon. Effectively, the Sun is pulling the plane of the Moon's orbit towards the ecliptic when the nodes of the Moon's orbit are at right angles to the Earth–Sun line. Looking at this another way the inclination of the Moon's orbital plane to the ecliptic is greatest (that is least effected by the gravitational pull of the Sun) when a line passing through the nodes also passes through the sun. The period of this $\pm 9'$ oscillation is therefore simply the time taken for the Sun to pass from one node to another, or half an eclipse year, and this period, 173.3 days, therefore separates the 'danger times' for lunar and solar eclipses.

*The shape of the Moon's orbit*
Although we shall never know whether prehistoric men ever did succeed in predicting eclipses, there are several factors which would no doubt have plagued their observations. Firstly, the Moon only rises once a day and, even when it does, the sky is often cloudy or the distant skyline may not be visible. Secondly, the light from the Moon is bent by refraction through the Earth's atmosphere, which makes the moon appear higher in the sky than it should be. The actual refraction angle may be at least $\frac{1}{2}°$ for the rising Moon and varies substantially with atmospheric conditions. And, thirdly, there is a parallax effect which makes the Moon appear to be lower in the sky than it really is by about $1°$. This parallax effect stems from the fact that the Moon is relatively close to the Earth and it is not being observed along a line passing through the centre of the Earth, but rather along a tangent to its surface.

But not only is the parallax angle appreciable, it also varies with the Earth–Moon distance. And the orbit of the Moon is elliptical, rather than circular. At perigee (that is at its nearest point to us) the Moon may be less than 227 000 miles from the Earth, whereas at apogee the Earth–Moon distance is more than 252 000 miles. The Moon's mean distance from Earth is 238 857 miles. It is this non-circularity of the Moon's orbit which sometimes results in annular solar eclipses (rather than total ones) when the Sun, Moon and Earth are perfectly aligned. For when the Moon is farthest from us its angular diameter may be insufficient to cover the disc of the Sun completely, thus making a total solar eclipse impossible. The Moon's shadow cone will just not reach the surface of the Earth under these conditions.

The time taken for the Moon to make successive passes through perigee is known as the 'anomalistic' month and, at 27.55 days, is slightly longer than the sidereal month as lunar perigee moves slowly around the Earth in an easterly direction (Figure 1.4). The net result of all this is that the most northerly rising of the Moon is once again affected, but this effect is very small and has a 93 year period.

One final aspect of eclipses which prehistoric man may have recognised was that 19 passages of the Sun through one of the lunar nodes (19 eclipse years) happens to contain exactly the same number of days as 223 synodic months (namely 6585). It follows that every 6585 days (or 18 years and 10 or 11 days depending on the number of intervening leap years) the Sun and the Moon will once again be in the same position in the sky relative to the lunar nodes. As a result, if there is a solar eclipse on a

Figure 1.4. *The anomalistic month.* The most distant point on the moon's elliptical orbit is known as apogee (*A*), whereas the nearest point is called perigee (*P*). The line *AP* is known as the line of apsides and rotates slowly eastwards, making one complete revolution of the Earth every 8.85 years. As a consequence of this rotation, the time between successive passages of the Moon through perigee (i.e. from *P* to *P'*) is slightly longer than the sidereal month. This period is known as the anomalistic month (27.55 days).

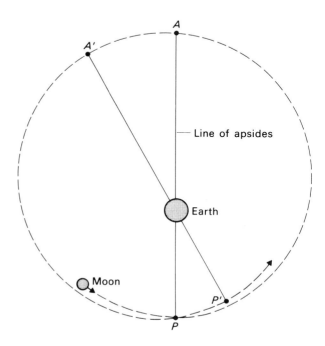

particular day, another solar eclipse will occur 6585 days later. Once this fact had been established, this period (known now as the Saros cycle) would have enabled eclipses to have been predicted for many centuries into the future and would have made further lunar observations quite unnecessary.

What was there left to know about the Moon and its motion? How long would the world have to wait until the next steps in lunar science were to be taken? Well there are certainly a number of minor aspects of lunar motion which we really cannot expect prehistoric man to have recognised. Some of the more complex of these will be briefly described in Chapter 5. But for him to have gone much further he would have needed a modicum of mathematics. How else could he possibly have computed the Moon's size and distance? For this we have to move forward to Classical Greece where scientists first started to investigate our Moon, not just in religious or even mathematical terms, but for what it really is, our sister planet.

## 1.3 The Moon in Ancient Greece

The tradition of astronomy in Greece was inherited not only from the Egyptians and the Babylonians, but also from the Chaldeans, the first known discoverers of the Saros eclipse cycle. The Greek astronomer, Thales, for example, predicted a solar eclipse on 28 May in 584 B.C., an event which brought the lengthy war between the Lydians and the Medes to a hasty conclusion. This eclipse is believed, incidentally, to be the earliest accurately dated event in human history.

Although the Greeks were accomplished mathematicians, mathematics in those days was synonymous with geometry because algebra and trigonometry had not yet been invented. So the Greeks had to try and measure the diameter of the Moon (and its distance from the Earth) by purely geometric means. This work was begun in the 3rd century B.C. by the Alexandrian, Eratosthenes, who made the first critical measurement, that of the Earth's circumference.

*The size of the Earth*
In the previous century, Aristotle had already summarised the principal astronomical arguments in favour of a spherical, rather than a flat, Earth. Firstly, he had pointed out that the shadow cast by the Earth on the Moon during a lunar eclipse is circular in outline. Secondly, he was aware that, when one travels north or south, a number of new stars appear in the sky, while others disappear below the horizon. Some of the stars which are visible from Egypt, for example, never rise at the latitude of Athens.

It was effectively this second argument which Eratosthenes used to measure the circumference of the Earth. He noted that at noon on midsummer day the Sun was directly overhead at Syene (now Aswan) in Upper Egypt, whereas at Alexandria to the north it was 7.2° south of the overhead point. As the Sun's distance is very large compared to the Earth's radius, the Sun's rays are all effectively parallel, and therefore

Alexandria must be 7.2° north of Syene in latitude (Figure 1.5). It follows that the circumference of the Earth must be 360/7.2 times the surface distance separating the two towns. Unfortunately, we do not know the precise length of the distance unit used by Eratosthenes (the stadium), but one reported value gives the Earth's radius (3960 miles) to an accuracy of 1 % and there is no reason to believe that such accuracy could not have been achieved by Eratosthenes.

*The Moon's distance and diameter*
The remoteness of the Sun, when compared to the Earth–Moon distance, was demonstrated by Aristarchus, another of the outstanding Alexandrian school of Greek astronomers. Aristarchus has the distinction of being the first man to propose that the Sun and not the Earth is at the centre of the Solar System. Unfortunately, this idea never caught on among his contemporaries and was not restated until Copernicus did so in the 16th century A.D.

As far as his lunar measurements were concerned, Aristarchus recognised that the angle between the Earth, the Moon and the Sun must be 90° when the Moon is at either first or last quarter (Figure 1.6). The angular separation of the Moon and the Sun as seen from the Earth at these times will then depend solely on the relative distances to the Sun and Moon. Aristarchus discovered that this angle is so close to a right angle that the Sun must be at least 20 times as distant as the Moon. But the precise moments of quadrature are difficult to ascertain (because of the Moon's irregular surface) and this explains why his distance ratio was still too small by a factor of 20.

Figure 1.5. *The size of the Earth*. The Sun is so far away from us that all rays of sunlight reaching the Earth can be considered to be parallel. At midsummer on the Tropic of Cancer the Sun is directly overhead. A vertical post set in the ground at Syene (*S*) will therefore cast no shadow at this time. But Eratosthenes noted that further north at Alexandria (*A*) the Sun was 7.2° south of the overhead point at noon. The distance between Alexandria and Syene must therefore be in the same proportion to the circumference of the Earth as 7.2° is to 360°. In other words the Earth's circumference must be equal to *AS* × 7.2/360.

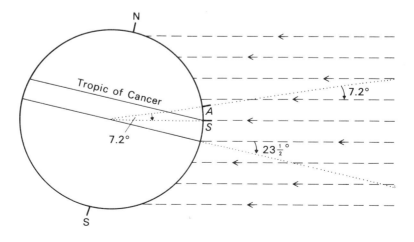

It is one thing to calculate the relative distances to the Sun and the Moon and quite another to determine these distances absolutely. But this was successfully achieved as long ago as the second century B.C. by Hipparchus, the greatest of all Greek astronomers. His most important contributions to astronomy were the invention of trigonometry, the discovery of the precession of the equinoxes, the introduction of the magnitude scale for star brightness and the compilation of the first reliable star atlas. But Hipparchus was also interested in the Moon and he calculated the lengths of the draconic and anomalistic months very precisely, using ancient eclipse records.

The method which he used to determine the Moon's size and distance was based on the size of the Earth's shadow during a lunar eclipse, the curvature of which had already been used as evidence for the Earth's sphericity by Aristotle. The diameter of this shadow depends on the radius of the Earth, and to a lesser extent on the relative distances of the Sun and the Moon. Using the results of Aristarchus for the ratio of the distances to the Sun and Moon, Hipparchus timed the passage of the

Figure 1.6. *The Moon's phases*. The phase of the Moon depends on the relative positions of the Moon, Earth and Sun. Half of the Moon is always illuminated by the Sun. But how much of this illuminated hemisphere is visible to us depends on the angle (*a*) between the Moon and Sun directions. At first or last quarter, the angle between the Sun and the Earth as seen from the Moon must, by definition, be 90°. The angle (*a*) at this time will therefore just depend on the relative distances to the Sun and the Moon. By measuring this angle (which is of course very close to 90°), Aristarchus estimated that the Sun must be 20 times more remote than the Moon. He was out by a factor of 20.

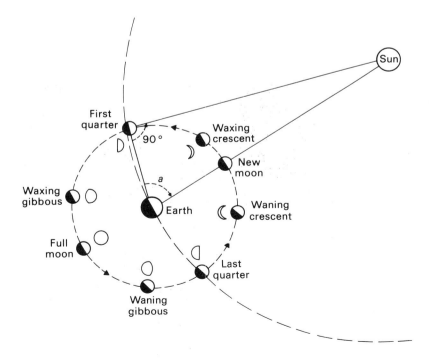

Moon through this shadow and concluded that the Moon's distance from the Earth must be nearly 59 times greater than the Earth's radius (Figure 1.7). If we take the radius of the Earth to be 3960 miles, and the angular diameter of the Moon to be $\frac{1}{2}°$, we can use Hipparchus's result to calculate a value of 233 000 miles for the Moon's distance and 1016 miles for its radius. The presently accepted mean values are 238 857 and 1080 miles, so this result represents one of the major achievements of Greek astronomy.

### Lunar parallax

One consequence of the relative closeness of the Moon is that a solar eclipse (one in which the Moon passes in front of the Sun) may appear to be total at one point on the Earth's surface, whereas elsewhere it is only partial (Figure 1.8). During the night this parallax effect allows us to determine the Moon's distance by triangulation, because the position of the Moon in the sky (relative to the stars at any particular time) varies from one place to another on the Earth's surface. And this triangulation method was in fact used to measure the Moon's distance by Ptolemy way back in the second century A.D. The method has only recently been superseded by radar and laser reflection techniques.

### Orbital motion

It was through Ptolemy that much of the Hipparchus astronomical work has survived. Indeed, a major part of Ptolemy's *Almagest* represents the work carried out by Hipparchus some 300 years earlier. But Ptolemy made some original contributions to lunar science himself, particularly when describing the Moon's orbital motion. In particular, he wanted to account for the variations in the angular velocity of the Moon around its orbit. And in this endeavour he was hampered by the widespread belief

Figure 1.7. *Lunar eclipses*. A lunar eclipse occurs when the Moon passes into the Earth's shadow. As the Moon's orbit is inclined to the ecliptic by about 5°, lunar eclipses can only happen when a full moon occurs at one of the nodes. The diameter of the Earth's shadow cone, when cast on the Moon during a lunar eclipse, depends on the radius of the Earth and, to a lesser extent, on the relative distances to the Sun and Moon. Hipparchus used Eratosthenes's value for the radius of the Earth, and Aristarchus's relative distance estimates, and then timed the passage of the Moon through the Earth's shadow to calculate the Moon's absolute diameter.

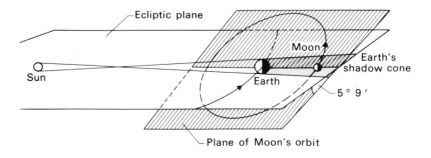

that the Moon and planets all move through space along circular paths and at uniform rates.

In order to explain the changes in the Moon's angular velocity, Hipparchus had already suggested that the Moon's orbit must be an eccentric, that is a circle with the Earth slightly displaced from its centre. But Hipparchus's distance measurements could only be carried out during lunar eclipses, whereas Ptolemy's triangulation method was applicable at all times. So Ptolemy was able to discover additional irregularities in the Moon's orbital motion and was forced to propose that the Moon must move around a small circle (an epicycle) the centre of which describes a larger circle (the deferent), the eccentric circle of Hipparchus (Figure 1.9). By adding still more epicycles Ptolemy could 'explain' the observed variations in the Moon's angular velocity quite accurately. But, unfortunately, he could not at the same time make his model compatible with measured Earth–Moon distances. In other words his models predicted variations in lunar angular diameter which conflicted seriously with observations.

The work of Ptolemy in the second century A.D. ended the second era in lunar science, an era in which the Moon had ceased to be treated as a god simply to be worshipped and had become an object amenable to scientific study. But not until the Renaissance was our planetary com-

Figure 1.8. *Solar eclipses*. The angular diameters of the Sun and the Moon are almost identical and therefore total solar eclipses are only observed for very short periods of time and in a very restricted area (*A*) on the Earth's surface. The Moon's elliptical orbit sometimes even results in the umbra failing to reach the Earth at all, in which case the best that can be seen is an annular eclipse. Inside the penumbral shadow (e.g. at *B*) only a partial eclipse is observed. This can be considered as just a parallax effect, in the same way as a pencil held at arm's length will change its position relative to more distant objects when viewed first with one eye and then with the other.

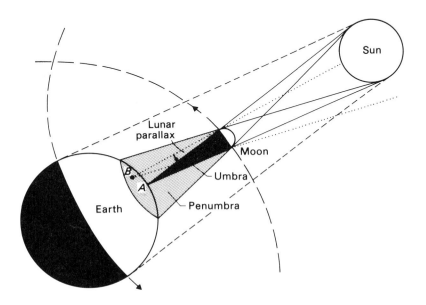

panion once more to be considered in these terms. And not until then were features on the Moon's surface described in some detail for the first time. In Greek times the Moon was generally believed to be a smooth orb.

## 1.4    The history of lunar cartography

Astronomy prior to the Renaissance was strictly limited to the study of the positions and apparent motions of the heavenly bodies. For more than a thousand years no serious attempts were made to try and discover the true natures of the Moon and planets and why they follow such complicated paths across the sky. Although astronomy thrived in India and the Arab countries during the Dark Ages of Western Europe, these eastern astronomers were solely concerned with improving observational techniques and with revising the astronomical tables compiled by the Greeks. They made very few fundamental advances. Indeed, the Arab astronomers' greatest contributions were not in observational astronomy at all, but in mathematics. Not only did they introduce our present decimal system of counting, they also invented another major mathematical discipline, algebra, without which further advances in astronomy would have been quite impossible.

Figure 1.9. *Circular versus elliptical orbits*. The Greeks believed that all heavenly bodies moved at uniform rates along circular paths. So, in order to account for the obviously non-uniform motion of the Moon, Hipparchus proposed that the Earth may not be at the centre of the Moon's orbit. Ptolemy went one stage further by demonstrating that the Moon's apparent motion could be better explained by introducing an epicycle. The Moon then moves uniformly around this epicycle, while its centre describes a circular orbit around the Earth. This conviction about circular orbits and uniform motions was not dispelled until as recently as the seventeenth century, when Kepler proved once and for all that the orbits of moons and planets are, in fact, elliptical, and that angular velocity is directly related to the distance from one focus of the ellipse. (The scale of the epicycle is much exaggerated in comparison with that of the deferent.)

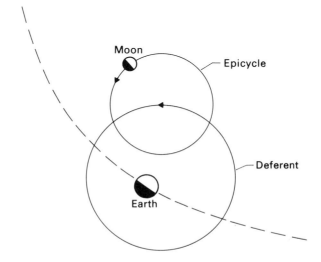

Fortunately, the precise planetary observations recorded by the Arab astronomers were not entirely wasted. They were put to good use by Copernicus in the sixteenth century in support of his Sun-centred theory for the Solar System. Copernicus was not the world's most accomplished observational astronomer, so the Arab measurements of the motions of the planets were the best that were available to him. But the same cannot be said of Tycho Brahe who was not satisfied until he had built an observatory which was capable of measuring angles to better than 1 minute of arc, or less than one-sixtieth of a degree. Without this accuracy, Johannes Kepler (a student of Tycho's) would probably never have been able to establish his celebrated laws of planetary motion.

It must have seemed to Kepler that further improvements in positional astronomy would be very difficult, if not impossible, to attain. But this would have been to ignore the pace of scientific and technological progress during the seventeenth century. While Kepler was struggling to find some pattern in the complex motions of the Moon and planets, an instrument was being invented which was to revolutionise positional astronomy and to open up many new and exciting fields for study. This invention was of course the telescope, first constructively pointed towards the heavens by Harriot in 1609 and Galileo in 1610.

Galileo used his primitive telescope to discover sunspots, the moons of Jupiter and the phases of Venus. But perhaps his most dramatic observations were made when he first looked at the Moon. The only map of the Moon which is known to date from the pre-telescope era was drawn only ten years earlier by Gilbert. Without optical aid, all Gilbert could do was to draw vague boundaries to separate what appeared to him to be dark and light areas on the Moon's surface. Galileo saw these two types of terrain very clearly through his telescope and describe them as *maria* and *terra* (sea and land) respectively, after their terrestrial counterparts. These unfortunate descriptions have survived to the present day, despite the fact that even Galileo recognised that there were probably no broad expanses of water on the Moon.

### Mare *nomenclature*

The system of *mare* nomenclature which we now use is largely due to Riccioli, whose map of the Moon published in 1651 accorded them mental or meteorological attributes using Latinised names. The *mare* which covers a large part of the western half of the Moon's visible hemisphere, for example, is called Oceanus Procellarum, or the Ocean of Storms. Most of Riccioli's names do little to reflect the harshness of the lunar environment and, if we had to rename the lunar *maria*, it is most unlikely that we would again choose such inapt descriptions as Serenity, Tranquillity and Fecundity. The Sea of Crises (Mare Crisium), on the other hand, is perhaps more appropriate when we look back on the rigours and achievements of the Apollo program. Nowadays we tend to invent names like Mare Cognitum, the Sea of Knowledge (Table 1.2).

Most of the smaller patches of dark terrain on the Moon are variously described as bays, lakes or marshes, again somewhat inappropriate

Table 1.2. *List of large lunar maria*

| Maria | Seas |
|---|---|
| Oceanus Procellarum | Ocean of Storms |
| Mare Aestatis | Summer Sea |
| Mare Anguis | Serpent Sea |
| Mare Australe | Southern Sea |
| Mare Cognitum | Sea of Knowledge |
| Mare Crisium | Sea of Crises |
| Mare Fecunditatis | Sea of Fertility |
| Mare Frigoris | Sea of Cold |
| Mare Humboldtianum | Humboldt's Sea |
| Mare Humorum | Sea of Moisture |
| Mare Imbrium | Sea of Rains |
| Mare Ingenii | Sea of Ingenuity |
| Mare Marginis | Marginal Sea |
| Mare Moscoviensae | Moscow Sea |
| Mare Nectaris | Sea of Nectar |
| Mare Nubium | Sea of Clouds |
| Mare Orientale | Eastern Sea |
| Mare Serenitatis | Sea of Serenity |
| Mare Smythii | Smyth's Sea |
| Mare Spumans | Foaming Sea |
| Mare Tranquillitatis | Sea of Tranquillity |
| Mare Undarum | Sea of Waves |
| Mare Vaporum | Sea of Vapours |

descriptions. The Apollo 15 mission, for example, landed not in Mare Imbrium (the Sea of Rains) but rather in the adjoining Palus Putredinis (the Marsh of Decay). This secondary nomenclature is not, however, all due to Riccioli. Sinus Medii (Central Bay), for example, was first named in a map drawn in 1645 by another prominent lunar cartographer of the century, Langrenus (Table 1.3).

Table 1.3. *List of small lunar maria*

| Small *maria* | Lakes, bays and marshes |
|---|---|
| Lacus Mortis | Lake of Death |
| Lacus Somniorum | Lake of Dreams |
| Lacus Veris | Spring Lake |
| Lacus Autumni | Autumn Lake |
| Lacus Aestatis | Summer Lake (same as Summer Sea) |
| Palus Somnii | Marsh of Sleep |
| Palus Nebularum | Marsh of Diseases |
| Palus Putredinis | Marsh of Decay |
| Palus Epidemiarum | Marsh of Epidemics |
| Sinus Aestuum | Seething Bay |
| Sinus Iridum | Bay of Rainbows |
| Sinus Medii | Central Bay |
| Sinus Roris | Bay of Dew |

*Lunar craters and mountains*

Galileo's drawings of the Moon bear little resemblance to the Moon as we know it today (Figure 1.18). But they depicted for the very first time the most conspicuous lunar features, the innumerable craters. The fine-grained debris from these impacts resulted in light-coloured ray systems which are particularly prominent at full moon. The ray system emanating from the bright crater Tycho, for example, traverses much of the nearside face of the Moon. This and other bright ray systems were quite faithfully recorded in 1645 by another early lunar cartographer, de Rheita, and indicate the youth of the craters with which they are associated.

But some impact ejecta is not so fine, and some of it may even be capable of producing craters. The first map to depict such secondary craters (those around the young crater Copernicus) was drawn as long ago as 1680 by Cassini, one of a very distinguished French astronomical family (Figure 1.10).

It was Riccioli who also initiated the practice of naming craters after

Figure 1.10. *Cassini's lunar map*. Drawn in 1680, this map shows secondary craters and ray systems around such fresh craters as Copernicus. Note in particular the long ray from Tycho and the looping one from Copernicus. (The British Library Board.)

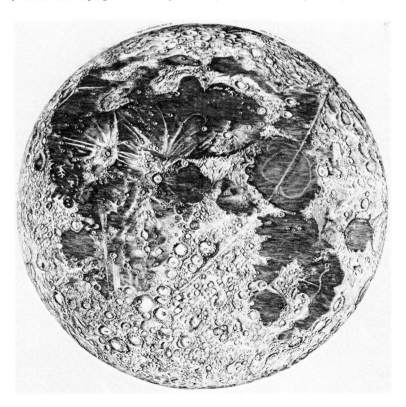

great scientists and philosophers, a practice which has survived to the present day. There are some inconsistencies, however, and the size of a crater does not always reflect the eminence of its namesake. Copernicus, Eratosthenes, Kepler, Aristarchus and Tycho are rightly all prominent features, but the craters named after Galileo and Newton are, by comparison, quite insignificant.

As the Moon's surface was revealed with ever increasing resolution, first by means of the telescope and more recently as a consequence of the space programme, the total number of lunar craters soon outstripped the number of suitable candidates for immortality. And a great many lunar craters, particularly those on the permanently invisible far side, still remain anonymous. But in 1969 craters were named after living people for the first time, namely Armstrong, Aldrin and Collins.

A number of very small craters at each Apollo site were given unofficial names for use during the missions. Some craters were dedicated to historical characters or places (e.g. Shakespeare, Camelot) while others were assigned more descriptive names (e.g. Cone, South Ray), or names of convenience (e.g. Plum, Buster). A tiny crater at the Apollo 16 landing site known as WC presumably belongs to the third category!

Unlike lunar craters, lunar mountain chains (most of which are really just the ramparts of very large craters) have always been named after their terrestrial equivalents. Around Mare Imbrium, for example, we have the Carpathians, the Apennines, the Alps, the Caucasus and the Jura, whereas the Pyrénées can be found around Mare Nectaris.

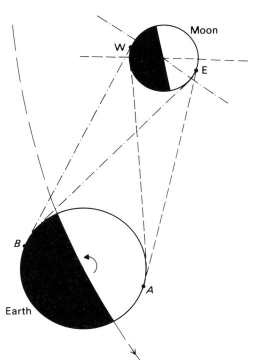

Figure 1.11. *Diurnal libration*. The relative proximity of the Moon (compared to the Earth's diameter) allows an equatorial observer to see more of the eastern limb of the Moon when it rises (*A*) than he will when the rotation of the Earth has carried him to *B* and the Moon is about to set.

## Optical librations

Another of Galileo's discoveries was that, although one hemisphere of the Moon is always turned towards the Earth (because of the synchronisation of the Moon's axial rotation with its orbital period), at one time or another it is possible to see slightly more than half of the Moon's surface. Only 41% of the Moon, in fact, is permanently invisible from Earth. Most of the occasional 9% is made visible by what are known as the Moon's optical librations. How then do these optical librations allow us to see more than half of the Moon's surface?

One libration was first explained by Galileo himself and is really just a parallax effect. When we observe the Moon, we do so not from the centre of the Earth, but rather from some point on its surface. The view which we obtain will then depend on our position on the Earth's surface, as the Moon is not really very distant when compared with the Earth's diameter. An observer at the Earth's North Pole, for example, will see slightly more of the Moon's north polar region than he would if he was on the equator. But to observe this parallax libration it is not necessary to do any travelling at all. We can simply make use of the Earth's rotation on its own axis to change our observing position. When the Moon rises it is possible to see slightly beyond its eastern limb, whereas its western limb is more clearly visible at moonset. The rotation of the Earth therefore gives rise to a diurnal (or daily) libration in longitude, the magnitude of which will be greatest at the Earth's equator (Figure 1.11).

The second optical libration follows from Kepler's first and second laws of planetary motion. The angular velocity of the Moon in its elliptic orbit around the Earth depends on its distance from us, being greater at

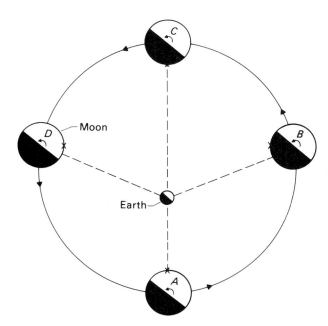

Figure 1.12. *Monthly libration in longitude*. An imaginary line joining the Earth and the Moon sweeps out equal areas in equal times. The time taken by the Moon to move from $D$ to $A$ is the same as that for it to move from $A$ to $B$, from $B$ to $C$, and from $C$ back to $D$. So the point which is in the centre of the lunar disc at perigee ($A$) and at apogee ($C$) will be displaced to the west of the centre at $B$ and to the east of the centre at $D$, by as much as $7°5'$. The ellipticity of the Moon's orbit, then, gives rise to a monthly libration in longitude.

perigee than it is at apogee. As a result, its revolution around the Earth sometimes exceeds and sometimes lags behind its axial rotation. The result is another libration in longitude, but this time one with a monthly period, rather than a daily one (Figure 1.12).

The third and final optical libration is a libration in latitude and is due to the 6 ° 41 ' inclination of the Moon's equator to its orbital plane. When the Moon is north or south of the ecliptic we can see slightly more of its south or north polar regions respectively (Figure 1.13).

The first cartographer to make some attempt to include the effects of the optical librations on a lunar map was Hevelius, in his drawings made in 1647 (Figure 1.14). These maps constitute man's first attempts to describe the far side of the Moon, a project which was not to be completed until the Moon was explored by spacecraft.

*Modern cartography*

It is interesting to compare the accuracy of seventeenth century lunar maps with maps of the Earth dating from the same period. In some ways the lunar ones were superior, despite the fact that terrestrial cartography had been established for more than 2000 years. The Moon's remoteness meant lunar cartography could not even begin until the invention of the telescope. But this very remoteness immediately enabled the entire nearside hemisphere to be observed and mapped, at least roughly. The Earth, on the other hand, had always to be surveyed from the ground, a much more complicated process.

But there is more to cartography than just recording visual impressions. It is necessary to define coordinate systems very carefully and to make extremely precise positional measurements. In the case of the Moon its remoteness again made progress difficult. The angular diameter of the full moon is only about $\frac{1}{2}$ ° and an angular error of only one minute of arc could correspond to a distance error on the Moon's surface of more than 70 miles. Advances in lunar cartography were therefore very closely linked to improvements in telescope design.

Strangely enough we have to return once more to Galileo to find the very first measurements of lunar distance. Although Galileo was not interested in the accurate positions of lunar craters he did make some attempts to calculate the heights of their ramparts. What he did was simply to measure the lengths of their shadows under a known angle of solar illumination. He was then able to conclude correctly that some lunar mountains rise as much as 5 miles above surrounding low lying areas.

Micrometer eyepieces were reportedly used on telescopes by several late 17th and early 18th century lunar cartographers, but the first really accurate lunar map (complete with a network of equatorial coordinates) was that compiled by Tobias Mayer in 1750. In this map the Moon's poles and equator were located by careful measurements of the Moon's librations, and the prime meridian was that passing through the mean centre of the lunar disc, in other words the central meridian after averaging out the effects of the Moon's librations.

Figure 1.13. *Monthly libration in latitude*. The Moon's equatorial plane is inclined by 1°32′ to the ecliptic and by 6°41′ to the plane of its orbit around the Earth. During the half of the month when the Moon is north of the ecliptic we can therefore see more of the Moon's south polar region (unshaded area on the right) than we can during the other half of the month, when it is the Moon's north pole which is tilted slightly towards the Earth. Even greater librations are observed from higher latitudes. An observer at A, for example, will see the hemisphere which is on his side of the line aa′.

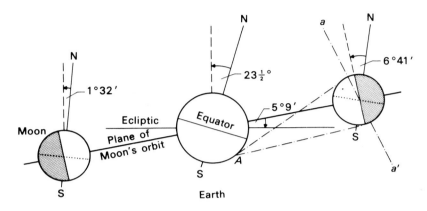

Figure 1.14. *Hevelius' lunar map*. The map of the Moon drawn by Johannes Hevelius in 1645 shows features which are only made visible through the Moon's optical librations. Note that the system of nomenclature is very different from that now in use, mainly due to Riccioli. (The British Library Board.)

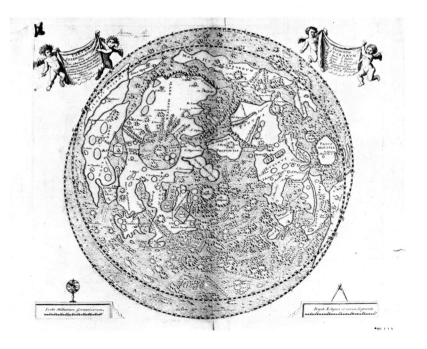

Since the publication of Mayer's map, lunar cartography has become more and more sophisticated and detailed. The monumental atlas compiled by Schmidt in 1878, for example, was divided into 25 separate sections and records the position of 32 856 individual features.

At around the beginning of the 20th century lunar mapping was revolutionised by the advent of astronomical photography, although for a while progress was hampered by poor quality photographic emulsions. Since then, however, a number of photographic atlases and maps have appeared. This mapping era culminated in a massive undertaking by the USAF in 1960 to cover the whole of the Moon's near side to a scale of 1:2 000 000, a project which was finally completed in 1964.

But this was certainly not good enough for Apollo, so the Lunar Orbiter programme led to a complete remapping of the Moon (including the far side) to the larger scale of 1:1 000 000. And now the Apollo programme itself has resulted in much higher quality photographic coverages of at least part of the Moon's surface and the Moon is now being mapped to the larger scale of 1:250 000 or about 4 miles to the inch. No doubt these projects will continue for many years (Figure 1.15*b*).

Figure 1.15. *Modern maps*. (*a*) This highly simplified drawing shows a number of the more prominent features clearly visible through a small telescope.

(*a*)

But what could be the point of mapping the Moon in such great detail? What can we possibly hope to gain from a knowledge of the precise positions of millions of tiny craters? Well first of all there was the technological one associated with landing the Apollo spacecraft. Maps had to be drawn at still larger scales (1:50 000 to 1:10 000) and models made of each landing area so that the astronauts could be made thoroughly familiar with their proposed site long before they ever reached the Moon.

The second and more long lasting objective however is to try and understand the geology and physics behind the processes which have shaped the Moon's surface. How does one area compare with another stratigraphically, for example, and what sort of rocks underlie a particular type of terrain? But before discussing these aspects of lunar geology, it is necessary to step back and review the means by which these high quality photographs were obtained in the first place. And this means discussing the exploration of the Moon by unmanned spacecraft.

## 1.5 Flybys, hardlanders and orbiters

Exploring the Moon from space turned out to be much more involved than had ever been imagined by even the most pessimistic writers of science fiction. Firstly, it was necessary to launch spacecraft through the Earth's atmosphere, and this meant the development of powerful, multistage rockets. There was no question of going to the Moon until

Figure 1.15 (*b*) Map based on Lunar Orbiter photography of Mare Tranquillitatis and Mare Serenitatis area. The Apollo 11 landing site is marked. (NASA.)

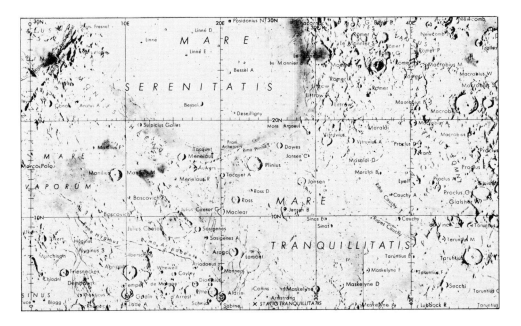

satellites had been successfully injected into orbit around the Earth. Secondly, there were some very serious navigational problems. In what direction should the rockets be directed, for example, in order to reach their target within an acceptable margin of error? Where is the spacecraft at any particular time and how is it orientated? How fast is it moving and in what direction? These were new and complex problems which could

Figure 1.16. *The first Russian probes*. (*a*) Lunik 1, the first spacecraft to reach the vicinity of the Moon. (*b*) Lunik 2, the first hardlander. (*c*) Lunik 3, returned the first pictures of the far side. (*d*) Zond 3, returned higher quality farside pictures.

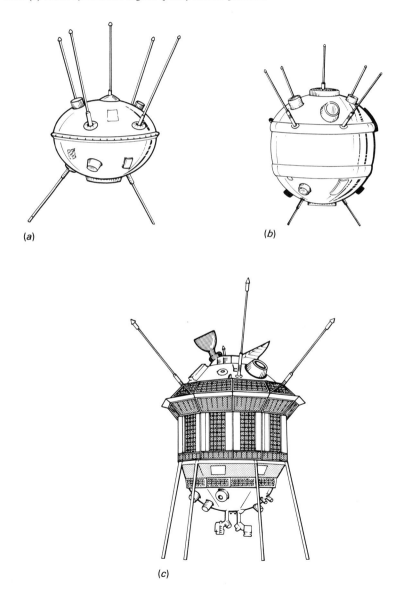

(*a*)

(*b*)

(*c*)

Table 1.4. *Russian unmanned lunar spacecraft*

| Year | Month | Spacecraft | Type | Accomplishments |
|------|-------|-----------|------|-----------------|
| 1959 | January | Lunik 1 | F | missed Moon by 3000 miles |
| | September | Lunik 2 | H | crashed near Archimedes |
| | October | Lunik 3 | F | first farside pictures |
| 1963 | April | Luna 4 | S? | missed Moon by 5000 miles |
| 1965 | May | Luna 5 | S | crashed in Mare Nubium |
| | June | Luna 6 | S | missed Moon by 10 000 miles |
| | July | Zond 3 | F | better farside pictures |
| | October | Luna 7 | S | crashed in Oceanus Procellarum |
| | December | Luna 8 | S | crashed in Oceanus Procellarum |
| 1966 | January | Luna 9 | S | first soft landing (Oceanus Procellarum) |
| | March | Luna 10 | O | first lunar satellite |
| | August | Luna 11 | O | experiments in orbit |
| | October | Luna 12 | O | pictures from lunar orbit |
| | December | Luna 13 | S | soft landing in Oceanus Procellarum |
| 1968 | March | Zond 4 | C | crashed on return |
| | April | Luna 14 | O | experiments in orbit |
| | September | Zond 5 | C | capsule recovered |
| | November | Zond 6 | C | capsule recovered |
| 1969 | July | Luna 15 | R | crashed in Mare Crisium |
| | August | Zond 7 | C | photographs returned |
| 1970 | September | Luna 16 | R | 101 gm soil (Mare Fecunditatis) |
| | October | Zond 8 | C | photographs returned |
| | November | Luna 17 | L (1) | landed in Mare Imbrium |
| 1971 | September | Luna 18 | R | crashed |
| | September | Luna 19 | O | experiments in orbit |
| 1972 | February | Luna 20 | R | 30 gm soil (Apollonius) |
| 1973 | January | Luna 21 | L (2) | landed in Mare Serenitatis |
| 1974 | May | Luna 22 | O | experiments in orbit |
| | October | Luna 23 | R | drill damaged on landing |
| 1976 | August | Luna 24 | R | 170 gm soil (Mare Crisium) |

F=flyby, H=hardlander, C=circumlunar flight, L=Lunokhod, R=sample return, O=orbiter, S=softlander

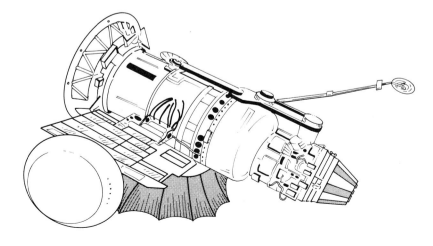

(*d*)

not be solved satisfactorily until major advances in telecommunications and computer technology had been made. Thirdly, it was necessary to design instruments which could be flown on a spacecraft without mal-functioning. Space is a harsh environment and delicate instruments such as television cameras can easily be damaged by the hard vacuum, extremes of temperature and intense radiation encountered there.

All of these problems took their toll of failed space probes, both American and Russian. Nevertheless, solutions were always found so that the exploration of the Moon could continue. But is it better to go all out for a manned landing or can the same results be obtained using unmanned craft? The Russians and the Americans both settled on a step-by-step approach, but it turned out that the ultimate objectives and achievements of each nation were very different.

Figure 1.17. *Farside photography*. Television pictures of the lunar far side taken by the Russian probes (*a*) Lunik 3 and (*b*) Zond 3. Note the absence of large dark *maria* and compare the resolution of these pictures with those taken by Zond 8 (Figure 1.24) and Lunar Orbiter 5 (Figure 1.22). (The Associated Press Ltd.)

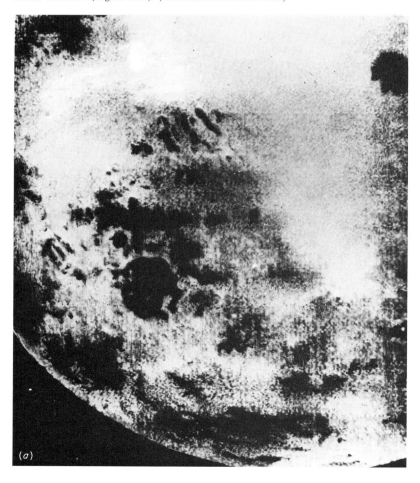

(*a*)

## Early Russian probes

The first step to the Moon is of course to escape from Earth orbit. And this goal was achieved by the Russians only two years after they had launched their first successful Earth Satellite, Sputnik 1. Between 1 January and 7 October, 1959, three Russian moon probes all reached the vicinity of the Moon (Figure 1.16). The first, Lunik 1, missed the Moon by only 3000 miles while the second was even more accurately directed, crash landing close to the crater Archimedes. Many will remember that moment (23 seconds after 10.02 pm GMT on 13 September) when signals being received from Lunik 2 by the radiotelescope at Jodrell Bank abruptly ceased. Never had such sudden silence been quite so awe inspiring.

The third Lunik probe was more successful still. Instead of crashing, Lunik 3 passed behind the Moon and, from a distance of about 3850 miles, took the first television pictures of the normally invisible far side (Figure 1.17) Strangely enough it did not look at all like the familiar near side. Instead, it was seen to be almost totally devoid of large dark *maria*.

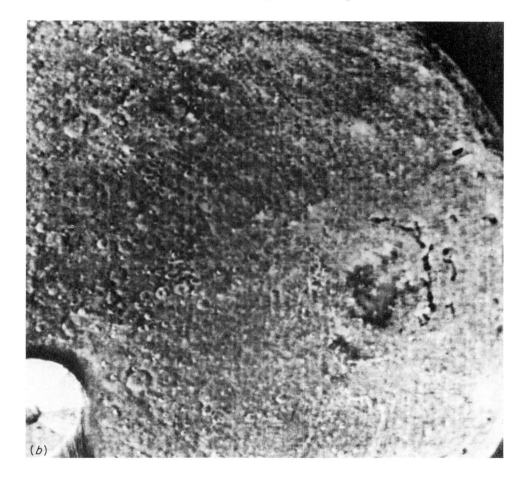

(b)

Figure 1.18. *Landing site map*. This map shows the landing points of most lunar spacecraft that successfully reached the Moon, including hardlanders, softlanders and sample return missions. (NASA.)

And even today, long after the end of the Apollo programme, we still do not have a totally convincing explanation for this contrast. In 1965, however, one of the Soviets' deep space probes (called Zond 3) gave us a clue by returning some more pictures of the far side of the Moon while *en route* to Mars. Its coverage was wider than that of Lunik 3 and the surface resolution was much improved. And what these new pictures showed for the first time was that, while there are indeed circular depressions on the far side, very few of them are filled with dark *mare* material. So it could be that the far side crust is just unusually thick (Figures 1.16 and 1.17). Figure 1.18 shows the landing positions of most of the spacecraft that have landed on the Moon.

Figure 1.19. *American Photographic Reconnaissance Missions*. The Ranger spacecraft (*a*) and Lunar Orbiter satellites (*b*) were equipped with both low resolution and high resolution optics and were powered by solar cells. (NASA.)

*Ranger*

But what of the Americans? How were they responding to the challenge being set by the Russians during the early sixties? Well their first main thrust took the form of Ranger, a programme which turned out to be not without its setbacks. The basic idea behind Ranger was to transmit pictures of the Moon from as near as possible to the lunar surface and so discover what the Moon looks like at really close range. This meant mounting television cameras on to crash-landing spacecraft (Figure 1.19).

Initially it had been intended to use retrorockets to hardland at 90 miles per hour, but this idea was dropped after the first three craft had failed so miserably. Rangers 3 and 5 missed the Moon completely, while Ranger 4 crashed on the far side out of radio contact (Table 1.5). The problems of interplanetary navigation were certainly proving to be more intractable than had originally been anticipated and something clearly had to be done.

The whole concept of the programme was duly upgraded, with the result that Ranger 6 was very nearly successful. This time in fact it was not the fault of poor navigation. The spacecraft arrived at its destination in Mare Tranquillitatis with pinpoint accuracy. But its cameras had failed soon after launch and therefore no pictures were ever returned. It was therefore left to its successor, Ranger 7, to do what no space probe had yet done, that is to televise a 'landing' on the Moon starting from an altitude of more than 1200 miles.

(b)

Table 1.5. *American unmanned lunar spacecraft*

| Year | Month | Spacecraft | Type | Accomplishments |
|------|-------|-----------|------|-----------------|
| 1959 | March | Pioneer 4 | F | missed Moon by 40 000 miles |
| 1962 | January | Ranger 3 | H | missed Moon by 22 000 miles |
| | April | Ranger 4 | H | crashed on far side |
| | October | Ranger 5 | H | missed Moon by 500 miles |
| 1964 | January | Ranger 6 | H | TV malfunction |
| | July | Ranger 7 | H | first close-ups (Mare Nubium) |
| 1965 | February | Ranger 8 | H | close-ups of Mare Tranquillitatis |
| | March | Ranger 9 | H | close-ups of Alphonsus |
| 1966 | May | Surveyor 1 | S | landed in Oceanus Procellarum |
| | August | Lunar Orbiter 1 | O | landing site photography |
| | September | Surveyor 2 | S | crashed in Sinus Aestuum |
| | November | Lunar Orbiter 2 | O | 205 pictures returned |
| 1967 | March | Lunar Orbiter 3 | O | 182 pictures returned |
| | April | Surveyor 3 | S | landed in Oceanus Procellarum |
| | May | Lunar Orbiter 4 | O | nearside mapping photography |
| | July | Surveyor 4 | S | crashed in Sinus Medii |
| | July | Explorer 35 | O | magnetic field studies |
| | August | Lunar Orbiter 5 | O | farside mapping photography |
| | September | Surveyor 5 | S | landed in Mare Tranquillitatis |
| | November | Surveyor 6 | S | landed in Sinus Medii |
| 1968 | January | Surveyor 7 | S | landed near Tycho |
| 1973 | November | Mariner 10 | V/M | pictures from 70 000 miles |

F=flyby, H=hardlander, S=softlander, O=orbiter, V/M=Venus/Mercury probe

Ranger 7 completed its mission on 31 July 1964 by returning progressively more and more detailed pictures of the Fra Mauro region, an area that was later to become the landing site of Apollo 14. The final picture taken by Ranger 7 shows an area comparable in size to a football pitch and was still being transmitted back to Earth when impact occurred. Rangers 8 and 9 were just as successful, returning thousands of close up pictures of Mare Tranquillitatis and the crater Alphonsus respectively (Fig. 1.20). The closest Ranger pictures were taken from an altitude of only 1800 feet.

So what did the Ranger pictures reveal about the Moon, apart from how best to get there? Well it was now possible to see that the lunar surface is pitted with much smaller craters than had previously been recognised. The smallest photographed were only a few feet in diameter. The pictures also showed large boulders resting on the surface. In some ways this discovery was comforting because some had predicted that the lunar *maria* might consist of vast seas of dust into which softlanding spacecraft might disappear on touch down. These boulders, however, clearly constituted additional hazards to be avoided when softlanding. But perhaps most important of all the Ranger pictures did prove that parts of the Moon were indeed sufficiently flat and boulderfree for a manned landing to be perfectly feasible. The question now was where?

## Lunar Orbiter

The search for a suitable Apollo landing site could clearly not be carried out with Ranger spacecraft. What was needed was a satellite in lunar orbit which could photograph large areas of the Moon on a single mission. A new American space programme was therefore initiated (appropriately named Lunar Orbiter) the main objective of which was to discover landing sites for Apollo. As it turned out the Lunar Orbiters of 1966–7 did far more than this. All five of them, in fact, succeeded beyond the wildest dreams of their designers.

The photographic hardware flown on Lunar Orbiter was much more sophisticated than the television cameras carried by Ranger. Ranger had to send pictures back to Earth as rapidly as possible. For, had the time taken to transmit a picture been greater than that prior to impact, the transmission of that picture would never have been completed. In contrast, photography from lunar orbit could be carried out at a much more leisurely pace using film, and could therefore be of considerably higher quality. The exposed films were developed on board and then divided into strips, each strip being scanned for transmission back to Earth using a light beam only 5 thousandths of a millimetre across. The 'joins' between these strips are still visible on all Lunar Orbiter photographs, despite extensive computerised 'cleaning up' procedures.

Figure 1.20. *Ranger photography*. This picture of the 40 mile diameter crater, Guericke, was taken by Ranger 7, shortly before it crashed into Mare Cognitum at the point shown. (NASA, Ranger.)

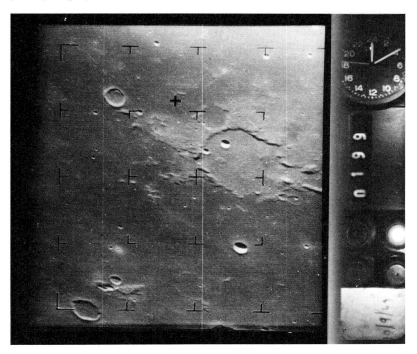

What was it possible to discover with Lunar Orbiter? The first three spacecraft (which all had orbits which took them to within 40 miles of the surface of the Moon) found several suitably flat landing areas for Apollo. This task accomplished, the remaining film could be used up on other areas of geological interest, such as the bright young crater Tycho and a volcanic province in Oceanus Procellarum known as the Marius Hills (Figure 1.21). By injecting the last two Lunar Orbiters into high altitude polar orbits, in fact, almost total coverage of the Moon was achieved (Figure 1.22). These later photographs still provide the basis for mapping those parts of the Moon which were not adequately covered by Apollo, particularly the polar regions.

Lunar Orbiter, then, was the first space programme to explore the Moon in a truly systematic way and the next step was to find out if the lunar surface really could support the weight of a spacecraft and to see what the surface of the Moon actually consists of. This meant sending a softlander, and the American softlander programme was called Surveyor. But that comes later – what were the Russians up to all of this time? Did they too have their orbiter and softlander programmes?

### Later Russian probes

After their initial successes the Russians also experienced problems. Lunas 4 to 8 were all failures. Between 1966 and 1968 however Lunas 10, 11, 12 and 14 went into lunar orbit successfully (Figure 1.23), and even returned television pictures of surface features. But the Russians were also continuing to explore the Moon as part of their circumlunar Zond programme. After the failure of Zond 4 in 1968 (when it crashed on Earth after returning from the Moon), they succeeded in recovering Zonds 5 and 6 intact, together with their live cargoes of turtles and fruit flies.

Figure 1.21. *The Marius Hills from Lunar Orbiter 2*. This young volcanic province was once highly favoured as an Apollo landing site. The large crater is Marius itself. Note the domes and ridges in the foreground. (NASA, Lunar Orbiter.)

And, in the next two years, Zonds 7 and 8 returned some excellent farside photographs, similar to those already taken by an American satellite (Lunar Orbiter 5) but exposed under very different illumination conditions. Such differences in lighting angle are frequently crucial to the recognition and interpretation of surface features. Subdued craters, for example, are only shadowed when the Sun is very low in the sky, whereas differences in reflectance efficiency (or albedo) are most apparent when the Sun is directly overhead. The ray systems around fresh craters are best observed when the Moon is full for example (Figure 1.24).

But what about the two missing Lunas? And what happened after Luna 14? Lunas 9 and 13 were, of course, softlanders, the Soviet equivalent of the larger, more sophisticated, American Surveyors. Lunas 17 and 21 on the other hand were automated roving vehicles (otherwise known as Lunokhods 1 and 2) whereas Lunas 16, 20 and 24 succeeded in

Figure 1.22. *The far side from Lunar Orbiter 5*. Note the almost complete absence of dark *maria*. The lava filled crater with the offset central peak is Tsiolkovsky, one of the more ambitious potential landing sites for Apollo. Note the avalanche to the bottom left of this crater. (NASA, Lunar Orbiter.)

bringing back small samples of lunar soil. In the next chapter the achievements of these softlanding spacecraft, together with the Apollo manned space programme, will be described in some detail. But, before that, the question to be asked is: what has it been possible to learn just from these high quality photographs? Are the dark *maria* dust bowls, for example, or could they be just seas of volcanic lava? Are the craters volcanic, like Vesuvius, or were they produced by impact, like the Barringer meteor crater in Arizona? The science of lunar geology must now take over from the technology of spaceflight.

## 1.6  Lunar geology

The task of the lunar geologist is not an easy one. He must always be objective and not interpret only those features which his own pet theory can most easily accommodate. But by the same token he must also beware of being too chauvinistic. Just because land movements here on Earth are now explained so well in terms of continental drift and plate tectonics, it does not necessarily follow that the surfaces of other planets

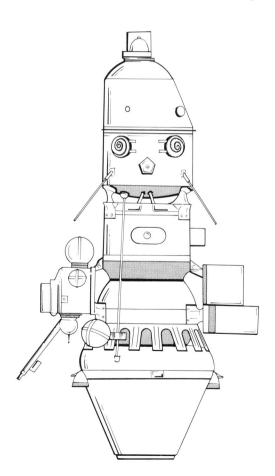

Figure 1.23. *Russian Orbiter*. Drawing of the first successful Russian lunar orbiter, Luna 10, which carried out gamma ray and micrometeorite studies.

are moulded by similar forces. Processes which may have negligible consequences here on Earth may predominate elsewhere in the Solar System.

But perhaps most important of all, the lunar geologist has a problem which his terrestrial counterpart does not have to contend with. For only very rarely is he given the opportunity to collect samples of rock. Field trips to the Moon have proved to be hugely expensive and are certainly not to be undertaken lightly. What, then, have geologists learnt about the Moon and where do the main bones of contention still lie?

### Uniformitarianism

Perhaps the most fiery controversy among geologists of the last century was that between the catastrophists and the uniformitarians. The advocates of catastrophism believed that the Earth was shaped by discrete violent events, one of which was supposedly recorded in the Bible, in the form of Noah's flood. Only by invoking such catastrophes could they

Figure 1.24. *The view from Zond 8*. This view of Mare Orientale is in marked contrast to that taken by Lunar Orbiter (Figure 1.29) under different lighting conditions. Note the difference between the two hemispheres. (Courtesy C. P. Florensky, Vernadsky Institute, Moscow, USSR.)

explain the occurrence of strange fossil animals in the most unlikely of places, such as on the tops of high mountains. According to the catastrophists, then, the ammonites were washed up by some huge flood and were then killed off by some catastrophic event, such as a drought.

The uniformitarians, on the other hand, contested that the rocks at the tops of mountains contain fossils simply because they were formed at the bottom of the sea. Only through the action of crustal forces over millions of years were they raised to their present locations. And the ammonites became extinct because, as time went by, they were no longer able to compete successfully with other forms of life in the sea.

Uniformitarianism, then, is best summed up by the phrase 'the present is the key to the past'. In other words, the processes which have operated in the past are exactly the same as those which are still going on today. It is just that they have been hugely magnified over periods of millions (or even hundreds of millions) of years. Certainly, this evolutionary approach to terrestrial geology has clarified many awkward problems. But it is still just possible that certain events, such as the extinction of dinosaurs, may have been truly catastrophic in character. But what about the Moon? Has its surface been moulded by a small number of catastrophic events, or can the features which we see today be explained better by gradual processes.

*Stratigraphy*

The first objective for lunar geology must be to establish some sort of relative time scale. It is not essential at this stage to know absolute ages, in millions of years. But it is important to define some sort of stratigraphic sequence, comparable in scope if not in detail, to our own geological column, with its eras, periods and epochs. Obviously the names would have to be different (a Lunar Cretaceous period might imply absolute ages between 70 and 130 million years and could conjure up visions of a Moon populated by dinosaurs) but the basic principle would be the same, namely that of the superposition of strata.

When it comes to terrestrial sedimentary deposits (such as river silt or windblown sand) a new sediment is always laid down on top of an older one. This simple relationship may subsequently be altered by folding and faulting and large gaps may develop in the stratigraphic record because of erosion. But, on Earth at least, the presence of fossils in sedimentary rock helps to clarify any ambiguities which may arise.

How then can stratigraphy be applied to lunar geology? How can an individual stratum on the Moon be recognised and related to some remote rock formation? Clearly palaeontology is not applicable here. For, even if there were fossils in lunar rocks, they would certainly not be recognisable from orbit. Nor can vertical sections of lunar crust be studied. For even where they might be expected to be visible, such as on the sides of high mountains, only rarely is it possible to see anything but thoroughly subdued slopes covered with thick layers of eroded debris (Figure 1.25). No, it is more a question of mapping the various types of terrain and then trying to discover which types are the oldest.

*Geological mapping*

This exercise was first carried out by Galileo, who saw that there are basically two types of terrain on the Moon, namely the *terra* (or highlands) and the *maria*. But the lunar geologist can now do much better than that. The high quality of the orbital photographs enables him to subdivide *mare* and highland areas. Some regions have a higher albedo than others for example, while some may only be brighter through a particular type of colour filter and must therefore have a different chemical composition. Similarly certain areas exhibit distinctive lineations or may appear to be hummocky or fractured.

Even if the origins of these different types of terrain are not immediately apparent they can still be mapped as being physically distinct from one another. Some might in fact have common origins, but this more detailed characterisation must come later. To begin with it is just the morphology, location and areal distribution of the different types of terrain which are important. But, where two terrain types meet one another, which one is the older? And by how much? Sometimes one formation clearly overlies another. The circular *maria*, for example, must be younger than the depressions in the highlands that they fill (Figure 1.26). And in the youngest *maria* it is sometimes possible to recognise flow fronts, where a fluid (which we now know to have been molten basaltic lava) clearly erupted over the older *mare* (Figure 1.27). But how much older is the underlying *mare* here and what about the situations where superposition relationships are not so obvious?

Figure 1.25. *Lineaments*. There are rarely more than subtle streaks to be seen on steep hillsides. But Silver Spur, shown here, exhibits broad scale layering. (NASA, Apollo.)

Well firstly it is possible to make use of the most characteristic features on the Moon's surface, the innumerable craters. For, whatever the true origin of craters, it is a reasonable working hypothesis that the more highly cratered a surface is, the older it must be. Thus, if two types of terrain are in contact, the less densely cratered formation must be younger than (and therefore stratigraphically above) the other. So what sorts of terrain are there on the Moon, how are they temporally related, and what pitfalls and uncertainties still remain?

Well firstly it is clear that the *maria* are generally younger than the highlands. True, some highland areas may be less densely cratered than the most ancient *maria*, but this deficiency in numbers does not extend to

Figure 1.26. *Geology of Imbrium area*. This highly simplified map of the Imbrium area shows features of widely different ages. The mountains defining the rings and ejecta deposits of the Imbrium Basin either pre-date the event, being uplifted blocks of crustal material (black, e.g. Apennines), or are contemporaneous with its formation (e.g. Jura mts). Young and somewhat older lava flows are discernible in Mare Imbrium and neighbouring Mare Serenitatis. The craters range in age from those similar in age to the Imbrium Basin (e.g. Sinus Iridum, Plato, Archimedes), through those of Eratosthenian age (e.g. Aristoteles, Eratosthenes) to the most recent, or Copernican age, craters such as Eudoxus, Aristillus, Autolycus and Timocharis. Note that Apollo 15 landed in the younger *mare* unit in Palus Putredinis.

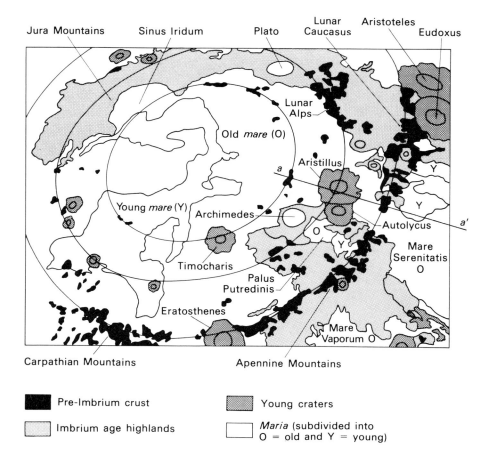

the larger craters, and must just reflect the fact that the smallest craters are more easily filled up in the highlands because of the steeper slopes there.

### *The* Maria

Between the various *mare* regions there are marked differences in age and physical appearance. Mare Tranquillitatis, for example, is bluish in colour, reflecting the high titanium content of the rocks collected from there by Apollo 11. Its highly cratered surface is consistent with these rocks being even older than the oldest known rocks on Earth, the 3750 million year old granite gneisses from eastern Greenland. The *maria* to the west, on the other hand, tend to be reddish in colour and are several

Figure 1.27. *A mare lava flow in Imbrium*. The very low Sun angle when this picture was taken highlights the flow fronts. In places they are more than 100 feet thick. (NASA.)

hundred million years younger. Based on remote geological studies, it appears, in fact, that the youngest *maria* of all are those in southwestern Mare Imbrium, and western Oceanus Procellarum. But quite how young these *maria* really are in absolute terms we can never know for certain without actually going there.

The problems of the origin of *maria* and the nature of the features which we can see there form a major part of the impact versus volcanism controversy and will therefore not be further elaborated here.

*Lunar basins*

The subdivision of highland formations, based on relative ages and origin, has proved to be a much more controversial area than the corresponding classification of the *maria*. The lunar highlands are predominantly light in colour and are generally very heavily cratered. So when were these areas last rejuvenated, if at all, and by what processes? And how is it possible to relate one highland unit to another when few, if any, of them can be the products of volcanic activity?

The crucial observation here is that the ejecta blankets from the largest craters of all, the circular *mare* basins, provide stratigraphic horizons which are, in some cases, recognisable over very wide areas. And the most widespread of these discrete deposits must be (as pointed out as long ago as 1893 by Gilbert) that derived from the largest and youngest such basin, Imbrium. The Imbrium Basin, then, provides the best starting point for lunar highland stratigraphy. But what is a basin? How does it differ from a crater and what does its ejecta blanket look like?

Lunar craters range in diameter from a few feet all the way up to real giants like Clavius, which is more than 140 miles from rim to rim. If we stood at the centre of Clavius, in fact, we would never know it, because the rim crest would be so far away that it would be well below the horizon (Figure 1.28).

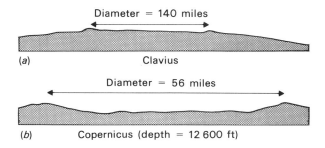

Diameter = 140 miles

(a)                    Clavius

Diameter = 56 miles

(b)          Copernicus (depth = 12 600 ft)

Figure 1.28. *Crater cross sections.* These typical cross sections show how shallow all but the smallest and freshest craters are, relative to their diameters. There is no vertical exaggeration in this diagram. Note how the largest craters are relatively less deep, and have flatter floors.

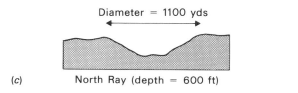

Diameter = 1100 yds

(c)          North Ray (depth = 600 ft)

But it is also possible to make out a number of even larger circular structures on the Moon. These structures consist of a series of concentric rings of mountains, which may be just a series of isolated peaks, or may instead be defined by continuous mountainous arcs. The Imbrium Basin, for example, has at least three of these concentric rings (Figure 1.26), the third one out from the centre being defined very clearly to the south and east by the Carpathian, Apennine and the Caucasus mountains, whereas the Alps define part of the second ring to the north. Like many other basins, this one is filled with dark materials, in this case to produce Mare Imbrium. And this dual nomenclature can be rather confusing when referring to the basin and not the *mare*, or vice versa.

The other basins on the near side include Crisium, Serenitatis and Nectaris (Figure 1.15). And it could be in fact that all other *maria* on the near side also mark the sites of former basins, the mountainous ramparts of which may have disappeared long ago. The irregular *maria* of Tranquillitatis and Fecunditatis, for example, were certainly formed where basins once existed. But the outlines of these basins are now only barely recognisable as such. Even the great Oceanus Procellarum may delineate a former basin. I was the first to propose the existence of such a basin and I preferred to call it Gargantuan, rather than Procellarum, because of its

Figure 1.29. *The Orientale Basin*. At least three concentric rings are clearly visible in this Lunar Orbiter picture of the Orientale Basin on the Moon's western limb. Unlike Imbrium, Orientale was not extensively flooded by *mare* lavas. Note the radial gouge to the south of the basin and how the ejecta blanket (known as the Hevelius formation) has mantled preexisting craters. (NASA, Lunar Orbiter.)

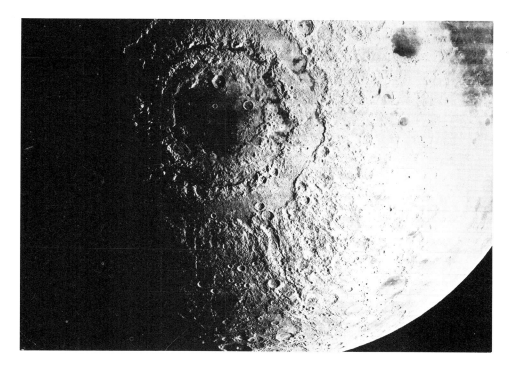

immense size. Support for the existence of the Gargantuan Basin is now steadily growing among lunar geologists. And, if it really does exist, then, at 1500 miles in diameter, it would certainly be a strong contender for the title of largest crater in the Solar System.

For that is all that basins are, craters which are so large that they must have been created by impacting asteroids several miles in diameter. And, instead of just a circular rim being formed, with possibly a central peak, the energy released during these events was so great that multiple ring structures were generated.

The prime example of such a multiple ring basin is Orientale, an even younger basin than Imbrium. Although first discovered from Earth, the Orientale Basin could not be studied seriously until after it had been photographed by Lunar Orbiter, because it is situated right on the Moon's western limb and can only be seen at all from Earth at times of very favourable libration. The Orientale Basin is so well preserved, in fact, that all three rings are almost completely intact. Furthermore, its interior is not *mare*-filled to any great extent, and the basin floor is therefore still exposed, revealing fractures and deposits which presumably date right back to the time of formation of the basin (Figure 1.29).

But, compared to Imbrium, Orientale is rather a small basin. Imbrium is nearly 600 miles in diameter and the material excavated from it must have been spread over a much wider area. Furthermore, as Imbrium is the second youngest basin on the Moon, it should be a relatively easy task

Figure 1.30. *Imbrium cross section*. A hypothetical cross section through Mare Imbrium, showing the stratigraphic relationships between the Apennine Mountains and the Imbrium ejecta blanket, the younger *mare* and the still younger crater Aristillus (see also Figure 1.26).

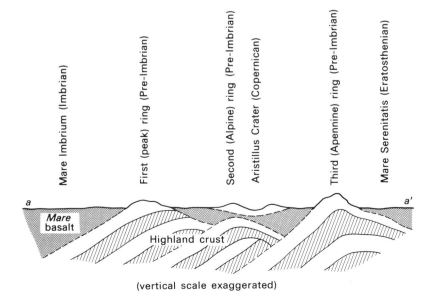

(vertical scale exaggerated)

to map its ejecta blanket, at least in highland areas where it has not been covered by *mare* materials (Figure 1.30). But what does the ejecta blanket of a basin such as Imbrium look like? We have no basins with ejecta blankets here on Earth, so how can we possibly expect to recognise one on the Moon? What is an ejecta blanket anyway? Is it produced by a shower of rocks and boulders travelling along independent parabolic trajectories, or was most of the material deposited from a thick cloud of fine dust and gas, a cloud which moved rapidly outwards from the impact point, closely hugging the surface contours?

*Basin and crater ejecta*

It now appears that both processes must occur during impact events. Around the young crater, Copernicus, for example, are a large number of small elongated craters and crater chains which must have been produced by secondary projectiles, large blocks of rock hurled out of the crater at low angles and below lunar escape velocity. Much older secondary craters are also visible on the Moon, and some of these may have been produced during basin-forming events. The Davy Crater Chain, for example, is certainly secondary in origin and must be very old (Figure 1.31).

Figure 1.31. *The Davy Chain*. This prominent feature is a good example of large scale secondary cratering. (NASA, Apollo.)

Around Orientale, on the other hand, is a well-preserved blanket of ejecta from the basin. At one time, this material must have behaved rather like a fluid, filling in craters and banking up in front of (and flowing around) any obstacle which happened to be in its path. In some places it is very thick indeed. Studies of the partly buried crater Hevelius, for example, suggest that it must have been mantled to a depth of about 300 feet by Orientale ejecta. Quite how far the Orientale ejecta blanket (or Hevelius formation) extends it is impossible to decide for certain. But there is a school of thought which believes that 'pools' of light-coloured material in the highlands (known as the Cayley formation, or highland light plains) may in fact be Orientale ejecta. And some of these pools are many hundred miles from the centre of the Orientale Basin (Figure 1.29).

As the Imbrium Basin is substantially larger than Orientale, its ejecta blanket must be even more extensive. And it is. Except where buried by more recent *mare* materials, the Imbrium equivalent of the Hevelius formation mantles much of the near side of the Moon. Usually referred to as the Fra Mauro formation (after the crater of the same name situated some 300 miles to the south of Copernicus), the Imbrium ejecta blanket was sampled by Apollo 14, the first manned mission to visit a non-*mare* area. This area has a characteristically hummocky appearance, with dunes and furrows radially aligned with respect to the centre of the basin (Figure 2.15).

But not all of this debris blanket may originate from the basin itself. Recent research suggests that much of it may be local in origin. This would certainly account for the observation that the walls of some old craters have disappeared on the side closest to the Imbrium Basin. They must have been broken down by the rapidly advancing debris shower and incorporated into it. It would also explain why the Apollo 14 rocks, which are some 3900 million years old, are such complex breccias.

*A lunar time scale*

The Imbrium event, then, provides a well-defined stratigraphic horizon for lunar chronology. Formations which are covered by Imbrium ejecta must be older than 3900 million years and are all assigned to what is known as the Pre-Imbrian system. One could liken them to Pre-Cambrian rocks here on Earth. The Imbrium Basin itself and its associated deposits (together with the Orientale Basin and the older *mare* units) are then assigned to the Imbrian system. The younger *mare* units, such as Palus Putredinis (visited by Apollo 15) and parts of Oceanus Procellarum (including the area sampled by Apollo 12) are assigned to a third system, known as the Eratosthenian. This Eratosthenian system includes most post-*mare* craters, including Eratosthenes of course, but excludes those with bright halos and rays, such as Tycho, Kepler and Copernicus. These young craters are assigned to a fourth system known as the Copernican which extends right up to the present time.

This then is the framework for the lunar geological time scale. And in the final chapter the entire history of the Moon will be built around it,

Figure 1.32. *Wrinkle ridges.* (*a*) Wrinkle ridges in southern Mare Serenitatis. Note the sharp colour boundary (see Figure 1.59) and the high abundance of rilles in the darker unit. Note also the concentric/radial alignments of the wrinkle ridges. (NASA, Apollo.) (*b*) Hypothetical cross section through a wrinkle ridge.

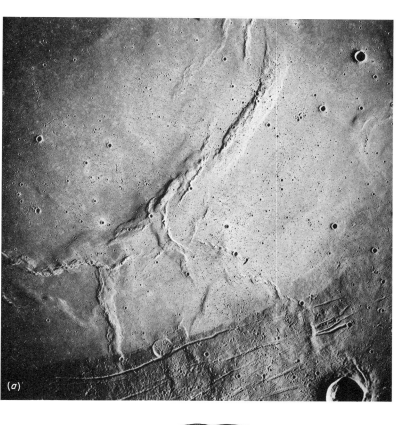

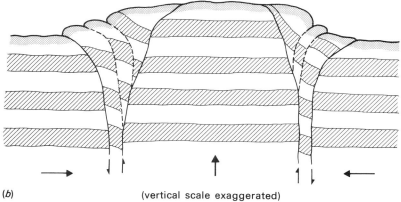

(vertical scale exaggerated)

with absolute ages assigned to events in each epoch. But we must now consider the origins of the various lunar features other than the basins. How were the *maria* produced for example? And are there really such things as lunar volcanoes, or were all lunar craters produced by impacting debris from space? In other words, which has contributed more to lunar geology, volcanism or impact?

## 1.7  Impact versus volcanism

The *maria* have been described as seas of volcanic lava and impacting asteroids have been held responsible for the formation of basins. But might not the *maria* instead be oceans of dust or impact melts? Could there be alternative explanations for the concentric mountain rings which define the basins? And what about the other characteristically lunar features, the smaller craters, the ridges and the rilles? Which of these formations are volcanic and which are impact-derived?

### *The* maria

The obvious place to begin this investigation is with the *maria* themselves. For, prior to the Apollo programme, there was certainly no

Figure 1.33. *High lava mark*. Below Mt. Hadley lavas have drained away leaving a high lava mark around the contact with the *mare*. (NASA, Apollo.)

shortage of mutually exclusive theories for *mare* origin, from dried up seas and lakes of asphalt to primordial deposits of unfractionated Solar System material. But the unequivocally basaltic rocks which were finally returned to Earth by Apollo 11 settled this controversy once and for all, by showing quite conclusively that the *maria* are indeed underlain by very extensive lava flows, similar in many ways to the rocks which are still building up the volcanic islands of Iceland and Hawaii.

But lava flows are not visible as such on the Moon. All those millions of years of exposure to meteorite bombardment have led to the development of a surficial debris layer (known as the regolith) which is several feet thick. So how do we know that volcanic bedrock really exists?

Well, the Apollo 15 astronauts visited Hadley Rille, a sinuous channel running across Palus Putredinis. And in the far wall of this rille, several hundred yards away, and several feet below the rille rim, were exposures revealing horizontal layers of basalt rock (Figure 2.26). The existence of volcanic lava at relatively shallow depths also accounts for the ubiquity of basalt boulders around even the smallest craters on the Moon (Figure 2.22).

A volcanic origin for lunar *maria* is, however, still not universally accepted. Indeed, it has been suggested that the basalts may have been derived from the regolith, rather than the other way round. But this hypothesis does not survive serious analysis. For the basalts are deficient in a number of elements which are very abundant in the regolith and which must therefore be largely extralunar in origin. Most of the nickel in *mare* soils, for example, comes from meteorites. So an impact-melted soil should contain abundant nickel which lunar basalts certainly do not.

### Wrinkle ridges

Arrested flow fronts in the *maria* have already been mentioned as evidence for the former existence of fluid lava on the Moon, but these are not the only positive relief formations visible there. When the Moon is illuminated at low Sun angle, it is also possible to see low relief wrinkles on *mare* surfaces (Figure 1.32). These wrinkle ridges may be discontinuous, are usually asymmetric in cross section and commonly extend across the *maria* for hundreds of miles. But what are they? Well, it looks as if some *mare* lavas solidified above their present level and then sank down, the wrinkle ridges forming by compression as the surface area of the *mare* decreased. The existence of 'high tide' marks around *mare* borders certainly adds support to this subsidence theory (Figure 1.33) as does the total absence of wrinkle ridges in the highlands.

But there are one or two other aspects of wrinkle ridges which must also be mentioned here. For one thing, they often delineate the buried rings of basins. The innermost rings of Imbrium and Gargantuan, for example, show up most clearly as wrinkle ridges. So maybe wrinkle ridges are not compressional features at all, but arose instead from the eruption or intrusion of lava along fissures associated with the underlying topography. Alternatively they may result from later crustal readjustments, in other words faulting. This latter explanation is supported by

Figure 1.34. *Linear rilles*. (*a*) The forked rille Hyginus in Mare Vaporum with its associated craters. (NASA, Lunar Orbiter.) (*b*) Network of graben rilles in the crater Goclenius extending beyond the confines of the crater. (NASA, Apollo.)

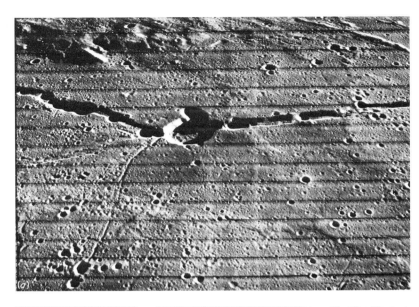

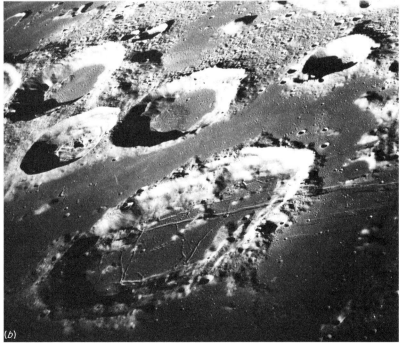

the fact that wrinkle ridges sometimes link up with simple faults in the highlands, such as the Lee–Lincoln Scarp (Figure 2.32). Whatever their true cause, however, it is generally agreed that wrinkle ridges are not due to impact.

### Sinuous rilles

The origins of lunar rilles are even less clear, largely as a result of their greater morphological diversity. There are two fundamentally different types, sinuous rilles like Hadley, and those which are more or less straight sided (Figures 1.34 and 1.35). Sinuous rilles invariably occur in the *maria* whereas the straighter variety do not always recognise highland–*mare* boundaries. What, then, are the important characteristics of these two rille types and how were they formed?

When Apollo 15 went to Hadley Rille it was difficult to know quite what to expect. It certainly looks like a dried up river bed. But it does not have any tributaries and is strictly confined to the *mare*, running along the base of the Apennine mountains for many miles before finally evaporating in Mare Imbrium. Its cross section is V-shaped, but this may just be the result of the impact-induced erosion of the walls and the accumulation of the eroded debris (or talus) in the bottom of the rille. That this erosion is still an active process today is evident from the thinness of the regolith on the rim of Hadley Rille. Bedrock here is probably only a foot or so below the surface, because any fine material created here is rapidly transferred to the bottom of the rille (Figure 2.25).

How then did sinuous rilles originate? Well Apollo 15 rocks contained no water whatsoever, so Hadley Rille certainly cannot be a true river bed. The most attractive theory is the 'collapsed lava tube' hypothesis in which a crust is presumed to have formed on top of the molten lava lake. The fluid lava then drained away from beneath this crust (out into Mare Imbrium in the case of Hadley Rille) leaving an underground tunnel. Such lava tubes, albeit on a much smaller scale, are well known here on Earth. But on the Moon they would have been easily ruptured by meteorite impact and so collapsed to produce sinuous channels. If this collapse was incomplete a row of coalescing craters rather than a continuous channel would have been the result (Figure 1.35b). Some sinuous rilles may never have been roofed in the first place but may instead have been open lava channels.

Perhaps the most dramatic sinuous rille of all is Schröter's valley on the Aristarchus Plateau in northern Oceanus Procellarum. Schröter's valley is unique because it consists of a sinuous rille inside another, much larger, sinuous rille. Furthermore, there is a large crater, aptly known as the Cobra's Head at the upper end of the rille (Figure 1.36). A number of other sinuous rilles (including Hadley) also originate in craters and this surely supports their volcanic origin. It also means that at least some craters on the Moon must be volcanic.

Figure 1.35. *Sinuous Rilles*. (*a*) A sinuous rille in the crater Posidonious. Note the break in the crater wall and the tortuous path of the rille, which eventually finds its way out on to the *mare*. (*b*) This chain of craters and elongate depressions is clearly a sinuous rille in the making. Note the elongated source crater and the wrinkle ridge at its lower end. (NASA, Apollo.) (*c*) Typical cross sections of a sinuous rille and a graben rille.

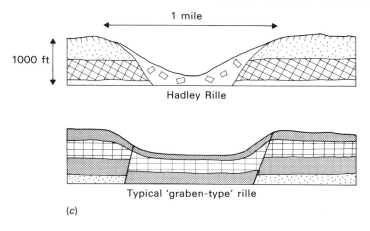

Hadley Rille

Typical 'graben-type' rille

(c)

Figure 1.36. *The Aristarchus plateau*. A volcanic plateau in western Oceanus Procellarum with two large impact craters, Herodotus and the much younger Aristarchus. Schröter's Valley, one rille inside another, originates in a volcanic crater, the Cobra Head, and disappears into the *mare*. (NASA, Apollo.)

*Straight rilles*

But what about the straighter sided rilles? The most popular theories for their origin involve faulting of some kind, and this may or may not have been associated with lunar volcanism. The formation of basins, for example, created radial lines of weakness which led, either immediately or at a much later date, to the formation of many such rilles (Figure 1.29). But by no means all rilles are radially orientated in this way. Around the Humorum and Serenitatis Basins, for example, there are many long, arcuate rilles (Figure 1.32). These would appear to be associated more with the *mare* lavas than with the formation of the basins themselves. The formation of Hyginus Rille may have been even more closely linked to *mare* volcanism as craters can be observed along its length (Figure 1.34*a*).

One notable feature of lunar geology is that there has been very little transverse movement within the lunar crust. Any 'continental drift' on the Moon should have made itself apparent in the form of distorted craters as well as in rift valleys. And no such craters are observed. So it is now generally accepted that these straight sided rilles (whether single or double) must represent vertical rather than horizontal displacements of the lunar crust. Some of those in the *maria* could well be slump features associated with the contraction of the *mare* surfaces, rather than with the formation of underlying basins. But although the majority of faults are vertical there are also some minor horizontal ones in the form of highland scarps. These scarps, however, are more akin to wrinkle ridges than to rilles and the horizontal displacements here are of very limited lateral extent (Lee–Lincoln Scarp, Figure 2.32).

*The lunar grid*

So what drives the vertical movements of the lunar crust? Well, it was pointed out many years ago that lunar rilles tend to be orientated in certain preferred directions. And, if such a lunar grid exists, tectonic forces on the Moon must presumably operate on a global scale. But, before rushing to conclusions about global tectonism, it should be remembered that there is no evidence whatever for continental drift on the Moon, so there may be less far-reaching explanations for the recorded rille alignments.

For one thing, a rille running north–south will always cast a more prominent shadow than one running east–west, a purely optical effect which also gives rise to preferred orientations for other lunar features, such as wrinkle ridges, mountains and craters. Laboratory experiments have shown in fact just how easy it is to simulate these apparent alignments under low angles of illumination. But the preferred directions may also be due to the predominant influence of one or two basin-forming events. A large number of rilles, for example, are found to be radial to the Imbrium Basin. So more compelling evidence is clearly required before the lunar grid can become an established fact.

*Domes, craters and cinder cones*

Turning now to some rather more controversial volcanic features, there is the question of lunar domes. These are low rounded hills (with convex upward profiles) which are frequently capped by small craters (Figure 1.21). Are these the long-sought lunar volcanoes then, or did impact craters just happen to be created on these particular mountain tops? A volcanic origin is certainly most likely for the Marius Hills, but we may have to wait until man goes there before we can be completely sure.

The nature of lunar craters as a whole has traditionally been a major topic for discussion and few controversies in the history of science have had such vociferous proponents. But since Apollo it has become clear that the vast majority of lunar craters are in fact of impact origin, despite evidence to the contrary. Arguments favouring volcanic origins based on the polygonal outlines, depth-to-diameter ratios and apparent alignments of craters have now been shown to be at best illusory and at worst positively misleading. Much of the effort on Apollo, in fact, was spent

Figure 1.37. *Alphonsus Crater*. The floor of Alphonsus is characterised by rilles and dark halo craters. This view is looking south of the flat floored 'walled plain' Ptolemaeus towards the central peak crater, Arzachel. (NASA, Apollo.)

chasing after what were thought to be volcanic craters and formations, but which subsequently turned out to be nothing of the sort.

Perhaps the best example of such a misinterpretation concerns the so called dark halo craters. Certain young lunar craters are surrounded by halos of unusually dark material rather than by bright ray systems. And it was suggested that they might be small volcanic craters, either cinder cones or fumaroles. The frequent association of dark halo craters with lunar rilles added credence to this viewpoint (Figure 1.37). But when such a dark halo crater (Shorty) was visited by Apollo 17, it turned out to be just a typical impact crater (Figure 2.32). Instead of being composed of volcanic cinders, the dark halo material here was found to consist predominantly of orange and black glass. Some would now maintain that this was not a typical dark halo crater, but support for the volcanic nature of dark halo craters as a whole was certainly much reduced after Apollo 17.

*Highland volcanics*

The previous mission, Apollo 16, also proved the lunar geologists wrong. But on this occasion the formations in question were in the highlands. The original feeling about the Highland Light Plains (or Cayley Formation) was that these apparently flat areas in the highlands were pools of volcanic lava which, in view of their high albedo, must be more acidic than those erupted in the *maria*. Similarly, highly silicic but more viscous volcanic lava was postulated to account for the so-called Descartes formation, an extremely light-coloured and hummocky area to the northwest of Mare Nectaris. But all that the Apollo 16 astronauts could see when they tried to sample these two formations (Figure 2.27) were breccias and impact melts. Not one highland volcanic rock was to be found. So if highland volcanics do exist on the Moon, they must be sought elsewhere.

Perhaps predictably, then, we have found strong evidence for both impact and volcanism on the Moon. The *maria* and its associated features (such as wrinkle ridges and sinuous rilles) were undoubtedly produced by basaltic lava, whereas the highlands were largely moulded by impact. But why are there so few volcanic craters in the *maria*? Where did all the lava come from if not through craters? Well it now appears that *mare* basalt viscosities were so low that the lava could flow very rapidly away from the fissures through which it was vented. Under these conditions it would have spread out in thin sheets and would never have piled up to produce volcanic caldera.

So what is now known about the mechanics of impact cratering and how can crater sizes and densities be used to improve our understanding of lunar chronology? How can the multiple ring structures of basins be explained, what are lunar rays and why do they disappear as a crater ages? In other words how exactly were lunar craters formed and how did they evolve?

## 1.8    Lunar craters

This discussion must begin with the question of how impact controls crater shape and how the resulting crater profiles vary with crater diameter, from the smallest depressions in the regolith all the way up to the vast multiring basins.

The tiniest craters lack any sort of raised rim or rocky ejecta blanket. And this is simply because the impacts responsible for them never penetrate down to hard rock. That these small hollows in the regolith are indeed of impact origin (rather than being volcanic fumaroles) is demonstrated by the fact that they are frequently lined with impact-melted glass (Figure 1.38). But this inference could not be drawn, of course, until man had visited the Moon with Apollo. The very existence of these tiny pits was only hinted at in the Ranger and Lunar Orbiter pictures.

Moving up the size scale a little there is a slight decrease in the relative depth of a crater when its diameter exceeds 100 feet or so. At this point the crater rim becomes noticeably raised above the surrounding area and

Figure 1.38. *Regolith craters*. Small (i.e. yard-sized) craters in the regolith are formed simply by the redistribution of materials within the fine-grained regolith itself. They tend to be rimless and have a rounded appearance. The freshest ones may be glass-lined. (NASA, Apollo.)

is mantled by rocks and boulders. So the impacts responsible for craters of this size must clearly have penetrated right through the regolith and started to excavate the more rocky substrata. And this discontinuity therefore provides an excellent method for finding out just how deep the regolith is at any particular site (Figure 1.39).

*Ejecta deposition*
As crater diameter increases further, deeper layers of the crust are exposed in the walls and deposited as ejecta, both inside and outside the

Figure 1.39. *Sub-regolith craters*. Craters more than a few hundred feet across were formed by impacts that penetrated through to the more rocky substrata. The crater Shorty, at the Apollo 17 site, is shown here, containing large blocks of *mare* basalt excavated from a depth of some 300 feet. (NASA, Apollo.)

Figure 1.40. *North Ray*. This crater was a prime target for Apollo 16 and its steep interior walls attest to its youth. This hypothetical cross section through the crater shows benches, talus and rocky outcrops in the far wall. (See Figure 2.29 for photograph.)

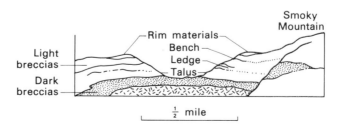

Figure 1.41. *Ejecta thickness distribution*. The thickness of the ejecta blanket from a crater decreases sharply with distance from the crater rim. The deepest materials are deposited closest to the rim, with shallower ejecta being deposited further out.

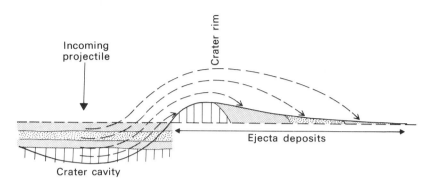

Figure 1.42. *Ejecta blanket characteristics*. The nature of the ejecta blanket around a crater changes with increasing crater diameter. As this diameter rises from about 300 yards (top left) to 13 miles (clockwise to bottom left), the abundance of blocks decreases and there is a corresponding increase in secondary cratering and the development of dunes and furrows. (From Oberbeck, V. R., Morrison, R. H., Horz, F., Quaide, W. L. & Gault, D. E. (1975). *Proceedings of the Fifth Lunar Science Conference*, pp. 111–36. New York: Pergamon.)

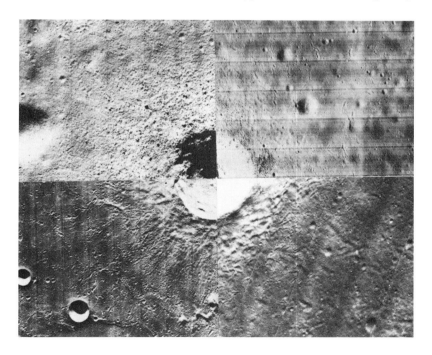

crater rim. The deposition of ejecta inside the crater itself usually obscures any of the preexisting stratigraphy but, in some cases, distinctive layers remain visible in the crater walls. A dark layer was clearly exposed in the bottom of North Ray Crater at the Apollo 16 site, for example, and stratification was also apparent in the far wall (Figures 1.40 and 2.29).

Theoretical calculations and laboratory simulations have both shown that the deepest materials excavated are deposited either inside the crater itself, or else very close to the crater rim, whereas the extremities of the ejecta blanket must consist of materials which were originally only at shallow depths (Figure 1.41). So the Imbrium rocks collected by Apollo 15 must be derived from deeper levels in the Pre-Imbrian crust than those collected from the Fra Mauro region (some 400 miles farther out from the centre of the basin) by Apollo 14. And the ejecta from small craters must presumably be distributed in much the same manner.

For craters more than a mile or so in diameter, secondary cratering starts to play a major part in the formation of the ejecta blanket, as rocks and boulders ejected from the crater thoroughly disturb the surrounding area. Only 20% of the ejecta blanket may actually come from the crater itself, the other 80% may be surface materials which have been turned over by, and mixed up with, the primary ejecta.

Among the characteristic features of a typical fresh ejecta blanket (already mentioned in the context of basins) are concentric dunes, radial ridges and chains of secondary craters (Figure 1.42). These features are characteristic not only of craters a few miles across but also of much larger ones such as Aristarchus (Figure 1.43).

Figure 1.43. *Profiles of young craters – Aristarchus*. Craters which are several tens of miles in diameter usually have terraced walls and complex floors. (NASA, Lunar Orbiter.)

*Crater morphology*

It is not just the ejecta blankets of craters which change as crater diameter increases. From being simple circular bowl-shaped depressions (Figure 1.44) larger craters on the Moon start to assume more complex forms. Slumping results in shallower floors (which sometimes exhibit swirl textures), terraced inner walls and more irregular outlines (Figure 1.45). The consequent reduction in depth – diameter ratio is enhanced by the existence of a less intensely fractured crust below a depth of two miles or so. As diameters exceed several tens of miles, depth–diameter ratios fall from as much as 1:5 down to as little as 1:30 or even less. This shallowness is surprising in view of the apparent depth of craters when seen near the terminator. Inner wall slopes may be as steep as 30°, but outer walls have a shallower gradient, generally less than 15°. Once again appearances are deceptive here. Contrary to popular belief, the Moon is not covered by towering pinnacles of rock, steep gorges and bottomless pits (Figure 1.46).

But slumping is not the only mechanism by which crater profiles are immediately modified. Features also appear which are due to the release of stresses built up during the impact event. As crater diameter increases still further it is possible to observe not only central peaks (Figures 1.22 and 1.43), but also peak rings and even multiple ring structures (Figure 1.29). The precise mechanisms by which the release of stress is expressed in these forms, however, are not yet clearly understood.

Figure 1.44. *Profiles of young craters – Mösting C.* Craters less than a mile or so across are circular bowl-shaped depressions with raised rims. (NASA, Lunar Orbiter.)

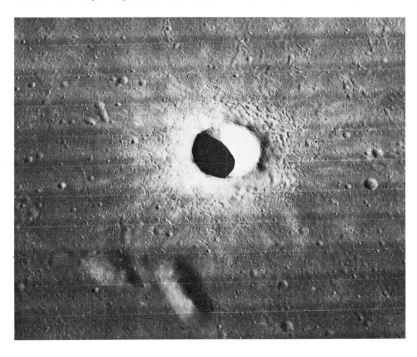

Figure 1.45. *Profiles of young craters – Dawes*. Intermediate-sized craters are frequently polygonal with slump features. (NASA, Apollo.)

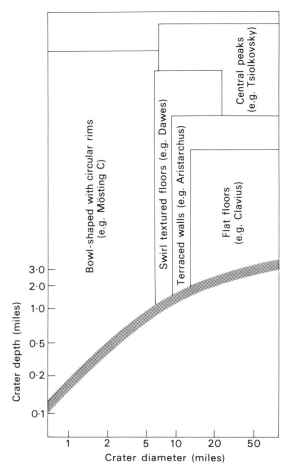

Figure 1.46. *Depth – diameter ratios*. There is a gradual decrease in the depth–diameter ratio for lunar craters with increasing diameter, partly due to wall slumping. (After Head, J. W. (1976). *Proceedings of the Seventh Lunar Science Conference*, pp. 2913–29. New York: Pergamon.)

Figure 1.47. *The young ones*. (*a*) Young lunar craters, such as Kepler (left) and Copernicus (right) are characterised by bright ray systems. (NASA, Lunar Orbiter.) (*b*) Giordano Bruno, just beyond the Moon's northeastern limb, may well have been formed in historical times. (NASA, Apollo.)

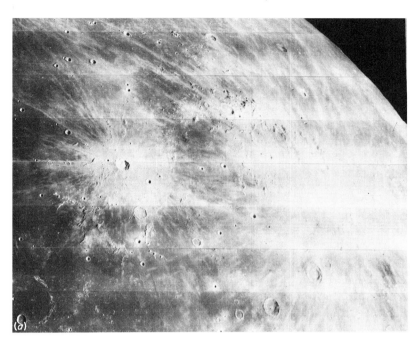

*The impact process*

One way to study impact cratering mechanics is through laboratory simulation. When tiny solid pellets are fired at cosmic velocities into suitable targets, the various stages of crater formation can be captured by high speed photography. But it is difficult to extrapolate the results of these experiments to craters several miles in diameter. The scaling problems are just too immense. In some ways this shortcoming is alleviated by being able to use terrestrial impact and artificial explosion craters as analogues. But there is still nothing on Earth to compare with the vast lunar basins.

What has become clear, however, is that the energy released by a meteorite impacting the Moon at several miles per second is quite sufficient to vaporise not only the meteorite itself but also many times its own mass of lunar crust. So it is certainly not surprising that some craters on the Moon contain what appears to be pools of impact melt. Others have floors exhibiting cracks and crenelations (Figure 1.52). At one time it was even thought that the *maria* themselves might be vast impact melts, generated during the excavation of the circular basins. But it is now known, of course, that the *mare* lavas are considerably younger than the

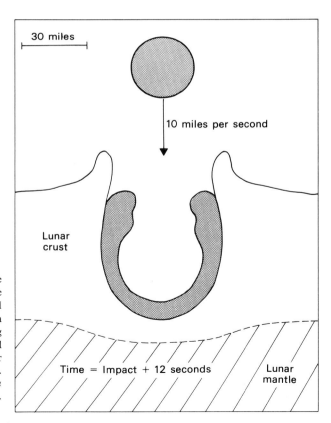

Figure 1.48. *The impact process*. The excavation of a lunar crater has more in common with an underground nuclear explosion than with, say, a splash into wet mud. The impacting projectile may penetrate several miles into the lunar crust. (After O'Keefe, J. D. & Ahrens, T. J. (1976). *Proceedings of the Seventh Lunar Science Conference*, pp. 3007–25. New York: Pergamon.)

basins which they flooded and they must have originated not within the crust but deep down in the Moon's interior.

One way of appreciating the true violence of crater excavation is to consider the impact of a basin-producing projectile, say a small asteroid 30 miles in diameter impacting at 10 miles per second. On impact, a shock wave is generated and this travels at some 5 miles per second through the projectile, reaching its trailing edge 6 seconds later. But by this time, of course, the trailing edge will already be 30 miles inside the Moon, because only then will it begin to register the shock wave and start to slow down (Figure 1.48). So meteorite impacts on the Moon are clearly more akin to underground nuclear explosions than to simple surface splashes. And the distribution of ejecta from lunar craters will therefore always tend to be more or less symmetrical, except in those extreme cases when the meteorite involved happened to strike the Moon at a grazing angle, as in the case of Proclus. The formation of the elongated crater Schiller may also have been due to a grazing impact, but this unusual feature may instead be a secondary crater associated with the formation of a major basin (Figure 1.49).

*Lunar rays*

So much, then, for the morphology of freshly formed craters and their ejecta blankets, but what happens to a crater with the passage of time? How does a crater age and is there any possibility of using its state of preservation to date it?

The youngest craters on the Moon are characterised by bright ray systems which are particularly prominent at full moon (Figure 1.24). As time passes, these bright rays are the first features of a crater to disappear. Precisely what they consist of is still not completely understood. But freshly exposed rock surfaces are certainly more reflective than fragments which have been subjected to radiation effects for millions of years. So lunar rays presumably fade when the initially light-coloured ejecta from a crater is impact melted or radiation darkened and is then thoroughly mixed up with the underlying regolith (Figure 1.47).

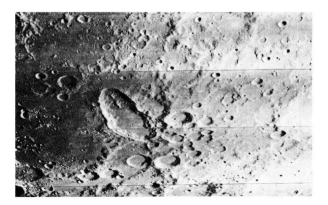

Figure 1.49. *Elongated crater*. The crater Schiller is very elongated and may either be a basin secondary or the product of a low angle primary impact. (NASA, Lunar Orbiter.)

*Crater evolution*

There are several factors which affect the evolution of a crater after the disappearance of its rays. The most important of these processes are obliteration, erosion, burial and isostatic readjustment. If a new crater is formed on the rim of a similar sized, or smaller, one, then much of the original crater structure may be obliterated. Two or three such critical impacts and the original crater will completely disappear.

The walls of a crater are also worn down by innumerable minor impacts. Such small scale erosion gradually fills up the crater interior and levels the rim, a process which gives the crater a progressively more subdued appearance. As time goes by, a lower and lower Sun is required for the rim to cast a shadow into the crater and so make it visible from orbit. Eventually, the crater becomes completely unrecognisable in the gently undulating lunar terrain.

If a crater is not destroyed by obliteration it may well be incorporated into (or filled in and buried by) *mare* lavas (Figure 1.50) or ejecta from nearby craters. The large *maria* and impact basins must have destroyed vast numbers of lunar craters in this way. And those still visible today must be only a small fraction of all craters which have existed since the formation of the Moon.

For the largest of these craters, the lunar crust did not have sufficient strength to support their high walls, nor to prevent their deeply depressed floors from welling up towards isostatic equilibrium. So very large craters have small depth–diameter ratios not only because their walls have been eroded away, and their interiors filled in, but also because of the slow upward readjustment of their floors and settling of their walls in response to lunar gravity. These stresses date back to the time of crater formation and have been exerted ever since, in other words for hundreds of millions of years (Figure 1.51). With these very large walled plains, as they are sometimes called, the floors take on the global curvature and the crater rims are then often below the horizon for an observer at the centre of the crater.

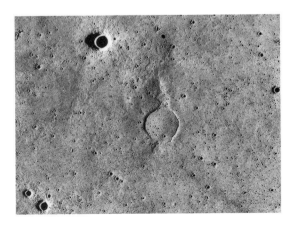

Figure 1.50. *A 'ghost' crater.* When *mare* lavas erupted in this area, some of the larger craters were not buried completely and their rims show up as rings. (NASA, Apollo.)

Figure 1.51. *The largest lunar 'crater'.* The largest circular feature on the Moon that is generally referred to as a crater is Bailly, near the south pole. Larger features are usually referred to as basins. (NASA, Lunar Orbiter.)

Figure 1.52. *Tycho.* With the possible exception of Giordano Bruno, Tycho is the youngest large crater on the Moon. Its rays traverse much of the near side. Note in particular the crenulated floor. (NASA, Lunar Orbiter.)

*Relative crater ages*

The superposition of one crater on another provides a direct means for classifying craters by age. The destroyed crater must obviously be older than the overlying one. But how much older it is only possible to decide after studying large numbers of overlapping craters, covering a wide spectrum of size and preservation (Figure 1.51). A table of features which are present (or absent) in a crater of a certain size and age may then be compiled. One such classification, involving three size groups, each subdivided into seven stages of crater preservation, has been used quite successfully to compare the age of one cratered area on the Moon with another. An area with large, poorly preserved craters, for example, is taken to be older than one in which all craters are in better states of preservation (Figure 1.53).

But this is not the only way to determine relative ages using craters. It is also possible to count the craters themselves. And crater counting, although inferior to the absolute dating of lunar rocks using radioisotopes, provides the best means for dating those areas of the Moon which are far removed from those visited by Apollo. So what are the principles of crater counting, how do the various methods compare, and what are the current uncertainties and shortcomings of this approach to lunar chronology?

Figure 1.53. *Crater classification*. All craters of Copernican age in this photograph that are larger than 250 feet in diameter have been classified according to their preservation state, with the freshest craters having the highest classification number. Those marked with an 'e' are Eratosthenian in age, but not all are shown. (From Swann, G. A. (1974). *Proceedings of the Fifth Lunar Science Conference*, pp. 151–8. New York: Pergamon.)

## 1.9   Crater counting

The basic principle behind crater counting is really very simple. The longer a particular area has been exposed to meteoritic bombardment, the more craters it will have accumulated. A region with a high crater density must clearly be older than one where the craters are more sparsely distributed. But when it comes to turning this principle into practice there are some practical difficulties. What are these problems then, how have they been overcome and what has now been learnt about the Moon just from counting craters?

Well, firstly, there is little point in counting all craters (irrespective of their size and state of preservation), and then simply dividing the total number by the area covered in order to obtain an age parameter. Small craters are invariably more abundant than larger ones and lighting conditions can be very different from one area to another. As a result, 100 yard diameter craters may be clearly recognisable in one photograph but completely unresolved elsewhere. The relative crater counts would clearly be completely meaningless under these conditions.

The craters must therefore be counted within a particular size range, or better still within a number of such size ranges. In practice, what is done is to construct a cumulative size distribution curve, in which the number of craters which are larger than a certain diameter is plotted against that diameter. Unlike a simple histogram of number against diameter, the cumulative size distribution curve will always have a negative slope. In other words if there are 10 craters per square mile with diameters greater than 10 yards, then there clearly cannot be more than 10 craters per square mile with diameters greater than 100 yards (Figure 1.54).

*Initial crater distributions*

So what does this distribution curve for lunar craters look like in practice? And how does it evolve with time? Well, the distribution of craters on a virgin surface is initially just a reflection of the relative frequency with which different sized meteorites are impacting the lunar surface. And the fact that there are many more small craters per unit area than larger ones simply means that small meteorites are relatively more abundant out there in space. This of course confirms terrestrial experience. Only very rarely have large meteorites (those large enough not to have lost all their cosmic velocity as they passed through our atmosphere) reached the surface of the Earth and produced impact craters. The same picture emerges at the lower end of the size scale when micrometeorites have been monitored by satellite-borne detectors (Figure 4.30).

What should then happen with the passing of time? Well at first the shape of the crater size distribution curve will not change very much, so long as the size distribution of meteorites stays essentially the same. But the absolute number of craters larger than any given diameter will of course steadily increase. As time passes, however, more and more craters will be formed on top of existing ones, while others will be buried by the ejecta from craters nearby. The craters which will be obliterated and buried most easily will obviously be the smaller ones. So there will start to be a relative deficiency of these smaller craters, a deficiency which will show up as a flattening in the size distribution curve (Figure 1.55).

*Relative ages*

The total number of these smaller craters may still continue to increase, of course, but there will come a time when they are being destroyed just as rapidly as they are being formed. The surface is then said to be saturated with all craters up to this particular size. As more and more craters of all sizes are produced, this saturation diameter gradually increases; the older the surface the larger it will be.

The saturation diameter will show up in the size distribution curve as a change in slope and its absolute value will reflect the exposure age of the area under study. The saturation crater diameter in the Fra Mauro formation, for example, is more than 400 yards, whereas in some *maria* it is less than 100 yards. What this means, in effect, is that the Fra Mauro formation just cannot accommodate any more 400 yard diameter craters, whereas 400 yard diameter craters in the *maria* are more widely spaced.

All other factors being equal, then, the crater distribution curve for a particular area will fall below that for an older area (and above that for a

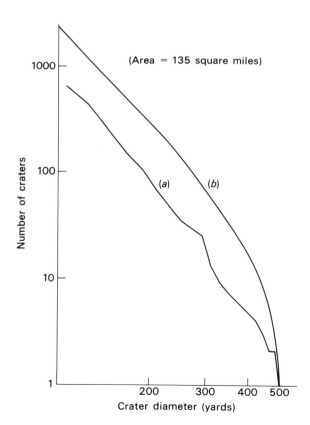

Figure 1.54. *Crater distribution by size.* It is more appropriate to display the crater size distribution as a cumulative curve (*b*) than as a histogram (*a*), because it is independent of the size interval chosen.

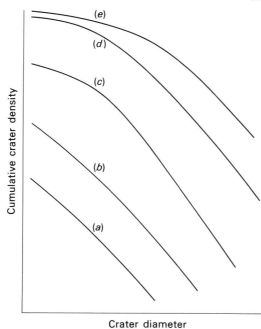

Figure 1.55. *Evolution of a cratered surface.* (*a*) The initial distribution reflects that of the impacting meteorites. (*b*) Crater numbers increase at a regular rate over all diameters. (*c*) Obliteration, erosion and burial start to deplete the smaller craters. (*d*) The surface becomes saturated with small craters. (*e*) The saturation crater diameter gradually increases.

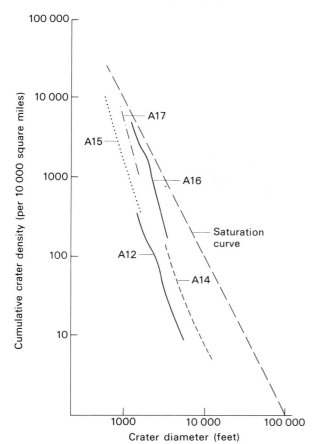

Figure 1.56. *Typical lunar curves.* As expected, the size distribution curve for the Apollo 17 *mare* lies between that for the young Apollo 15 *mare*, and the older highland plains as visited by Apollo 16. (After Neukum, G., König, B., Fechtig, H. & Storzer, D. (1975). *Proceedings of the Sixth Lunar Science Conference*, pp. 2597–620. New York: Pergamon.)

younger one) across the entire diameter range. So, if a crater size distribution curve can be bracketed by curves from two areas of known age then the undated area must be of intermediate age. The position of the crater size distribution curve therefore complements the saturation crater diameter as a relative age indicator. And this is what happens in practice. The curve for the Taurus–Littrow valley, for example, is found to lie between those for the younger *mare*, Palus Putredinis, and the somewhat older Cayley Plains (Figure 1.56). And this intermediate age has now in fact been confirmed by dating rocks from all three areas.

### Illumination effects

But this is not the end of the story. Not only are the results from different laboratories sometimes conflicting between themselves, they frequently do not match the radiometric ages of rocks recovered from the areas under study. The Apollo 17 site, for example, was once thought to be much younger than it turned out to be. So what can be done to improve the reliability of crater counting?

Well, firstly, there is the question of size. In other words, what exactly is meant by the diameter of a crater? Because crater numbers drop sharply with increasing diameter (there may be 100 times more craters larger than 300 yards than there are larger than 500 yards), a precise definition of crater diameter is clearly required, if only for the sake of consistency among results from different laboratories.

This may seem to be a trivial question, the trivial answer being simply the distance from rim to rim across the centre of the crater. But how is this rim to be defined? Well, it turns out that its position appears to change throughout the day. The profile of a lunar crater may be such that only at very low Sun angles does the highest point on the rim cast a shadow into the interior. At other times we might be underestimating the crater diameter (Figure 1.57). One way around this problem has been to make use of pairs of photographs which overlap and can therefore be viewed stereoscopically. The position of the true rim of the crater (in

Figure 1.57. *Crater rims – where are they?* The highest point of a crater does not always define the shadow in the crater. So the apparent diameter varies with illumination angle.
(*a*) Grazing angle (rim appears at *r*).
(*b*) Angle from rim to centre (rim appears at *r'*).
(*c*) Critical angle = interior slope angle *x*.

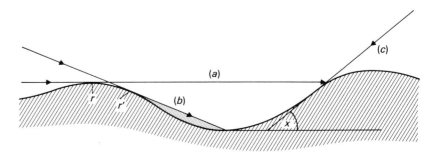

other words its highest point) then shows up clearly and unambiguously, even under a very high Sun angle.

But the angle of solar illumination not only affects the measurement of crater diameters, it can also change their total numbers. For the very oldest craters are so shallow that their interiors are only shadowed at all when the Sun is just above the horizon. At other times the crater may be completely unrecognisable. And there certainly seem to be many more craters on the Moon when the solar illumination angle is small than when the Sun is well up (Figure 1.58). What is obviously a subdued crater in the early morning, for example, may disappear when the Sun has risen only a few degrees. So some of the inconsistencies between crater counting laboratories may just be due to differences in solar illumination angle.

By counting all craters, irrespective of their preservation state, we are, however, throwing away some valuable age information. For it takes time for the rim of a crater to be worn down so much that it only casts a shadow at all when the sun is less than a degree or so above the horizon. And, the larger the crater, the longer this erosion will have taken. So the relative age of an area can be readily established by simply finding the diameter of the largest crater that is only just shadowed (or is shadowed halfway across the crater floor) under an illumination angle of, say, 1° (Figure 1.59b).

This procedure, like crater counting itself, is however not quite that simple. For, while there are large numbers of near terminator lunar photographs, only a few were exposed under exactly the same solar illumination angle. So the critical diameter has to be corrected mathematically, using a theoretical model for crater erosion. But, having made this correction, the differences in age between one part of the Moon and another are quite apparent. Much more so, in fact, than when one just counts the craters. In Mare Imbrium, for example, it is possible to recognise four chronologically distinct periods of volcanism, the youngest of which resulted in lava eruptions in the southwest of the *mare* (Figure 1.27). Here, the largest craters eroded down to an angle of 1° are only 200 yards across. Elsewhere in Mare Imbrium, however, it is possible to find 350 yard craters with 1° walls. So this method certainly seems to be very sensitive to what could be really quite small differences in absolute age (Figure 1.59).

*Buried craters*
Having described the techniques for using craters to establish age relationships, it is now appropriate to discuss what these ages really mean. How can they be correlated with absolute radiometric ages? Well, one of the problems here is that a particular area can effectively have more than one crater age. And this is because the lunar surface is only rarely completely rejuvenated as far as craters are concerned. An area which is virgin with respect to craters a mile or so in diameter, may still exhibit the partly buried remains of much larger craters formed during an earlier epoch.

In southern Oceanus Procellarum, for example, all of the smaller craters must have been formed since the formation of the *mare* itself. But the basaltic lava flows were of insufficient thickness there to bury completely the largest craters which already existed. As a result, a number of 'ghost craters' are visible which must be older than the lavas which partly buried them and possibly date back to the Imbrium event (Figure 1.50). So when it comes to crater counting, it is clearly necessary to exclude such ghost craters wherever possible. In the *maria* this is no great problem because the eruption of basalt lavas there was a well-defined event. But in the highlands, the light plains may well consist of debris which accumulated from the surrounding hills over periods of many millions of years. So the significance of ages based on crater counting would seem to be less easy to appreciate in the highlands than it is in the *maria*.

Figure 1.58. *The effects of illumination*. Two pictures of the same area under different illumination conditions are very different. The near terminator picture (*a*) seems more densely cratered than that taken under a high sun (*b*). (NASA, Apollo.)

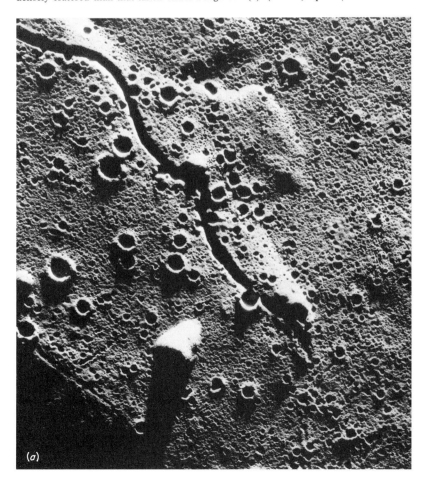

(*a*)

But the existence of partially buried craters can at least be put to some good use. For they enable us to establish just how deep some of the *maria* really are. From topographic studies of fresh craters it is possible to establish how their diameter–depth ratios vary with increasing diameter. So, on finding a 20 mile diameter crater which was very nearly submerged by lava, for example, it might be possible to say that the *mare* at that particular point is, say, 5000 feet thick. A number of *maria*, including Tranquillitatis and Nectaris, have been 'sounded' in this way and only the large circular *maria*, such as Imbrium, in fact, have not succumbed to this approach. Even here, however, it is most unlikely that the *mare* lava flows are much more than a few miles in thickness.

*Substrate effects*

The number of craters within a particular size range should theoretically increase until the craters are being destroyed just as rapidly as they are being formed. And therefore no two crater size distribution curves should ever cross one another. All such curves should gradually move up until they finally coincide with the theoretical saturation curve. But this is not what is always found in practice. In Mare Tranquillitatis, for

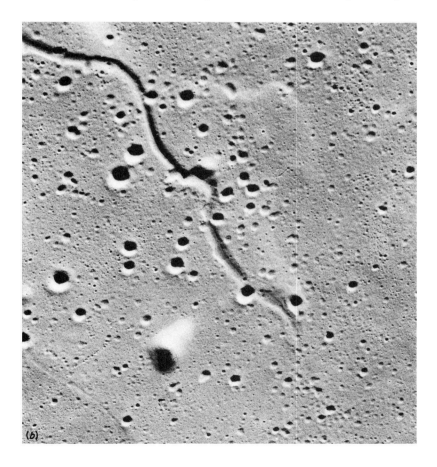

(b)

example, we find fewer 100–200 yard craters per square mile than in the clearly much younger Mare Imbrium. So it would appear that the numbers of small craters actually decrease, rather than increase, after the saturation level has been reached.

Why should this be the case? Well it is now believed that the survival of small craters may also be affected by seismic modification. Craters in a thick mature regolith may be more susceptible to collapse (when the regolith is vibrated by nearby impacts) than those which were formed in a

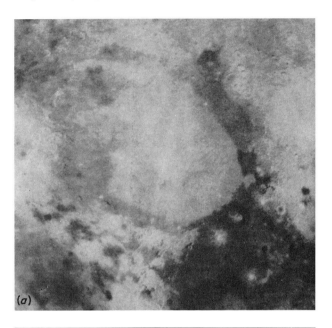

(a)

Figure 1.59. *Age/colour relationship*. (a) A colour difference map of the Mare Serenitatis area. Note in particular the dark unit to the southeast that includes the Apollo 17 landing site. (b) A map showing the relative ages of the lava units in Mare Serenitatis based on the maximum size of crater whose walls have been eroded down to such an extent that the rims just cast a shadow to the centre of the crater under a solar illumination angle of 1°.

A (youngest)    $d = 180$ yards
B               $d = 210$ yards
C               $d = 270$ yards
D               $d = 300$ yards
E (oldest)      $d = 380$ yards

(After Boyce, J. M. (1976). *Proceedings of the Seventh Lunar Science Conference*, pp. 2717–28. New York: Pergamon.)

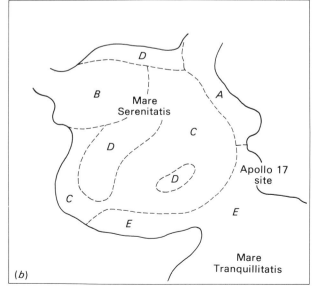

(b)

more coherent substrate. The thicker the regolith, then, the shorter the life expectancy of small craters. It may sound a little far fetched, but at least it fits the facts.

*Other complicating factors*
Before leaving the subject of crater counting, it is worth while to mention one or two complicating factors, if only to demonstrate just how difficult this technique can be.

Firstly there is the question of secondary craters, in other words, craters produced not by the meteorite itself but by its primary ejecta. These secondary craters must clearly be excluded, but how are they to be recognised in the first place? It is certainly true that secondary craters are generally elongated and are often associated with one another in chains and clusters. But there are doubtless many individual ones as well and these could quite easily pass unnoticed.

Secondly, some lunar craters may be volcanic. So an area with volcanic craters will appear older than a contemporary formation elsewhere. As it is difficult to recognise volcanic craters in the first place, this secondary source could well constitute a serious shortcoming of all crater counting methods.

Then there is the problem of measuring distances on the lunar surface. It must be possible to calculate the positions, altitudes and orientations of spacecraft very precisely in order to calculate accurate diameters from photographic measurements. The controlled metric photography of the later Apollos, combined as it was with improved tracking facilities and laser altimetry, went a long way towards alleviating some of the earlier uncertainties (see Chapter 5).

How useful has crater counting really been, then, and what more can still be done? Well, looking back, the earliest studies did little more than define the inherent limitations of the technique and sometimes led to predictions which proved misleading. But this situation is improving all the time. And this is indeed fortunate. For now that there is little prospect of more rock samples being returned from the Moon before the end of the century (at least by manned space missions) it is perhaps just as well that the various groups still working in this field are beginning to obtain consistent results.

What has been established beyond all doubt, however, is that there are indeed regions on the Moon which are younger (and older) than those sampled by Apollo. Western Oceanus Procellarum and Mare Imbrium, for example, contain some remarkably young lava flows. And some craters, such as Copernicus and Tycho, are much younger still. But appearances can be deceptive and just how young these lava flows and craters really are, in absolute terms, we cannot yet say. Nor can we even guess at the ages of certain highly cratered highland areas far removed from the rejuvenating effects of the Imbrium Basin. In order to do that it is necessary to obtain some absolute dates. And that meant collecting rocks with Apollo and dating them using naturally occurring radioisotopes.

# 2. The Moon landers

## 2.1 Surveyor

Before setting foot on the surface of the Moon, we had to send a series of unmanned spacecraft to see if a soft landing really was feasible. Retro-rocket and landing-gear performance had to be evaluated in the low gravity and high vacuum of interplanetary space and the descent engines had to be controlled successfully using an onboard radar system. The Apollo program could not be jeopardised simply because we did not fully appreciate the environmental constraints and engineering problems associated with making a soft landing on the Moon.

But these were not the only objectives of the American Surveyor program. Ranger had already photographed a few small areas of the Moon in great detail, providing ample proof that the Moon's surface

Figure 2.1. *Surveyor 1 view.* This picture shows the terrain in the immediate vicinity of Surveyor 1. (NASA, Surveyor.)

really is strong enough to support the weight of large boulders. But the idea that the *maria* might consist of unconsolidated dust was firmly entrenched and we were also, of course, very keen to see what the Moon's surface looks like at really close range. Just how fine grained is the soil, what does it consist of and what are its mechanical properties? These were just some of the questions that Surveyor set out to answer (Figure 2.4).

### The first landing

The Russians were, in fact, the first to succeed with a soft landing on the Moon, with Luna 9 pre-empting Surveyor by just a few months. But Surveyor employed more sophisticated concepts than the early Lunas and five out of the seven spacecraft reached their appointed targets without serious mishap. Only Surveyor 2 (which went into an uncontrollable spin during a midcourse manoeuvre) and Surveyor 4 (with which radio contact was lost just before touchdown) failed to complete their missions (Figure 1.18).

Surveyor 1 landed close to the crater Flamsteed in southwestern Oceanus Procellarum, where it was later to be photographed by Lunar

Figure 2.2. *Boulder close-up*. Surveyor 1 landed close to a 3 foot long boulder. From its appearance, it looked like basalt, a conclusion that was confirmed by Apollo. (NASA, Surveyor.)

Orbiter 3. Thousands of television pictures were transmitted back to Earth, showing us that the *mare* surface there is essentially flat, apart from in the immediate vicinity of small craters (Figure 2.1), and that it is composed predominantly of firm, but fine-grained, material. The occasional larger rocks and boulders looked as if they had a volcanic origin (Figure 2.2). In most directions the horizon was only about a mile distant, but mountain tops as far away as 12 miles were also visible, despite the Moon's strong curvature (Figure 2.3).

Surveyor 1 (Figure 2.4) continued to function for some time after the Sun had set, returning good pictures of the solar corona (the outer part of the Sun's atmosphere which can normally only be seen during a total solar eclipse) and of the Moon's surface illuminated by earthshine. After being switched off for the rest of the long lunar night, the spacecraft was successfully reawakened and, despite damage to its solar panels, it continued to function perfectly well for another eight months.

*Regolith mechanics and chemistry*
After bouncing a couple of times Surveyor 3 also landed in Oceanus Procellarum and was later visited by the crew of Apollo 12 (Figure 2.14*b*). A trenching device on a telescopic arm was used to test the bearing strength of the topmost layer, to expose the subsurface and to pick up soil fragments and deposit them on a landing footpad for closer inspection. Although what appeared to be coherent rocks sometimes broke up quite easily, like clods of wet sand, the bearing strength of the regolith below a depth of an inch or so, was clearly quite sufficient to support the weight of a much larger spacecraft. So all now seemed to bode well for Apollo.

Figure 2.3. *Distant mountains.* Surveyor 1 landed 12 miles inside a large, partially buried, crater, the rim of which can be seen in this composite picture. (NASA, Surveyor.)

Surveyor 5 landed in Mare Tranquillitatis, an area already closely studied by Ranger 8 and later to become the target for Apollo 11. But instead of using an excavator, Surveyor 5 lowered a box-like instrument on to the top of the regolith to perform the first ever local chemical analysis of another planet (Figure 2.5). A similar instrument was later taken by Surveyor 6, which landed close to a 100 foot high wrinkle ridge in Sinus Medii. This device incorporated a small quantity of curium-242, a transuranic isotope that acts as a powerful radioactive source of alpha particles, or energetic helium nuclei. When directed towards the lunar surface, these alpha particles may suffer one of two fates. Some will just be scattered and eventually slow down and pick up electrons to become normal helium atoms. But a few will initiate nuclear reactions, transmuting one element into another and releasing protons (and other particles) as byproducts.

Figure 2.4. *Surveyor spacecraft.* Mock-up of Surveyor. (NASA.)

Counting the energetic backscattered alpha particles was, however, the main objective of the experiment. So what principle is involved here? Well, if the target nuclei in the soil were of infinite mass then the reflected alpha particles would retain all of their original energy. But real atoms are not so heavy and will therefore recoil when struck. And, the smaller the atom, the more energy they will remove from the impacting alpha particle. So, by analysing the energies of the alpha particles arriving back in the instrument, the distribution of atomic masses within the lunar regolith could be readily determined. A very simple principle, but a most effective one. Complementary chemical information was supplied by analysing the protons as well. Because these proton energies depend on the nature of the particular target isotope, not just on its atomic mass.

What then did the alpha particle experiments flown on Surveyors 5 and 6 tell us about *mare* chemistry? Well the local soils were clearly rich in iron, magnesium, calcium, aluminium and titanium, and their silicon and oxygen contents were quite consistent with the popular idea that the lunar *maria* consist of basaltic lavas (Table 2.1). Potassium was scarcer than in continental Earth rocks (such as granite) and there was no trace of carbon, that all important element of life.

### A better view

The television camera systems flown on Surveyor became gradually more sophisticated as the program progressed, colour and polarising filters greatly increasing the information content of the photographs, and the varying illumination conditions being another critical factor.

But, until Surveyor 6, our view was essentially two dimensional, because focus-ranging was impractical at distances beyond a few yards.

Figure 2.5. *Chemical analyses.* The lunar regolith was analysed by Surveyors 5, 6 and 7 using this box-like instrument, an alpha particle back-scattering experiment. (NASA, Surveyor.)

Table 2.1. *Results of the chemical analyses of lunar soil performed by the alpha scattering devices on Surveyors 5, 6 and 7. Note how the highland soil is richer in calcium and aluminium and poorer in iron and titanium than the mare soils. The relatively higher titanium content of the Mare Tranquillitatis soil compared to that from Sinus Medii is also apparent*

| Spacecraft | Surveyor 5 | Surveyor 6 | Surveyor 7 |
|---|---|---|---|
| Landing Site | Mare Tranquillitatis | Sinus Medii | Tycho |
| Element | per cent by atom | | |
| Oxygen (O) | 61.1 ± 1.0 | 59.3 ± 1.6 | 61.8 ± 1.0 |
| Sodium (Na) | 0.5 ± 0.2 | 0.6 ± 0.2 | 0.5 ± 0.2 |
| Magnesium (Mg) | 2.8 ± 1.5 | 3.7 ± 1.6 | 3.6 ± 1.6 |
| Aluminium (Al) | 6.4 ± 0.4 | 6.5 ± 0.4 | 9.2 ± 0.4 |
| Silicon (Si) | 17.1 ± 1.2 | 18.5 ± 1.4 | 16.3 ± 1.2 |
| Calcium (Ca) | 5.5 ± 0.7 | 5.2 ± 0.9 | 6.9 ± 0.6 |
| Titanium (Ti) | 2.0 ± 0.5 | 1.0 ± 0.8 | 0.0 ± 0.4 |
| Iron (Fe) | 3.8 ± 0.4 | 3.9 ± 0.6 | 1.6 ± 0.4 |

Figure 2.6. *Highland panorama.* In this Surveyor 7 composite, the rocky and undulating terrain near Tycho is clearly apparent. (NASA, Surveyor.)

So, after completing its high resolution panoramic survey, Surveyor 6 became the first spacecraft to take off from another planet. But it was not up for long, touching down barely 8 feet away. This journey, albeit short, provided an opportunity for studying the erosional effects of the exhaust gases and to get a good look at the imprints made by the three landing pads. But, more important still, it meant that a new photographic sequence could now be obtained to provide stereoscopic coverage of much of the surrounding area. The same objective was achieved on Surveyor 7 by the simple expedient of incorporating extra mirrors.

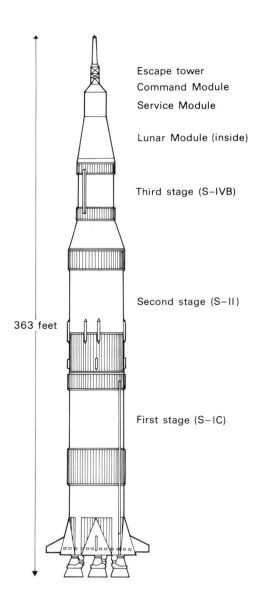

Escape tower
Command Module
Service Module

Lunar Module (inside)

Third stage (S–IVB)

Second stage (S–II)

363 feet

First stage (S–IC)

Figure 2.7. *Saturn 5*. This was the rocket that sent Apollo to the Moon, only the uppermost stage (SIVB) reaching the lunar surface, crashing to provide an artificial seismic source.

Figure 2.8. *CSM*. The command and service module, as viewed by the Apollo 15 LM crew (*a*), and in diagrammatic form (*b*), showing the main features. (NASA, Apollo.)

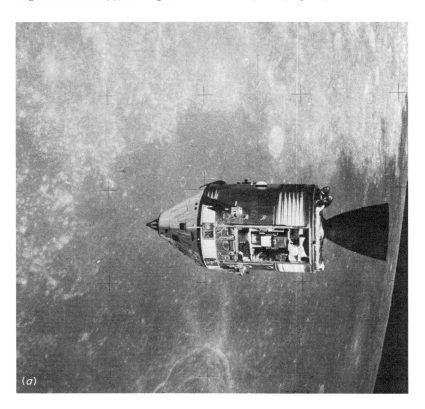

(*a*)

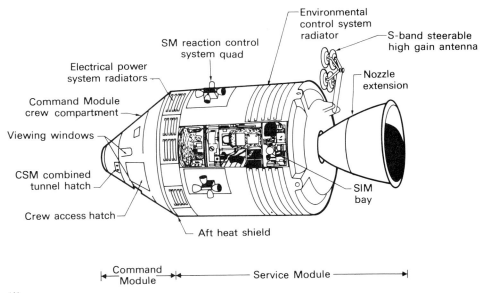

(*b*)

*The highland mission*

The success of Surveyor 6 meant that we now knew of four *mare* areas that were eminently suitable for the first manned landing. All four sites were flat enough and sufficiently free of boulders and craters for Apollo to be able to land safely. So it made sense now for the last of the series, Surveyor 7, to be a strictly scientific mission and to go to the highlands for the first time. Even if this last flight failed, the program as a whole would still have been an unqualified success.

In the event, Surveyor 7 was not a failure. But it came very close to being one, narrowly missing a field of boulders where it could quite easily have toppled over and crashed. It performed the first chemical analysis of the highlands near the crater Tycho, where the terrain was found to be very different from the three *maria* visited so far. The landscape was undulating and rocky (Figure 2.6) and the soil was chemically poorer in iron and titanium and correspondingly richer in calcium and aluminium (Table 2.1). All in all it looked as if the lunar highlands must consist predominantly of calcium–aluminium silicates rather than of the iron–magnesium ones that are so characteristic of the *maria*. This conclusion would certainly explain their lighter colour.

The last three Surveyors laid the foundations for experimental science on the Moon, an activity soon to be followed up by Apollo. As a final example of this, one of Surveyor 7's last tasks was to attempt to detect a laser beam directed at it from Earth in preparation for Apollo laser ranging. The success of this experiment meant that in the unlikely event

Figure 2.9. *SIM bay.* A sketch of the Scientific Instrument Module (SIM) bay, a part of the Service Module on the last three Apollos, showing the experiments housed there. (NASA.)

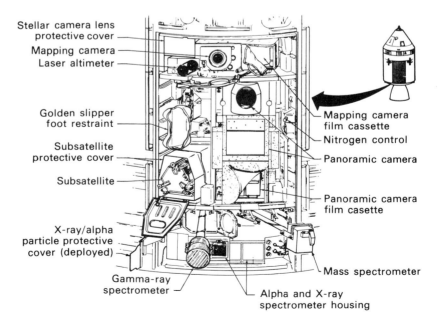

Stellar camera lens protective cover
Mapping camera
Laser altimeter
Golden slipper foot restraint
Subsatellite protective cover
Subsatellite
X-ray/alpha particle protective cover (deployed)
Gamma-ray spectrometer
Mapping camera film cassette
Nitrogen control
Panoramic camera
Panoramic camera film casette
Mass spectrometer
Alpha and X-ray spectrometer housing

of astronauts ever becoming stranded on the Moon without a radio, there would at least be some primitive form of one way communication from Earth available.

## 2.2  The early Apollos

By the time that Surveyor had shown that the Moon's surface was really not so inhospitable after all, the American manned space effort was already well advanced. The feasibility of sending men into space had been well proved by the Mercury program, and Gemini was now demonstrating that man could stay in space for long periods of time, manoeuvre his spacecraft precisely and undertake operations out in space clad only in a spacesuit. But Mercury and Gemini were of course just preliminaries. The most important program as far as the Moon was concerned was Apollo, America's plan to land a man on the Moon and return

Figure 2.10. *The Lunar Module*. This diagrammatic representation of the LM shows the main features of the craft in which lunar landings were made. The shaded descent stage was left behind on the Moon after the mission. Only the ascent stage redocked with the Command Module. (NASA.)

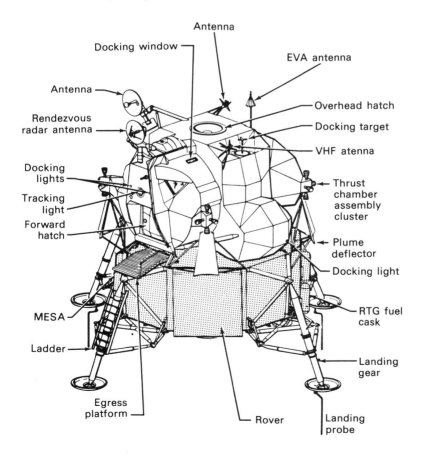

him safely to Earth before 1970, thus fulfilling the national goal set by President Kennedy way back in May 1961.

Apollo was a three man spacecraft, launched by the largest rocket ever built, the 363 foot high, three stage Saturn 5, weighing more than 3000 tons (Figure 2.7). The spacecraft itself also consisted of three parts, the Command Module (Figure 2.8), for travelling to and from the Moon, the Service Module, which contained the rocket engines and, in the later Apollos, the Scientific Instrument Module (Figure 2.9), and the Lunar Module (LM), itself in two parts, in which two of the three astronauts were destined to make the actual lunar landing (Figure 2.10).

One reason for having a separate lunar module was that a single spacecraft cannot carry enough fuel to take it to the Moon and back again. There had to be an orbital rendezvous somewhere and, on balance, a lunar one seemed preferable. Similarly, there was no point in having a reentry vehicle which consisted mostly of empty fuel tanks. So the

Figure 2.11. *The way to the Moon.* This diagram shows the 14 stages involved in making a lunar landing and returning to Earth. Apollo was on its way 2½ hours after liftoff and Earth Orbit Insertion. It reached the Moon 3 days later and the landing phase began after a further 24 hours. The return journey consisted of liftoff, redocking, trans-Earth injection, reentry and, finally, splashdown.

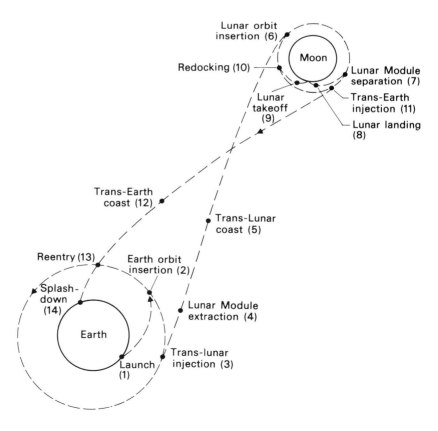

service module was designed to be jettisoned just before reentry, leaving the command module to bring the astronauts safely back to Earth (Figure 2.11).

*The test flights*

The earliest Apollos were designed to test the various components of the Apollo system in progressive stages. The launch vehicle had to have at least one successful test flight and the Apollo capsule had to be checked out fully on the ground. It was during one of these ground trials, in fact, that three astronauts lost their lives in a fire, an event which led to major improvements being made to the design of the command module. There was some delay, then, before the first manned orbital test flight of the command module could be made, in the form of Apollo 7 (Table 2.2).

Table 2.2. *The 11 manned Apollo missions, their astronauts and accomplishments. Astronauts marked with an asterisk made a lunar landing. Note the greater return payloads and traverse distances of the last 3 missions*

| Apollo mission | Astronauts CDR LMP CMP | Year | Month | Accomplishments | Sample wt (kg) | EVA time (hr) | EVA distance (km) |
|---|---|---|---|---|---|---|---|
| 7 | Schirra Eisele Cunningham | 1968 | Oct | First test of CM in Earth orbit | — | — | — |
| 8 | Borman Lovell Anders | 1968 | Dec | First trip around the Moon | — | — | — |
| 9 | Mcdivitt Scott Schweickart | 1969 | Mar | First test of LM in earth orbit | — | — | — |
| 10 | Stafford Young Cernan | 1969 | May | First test of LM in lunar orbit | — | — | — |
| 11 | Armstrong* Aldrin* Collins | 1969 | Jul | First lunar landing (Mare Tranquillitatis) | 21.7 | 2.2 | 0.5 |
| 12 | Conrad* Bean* Gordon | 1969 | Nov | Surveyor 3 reunion (Oceanus Procellarum) | 34.4 | 7.6 | 1.3 |
| 13 | Lovell Swigert Haise | 1970 | Apr | Mission aborted due to explosion in SM | — | — | — |
| 14 | Shepard* Mitchell* Roosa | 1971 | Feb | MET used at Fra Mauro (Cone Crater) | 42.9 | 9.2 | 3.4 |
| 15 | Scott* Irwin* Worden | 1971 | Jul | First test of Rover (Hadley–Apennine) | 76.8 | 18.3 | 27.9 |
| 16 | Young* Duke* Mattingley | 1972 | Apr | First highland trip (Cayley–Descartes) | 94.7 | 20.1 | 27.0 |
| 17 | Cernan* Schmitt* Evans | 1972 | Dec | First Scientist (Taurus–Littrow) | 110.5 | 22.0 | 30.0 |

Apollo 8 was the first manned spacecraft to go to the Moon, but it could not do very much when it got there because the lunar module was still far from ready at that time. The crew could, however, describe the experience of being behind the Moon and they brought home some really excellent pictures from orbit of its monotonously grey surface (Figure 1.34*b*).

Figure 2.12. *Tranquillity Base.* (*a*) The Apollo 11 landing site as the astronauts approached it from the west. (NASA, Apollo.) (*b*) The terrain in the immediate vicinity of the Lunar Module. Note the rocky 'ray' from West crater. (NASA, Apollo.) (*c*) A map of the site, showing the Lunar Module and sampling areas.

The next critical step was to test the lunar module itself, first in Earth orbit (with Apollo 9) and then in the vicinity of the Moon (with Apollo 10). The Apollo 10 LM separated from its command module and descended to within ten miles of the lunar surface, whereupon the astronauts had the strange sensation of being 'below' the distant mountain peaks. There really is very little by which to judge distances on the Moon.

After a successful redocking and return to Earth, there was nothing for it now but to try a landing, and this was of course accomplished in Mare Tranquillitatis on 21 July 1969 by the crew of Apollo 11, Neil Armstrong, Buzz Aldrin and Michael Collins, an event which heralded an exciting new era in lunar science.

### Tranquillity Base

The first landing site for Apollo was not, of course, chosen for purely scientific reasons, although we were interested to see if the *maria* did indeed consist of basalt lava flows. It had to be close to the equator, to make getting home easier, and as flat as possible, just in case difficulties were experienced during landing and a new site had to be found quickly (Figures 1.18 and 2.12). In the event, the LM did overshoot slightly, but the astronauts managed to find an alternative landing site without too much difficulty. Based on the experience gained here, the descent profiles on later missions were made much steeper.

Once equipped for going out on to the lunar surface (Figure 2.13), the crew of Apollo 11 had some well-defined scientific tasks to perform, the first of which was to collect a contingency dust sample, just in case the mission had to be aborted prematurely. It certainly would have been tragic to land on the Moon successfully and then fail to bring back even the smallest quantity of lunar dust.

Several experiments then had to be deployed, including a seismometer (Figure 5.35), a laser reflector (Figure 5.13) and a sheet of aluminium foil

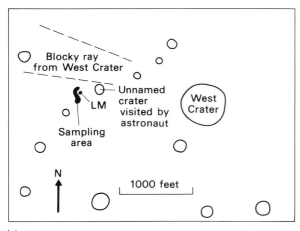

(c)

for collecting atoms from the Sun (Figure 4.8). The details of these and other Apollo 11 experiments (and of the more comprehensive packages carried by subsequent Apollos) will be described in later chapters. The astronauts then continued their collection of rocks and soils and packed them up in preparation for their return to Earth (Figure 2.40b).

The time available for extravehicular activity (EVA) on Apollo 11 was, of course, necessarily rather restricted, this being the very first landing mission. So the required samples were collected as quickly as possible, with quantity being considered more important than quality (Table 2.2). On later Apollos rock collection and documentation was to become a much more sophisticated process.

### The Surveyor 3 reunion

Having proved that man really could land on the Moon and then work efficiently in one-sixth of Earth gravity (despite the fact that lunar dust gets everywhere), it was the objective of Apollo 12 to repeat the performance in the Moon's western hemisphere. The target selected for this second landing mission was just a few hundred yards from Surveyor 3, which had come to rest in southern Oceanus Procellarum some $2\frac{1}{2}$ years earlier (Figure 2.14b). The idea here was to bring certain pieces of hardware from the old spacecraft back to Earth to see how they had suffered from their long exposure to the harsh interplanetary environment.

Figure 2.13. *Lunar astronaut*. This drawing shows some of the equipment worn, or carried, by astronauts during their periods of Extra-Vehicular Activity (EVA). (NASA.)

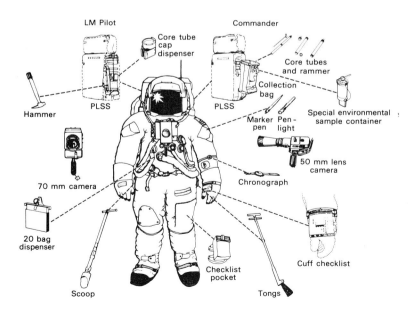

But the object of Apollo 12 was not just to recover a few parts from a long-dead spacecraft. That was just an added bonus. Its primary objective was to sample what was clearly a younger, and distinctly redder, *mare* unit (Figure 5.18). So how much younger would the rocks be in absolute terms and of what did the light-coloured rays from Copernicus (one of which crosses the landing site) consist? There was a strong possibility here that we might be able to sample Copernicus ejecta at the proposed Apollo 12 landing site (Figure 1.47*a*).

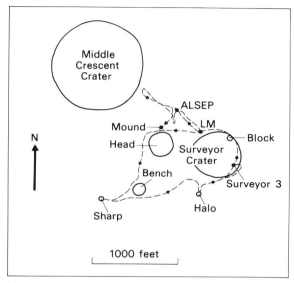

(*a*)

Figure 2.14. *Apollo 12 traverses.* (*a*) This map shows the paths taken by the Apollo 12 crew. (*b*) During their EVA, the Apollo 12 crew visited Surveyor 3. Note the bounce prints and trenches. (NASA, Apollo.)

In the event, Apollo 12 succeeded in making a rather more precise landing than its predecessor had done. Not only did this mean that the Surveyor 3 reunion meant a walk of only a few hundred yards (a short enough distance now that the EVA time had been extended somewhat, Figures 1.18, 2.14) it also paved the way for missions to the highlands, where a landing might require considerably more skill on the part of the Lunar Module pilot in order to avoid crashing into mountain peaks.

So what did the Apollo 12 crew manage to accomplish, apart from collecting a few bits from Surveyor? Well, as far as the rocks in Oceanus Procellarum were concerned, they found a much higher proportion of crystalline basalts than their predecessors had done at Tranquillity Base. At least half of the large Apollo 11 rocks had been lumps of shock-welded soil, otherwise known as regolith breccias. There were very few Apollo 12 breccias, although one of them (called rock 13) did become the best known of all lunar rocks, because of its very high content of radioactive elements.

Figure 2.15. *Fra Mauro*. Photograph of the Fra Mauro region taken by Apollo 12, showing the Apollo 14 landing site. Note the ridges and furrows, radial to the Imbrium Basin. (NASA, Apollo.)

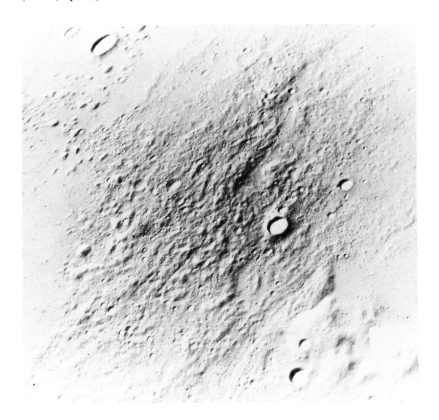

There was also more variety among the soil samples. In particular, there were some patches of light-coloured soil that could well have been that sought-after Copernican ray material. So here was an opportunity to date this most prominent of lunar craters.

All in all the sampling of the Moon by Apollo 12 was rather more comprehensive than had been possible on Apollo 11. But the time had now come to extend the sampling capabilities of Apollo much further by making a trip to a non-*mare* area. And the spot chosen for Apollo 13 was on the Imbrium ejecta blanket, not far from the crater Fra Mauro (Figure 1.18 and 2.15).

*Failure then success*

Some would say that Apollo 13 was ill fated from the start. Be that as it may, an explosion in the Service Module on the way to the Moon put paid to any possibility of a landing on that occasion. And it said a great deal for the flexibility of the Apollo system, in fact, that the astronauts managed to get back to Earth at all.

The Fra Mauro formation was duly taken over as the target for Apollo 14, its precise objective being a small crater on top of one of those ridges which are radially orientated with respect to the Imbrium Basin. The idea here was to make use of the fact that this crater, known as Cone, must have excavated materials from well beneath the regolith, materials that must themselves consist of ejecta from the Imbrium Basin. A suitable crater is clearly much more effective than an electric drill when it comes to sampling subregolith bedrock.

As it turned out it was easier said than done to make the trip on foot from the Lunar Module down in the valley up a 15° slope to the 1000 foot

Figure 2.16. *Apollo 14 traverse map*. This photograph shows the paths taken by the crew of Apollo 14 during their EVA. Note how close they went to the rim of Cone Crater. The letters indicate sampling stops. (NASA, Apollo.)

diameter crater on top of the hill (Figure 2.16). Despite having the use of a handcart, or Modular Equipment Transporter (MET), the astronauts only just managed to reach the boulder-strewn rim of the crater and, at the time, did not realise just how close they had come (Figure 2.17). The rocks they sampled there were in fact only some 60 feet from the rim.

These 'white rocks' were clearly very different from anything brought back from the Moon so far (Figure 2.18). No one had known quite what to expect from Fra Mauro but we had all become very used to the idea of crystalline volcanic rocks on the Moon and practically all of the large samples returned by Apollo 14 were complex breccias, rock mixtures welded together at high temperatures. Indeed the complex nature of these impact breccias was to pose a real challenge to petrologists for several years to come (Figure 3.25).

In some ways Apollo 14 can be considered as the intermediate mission. Its scientific objectives were clearly more ambitious than those of Apollo 11 and 12, but the attainment of these objectives was hampered by serious restrictions on astronaut mobility, despite the use of the MET. This was about to change however. On each of the final 3 Apollo missions a Lunar Roving Vehicle (LRV or Rover) was taken to the Moon, the return payload was increased and more EVA time was available for site exploration (Table 2.2). As a result, complex sites could be investigated thoroughly and comprehensive suites of samples returned to Earth. It could be argued, in fact, that more science was accomplished on each of the last three missions than had been possible on all of the first three combined.

Figure 2.17. *The Modular Equipment Transporter.* The walk up to Cone Crater was made possible by the use of a handcart, or MET. (NASA, Apollo.)

## 2.3 Hadley–Apennine

Having designed and built a Lunar Roving Vehicle in order to improve astronaut mobility and increase the scientific return of Apollo, the question now was: where to go with Apollo 15? It was clearly pointless to use it to go where we could just as well have landed the Lunar Module in the first place, so from now on there would have to be multiple objectives. Having travelled all that way, the Apollo 15 crew would have to achieve much more than their predecessors had done.

*The aims of Apollo 15*
The target finally chosen for Apollo 15 was in Palus Putredinis (the Marsh of Decay) to the southeast of Mare Imbrium, at a point just inside the third (i.e. Apennine) ring of the Imbrium Basin (Figure 1.18). By landing here the astronauts could sample highlands and *mare* on a single mission. But that was not all. For winding across the *mare* floor and along the highland–*mare* contact was Hadley Rille, in which boulders and

Figure 2.18. *The White Rocks*. The rocks in the vicinity of Cone Crater were complex breccias containing large pale clasts. This unfilleted boulder was called Contact Rock. (NASA, Apollo.)

stratification had been observed from lunar orbit (Figure 2.19). So here at last was an opportunity to investigate one of these enigmatic sinuous rilles directly. Were they created by running water or carved out by molten lava? Apollo 15 was going to answer this crucial question once and for all.

Our principal aim on Apollo 15, though, was to discover what rock types occur along the Apennine Front. Because there at the foot of Hadley Delta, a mountain rising more than 10 000 feet above the *mare* plain (taller than the highest peak in the terrestrial Pyrénées), crustal material brought up from at least that depth by the Imbrium event should have been exposed. So on Apollo 15 we would be using a basin rather than just a minor crater to sample the lunar subsurface.

Figure 2.19. *Hadley–Apennine.* Palus Putredinis and the lunar Apennines in the vicinity of Hadley Rille. Note the secondary craters from Aristillus and Autolycus. Note also that the rille itself originates in an elongate depression. The area enclosed in a box is that shown in Figure 2.20. (NASA, Apollo.)

*The first EVA*

In the event, Apollo 15 accomplished practically all of its planned objectives, bringing back a larger and more comprehensive selection of rocks and soil than any of its predecessors. Furthermore, the EVA time on Apollo 15, distributed as it was over three sampling traverses, was equivalent to the total that was available on the first three missions put together (Table 2.2).

After setting up the Apollo Lunar Science Experimental Package (ALSEP), the first EVA period was taken up with a trip across the *mare* to an ancient crater on the lower slopes of Hadley Delta known as St George (Figure 2.20). We wanted to visit St George in order to find and sample boulders of excavated bedrock on its rim. We had to find large rocks because small ones lying elsewhere on the regolith could have come from anywhere, particularly at a site as complex as Hadley–Apennine.

The journey southwards from the LM was no problem at all for the first Rover, a journey monitored from Earth by means of a remote-controlled television camera (Figure 2.21). On the way, the astronauts stopped to collect samples at Elbow, a small crater close to the lip of Hadley Rille. But St George Crater itself was somewhat disappointing. The astronauts did indeed find one large boulder there, but this is now thought to have originated from a *mare* area rather than from the underlying bedrock.

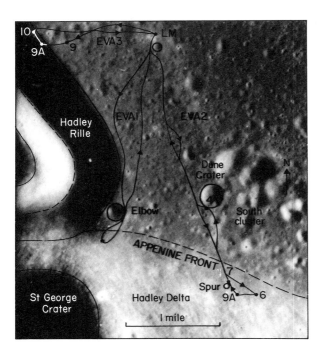

Figure 2.20. *The Apollo 15 landing site.* The 3 EVA traverses made by the Apollo 15 crew are shown in this photograph of the immediate vicinity of the LM. (NASA, Apollo.)

Figure 2.21. *The Lunar Roving Vehicle (LRV)*. (*a*) Saluting the flag with the LM, LRV and Hadley Delta in the background. (NASA, Apollo.) (*b*) Schematic representation of the LRV showing its principal components. Note the remote controlled TV camera mounted at the front. (NASA.)

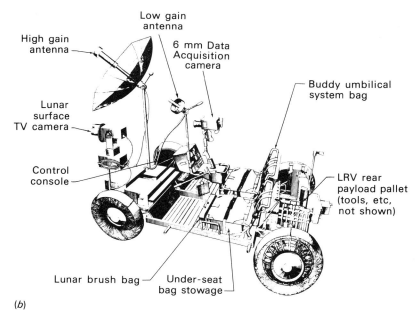

(*b*)

*South Cluster*

The second EVA traverse was more ambitious than the first, again taking the astronauts to the Apennine Front, but this time to a point farther east by way of a group of secondary craters known as South Cluster (Figure 2.20). It could well be that the secondary projectiles responsible for this group originated from the Copernican age craters to the north, Aristillus and Autolycus. Much of the landing site, in fact, is believed to be mantled by ray materials from these two young craters.

Our main reason for visiting South Cluster, however, was to investigate the *mare* stratigraphy in Palus Putredinis. We wanted to know what basalt types exist beneath the regolith at various depths. For by studying chemical variations within a deep sequence of lava flows we might be able to learn something about the chemical evolution of the lunar interior from which their parent magmas were derived.

The basalt rocks collected in the vicinity of South Cluster did indeed make up a varied suite, some boulders containing large holes several inches in diameter (Figure 2.22). Quite what was responsible for this

Figure 2.22. *Dune Crater boulder*. This boulder on the rim of one of the craters comprising South Cluster is unusual because of the large size of vesicles in it. (NASA, Apollo.)

vesiculation is still somewhat of a mystery (Figure 2.23). Other basalts were almost glassy in texture, indicating that their parent lavas must have cooled very quickly indeed on eruption.

### Genesis rock

When the astronauts finally reached the Apennine Front for the second time they were amply rewarded. Because one of the very first rocks examined there was almost pure white in colour, and so was immediately hailed as Genesis Rock, the rock the mission planners had very much hoped to find (Figure 2.41). For it had been predicted that the Apennines should consist of ancient highlands. And the highland crust was thought to consist of a rock type known as anorthosite, pale fragments of which had been turning up in soils ever since the first landing mission.

But, exciting as Genesis Rock was, this particular sample was by no means typical of the rocks of the Apennine Front. Indeed, Genesis Rock itself was just a clast in a breccia of anorthositic gabbro. Another anorthositic gabbro collected from this area came to be known as the Black

Figure 2.23. *Vesicular basalt.* This Apollo 15 basalt must have crystallised from a lava that frothed on eruption. (NASA, JSC–LRL.)

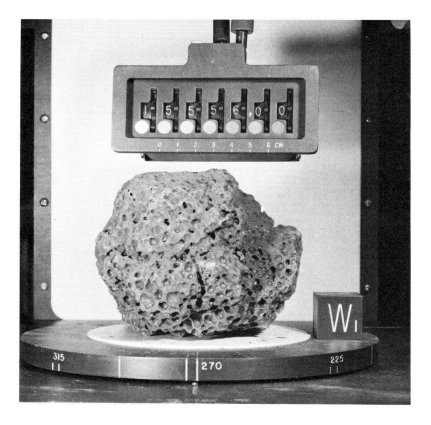

and White Rock, for obvious reasons. Most of the local rocks, however, were breccias welded together with brown glass, breccias not too dissimilar from those returned by Apollo 14 and Apollo 12. So hopes for learning much about the lunar highlands on this occasion were dashed.

*Green glass beads*

The Apennine Front did however have another surprise in store, for on the edge of a small crater called Spur, the Apollo 15 crew discovered clods of green soil! Far from being the first evidence for plant life on the Moon, these clods were, in fact, composed almost entirely of tiny glass spheres, more or less broken and devitrified and presenting a truly remarkable sight under the microscope at low magnification (Figure 2.24). A few of these spheres were as much as half an inch in diameter and they were not just confined to the immediate vicinity of Spur Crater. Indeed, glass beads of various shapes and colours are ubiquitous components of all lunar soils. It was just the vivid colour of those at Spur, combined with their high concentration there, which created so much excitement during the mission. The most popular theory for the origin of these particular spherules is that they were once droplets in a lava fountain, which froze in midflight, accumulated as a pyroclastic deposit, and were later exposed by the minor Spur Crater impact. Their ages are certainly consistent with that of the *mare* volcanism in the area.

Figure 2.24. *Green glass.* Clods of green soil, collected at Spur Crater by the Apollo 15 crew, were composed entirely of bottle-green glass droplets. The largest beads here are about 0.1 mm in diameter. (NASA, JSC–LRL.)

*Hadley Rille*

The third EVA period took the astronauts westwards to visit Hadley Rille for the second time. It had also been intended to visit North Complex, some low hills to the north of the LM, but this had to be abandoned because time was short (Figure 2.20).

The floor of the rille, which has a V-shaped cross section along most of its length, had clearly been filled in by debris (or talus) eroded from its slopes by minor meteorite impacts. Some of this debris was very massive. Boulders as large as 100 feet across could be seen below Elbow Crater.

This active mass wastage meant that the soil layer on the rim of the rille, instead of being several metres thick as it was elsewhere, was essentially non-existent. Some of the basalt boulders jutting out of the rille wall just below the rim, in fact, were the nearest we ever came to sampling lunar bedrock during the entire Apollo program (Figure 2.25). Elsewhere on the Moon bedrock is usually rendered totally inaccessible by that thick covering of fine dust known as regolith.

So what caused the rille to be formed in the first place? Well the area was certainly flooded to a considerable depth by thick basalt lava flows. Indeed, in the far wall of the rille, nearly a mile away, 60 foot thick

Figure 2.25. *Bedrock*. These boulders of *mare* basalt outcropping near the lip of Hadley Rille were essentially undisturbed blocks of bedrock. (NASA, Apollo.)

layered outcrops of *mare* basalt were clearly visible (Figure 2.26). And the fact that none of the basalts returned from the vicinity of the rille contained any traces of water whatsoever rules out the possibility of there ever having been a river there in the past. So the only question remaining now is how, and why, did flowing lava create such a channel?

The most popular idea is that this part of Palus Putredinis was once flooded to a high level with molten basalt, which managed to break through the so-called Fresnel Ridge to the north (Figure 2.19). The lava then drained out into Mare Imbrium underneath a chilled surface crust. This drainage would then have produced an underground channel, known on Earth as a lava tube. Subsequent meteorite bombardment would have caused the *mare* surface to subside and the lava tube to collapse to form a sinuous rille.

The collapsed lava tube hypothesis is certainly supported by the fact that the rille is deepest where it is widest and there is some evidence for a high lava mark along the foot of Mt Hadley. The North Complex Hills were also apparently once covered by *mare* lavas. Closer to the ridge it may be that Hadley Rille was never roofed in the first place.

Figure 2.26. *Basalt lava flows.* This telephoto picture of the far wall of Hadley Rille shows layered and vertically jointed outcrops of *mare* basalt, where the rille wall is so steep that a regolith cannot be supported. (NASA, Apollo.)

## The future for Apollo

The considerable achievements of Apollo 15 showed us for the first time just how important it is to be able to send men to the Moon in order to perform lunar science. Unmanned rovers and return capsules can achieve a great deal within their limitations, but, when it comes to making decisions in the light of experience, there really is no substitute for having a man (or woman) on the spot. The wealth of experimental data returned by this, the first of the more sophisticated Apollos, showed just how much can be achieved on a single mission.

But manned lunar exploration is hugely expensive. And with Apollo 15 we were only just beginning to scratch the surface of the Moon, scientifically speaking. Unfortunately financial cutbacks now meant that there could only be two more Apollos. So where should the last two go in order to maximise the scientific return from the program?

Figure 2.27. *The Apollo 16 site* (*a*) The Apollo 16 mission was to the Southern Highlands. The area in part (*b*) is enclosed in a box. (NASA, Apollo.)

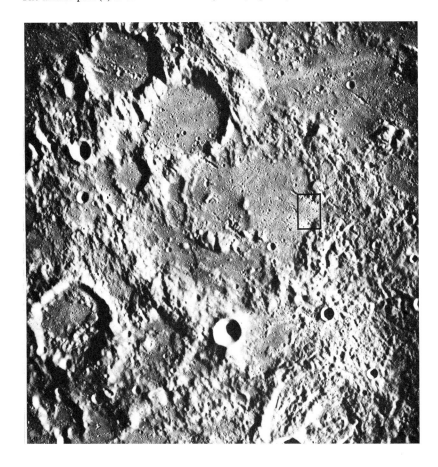

Well the site for Apollo 16 had already been chosen by this time, it had been selected on the basis of Apollo 12 pictures and studied in some detail by Apollo 14. There really was no question about it, we just had to find some typical lunar highlands.

## 2.4 Cayley–Descartes

The Cayley–Descartes area was, in fact, only one of many candidate sites shortlisted for Apollos 16 and 17. Among the others (and one which the volcanologists had always favoured) was the Marius Hills in western Oceanus Procellarum, thought to be one of the youngest volcanic provinces on the Moon (Figure 1.21). But we had already sampled *maria* on 3 occasions so we had to find at least one typically highland area. The most ambitious alternative was the crater Tsiolkovsky (Figure 1.22), one of the very few farside craters to have been flooded by dark-coloured lavas. The communications problems associated with a farside landing, however, although clearly not insuperable, did seem unnecessarily risky in view of the fact that there were still exciting places to visit on the near side.

(b) The 3 EVA traverses are shown on this photograph, together with the main features and numbered sampling stops. (NASA, Apollo.)

There also remained considerable support for a mission to the interior of a large young impact crater. Not only would this have enabled us to sample the deep highland crust, it would also have shed more light on the meteorite impact process itself. But one of our prime considerations when selecting a possible landing site for Apollo was that it should provide several sampling objectives and a crater such as Tycho (Figure 1.52) did not seem to offer much in the way of variety. A more promising alternative was Alphonsus, in which Ranger 9 had crash landed (Figure 1.37) back in 1965. With its rilles and dark halo craters, the floor of Alphonsus certainly had potential as an Apollo site. But what we really wanted was an area of typical highlands and so Alphonsus was discarded on the grounds that it was rather close to the Imbrium Basin and could always be reconsidered for Apollo 17.

### The landing site

The site finally chosen for Apollo 16 was in the Southern Highlands, further removed from Imbrium than the Fra Mauro site and closer, in fact, to Mare Nectaris (Figure 1.18). The landing point itself was right on the boundary between some unusually bright coloured, hummocky terrain (known as the Descartes formation after a local large crater) and a typical patch of highland light plains (or Cayley formation), a unit which accounts for several per cent of the lunar near side (Figure 2.27a).

Now the general consensus of opinion at the time of Apollo 16 was that the Descartes formation consisted of volcanoes built up from viscous, silicic lavas. The Cayley, on the other hand, was believed to be composed of highland basalt lava flows, which had flooded the hollows in, and between, those innumerable highland craters. So, all in all, we had been led to expect to find some new and interesting volcanic rock types at this Cayley–Descartes site. As it turned out, however, nothing could have been further from the truth.

The two most prominent features of the landing area (the appropriately named North Ray and South Ray Craters) were what had dictated the exact landing point for the Apollo 16 LM (Figure 2.27b). For, just as Cone Crater had been used to sample Fra Mauro, so South Ray Crater should have excavated the light plains unit and North Ray the Descartes formation. The traverses on Apollo 16 therefore reflected our primary interest in these two craters and their respective ejecta blankets.

### South Ray

During their first short excursion westwards, the astronauts noticed that the critical diameter above which fresh craters tend to have blocky rims was unusually high, implying that the highland regolith must be at least 30 feet thick. This value, later confirmed by seismic studies, must reflect the greater age of highlands when compared with *maria*.

The second EVA traverse took the Apollo 16 crew southwards, in order to collect ejecta from South Ray. But there were no plans to visit the crater itself, because of the roughness of the terrain in its immediate vicinity. Instead, the idea was to stop on one of the very prominent rays crossing the southern part of the landing area (Figure 2.27). These rays turned out to be much blockier than the surface nearer to the LM, and some of the rocks from these rays are now known to have been on the lunar surface for only two million years, making South Ray Crater very young indeed by lunar standards.

In order to sample one ray, the astronauts put the Rover through its paces (the so-called Grand Prix) and sped up the side of a low hill (known as Stone Mountain), which defines the boundary of the hypothetical Descartes formation. But it is now believed that this ridge, curving northwards towards Smoky Mountain and North Ray Crater, must be the broken rim of a large, old and very subdued crater. The total lack of unambiguously volcanic rocks on Stone Mountain (or anywhere else at the Apollo 16 site for that matter) now makes a volcanic origin for the Descartes formation (and its very existence as a discrete geological unit) highly improbable.

From their vantage point high up on Stone Mountain, however, the astronauts had a superb view across the valley, and had no difficulty at all in tracing the ray which they had just sampled out over the plain and up on to the rim of South Ray Crater. There could now be absolutely no doubt about the provenance of the rocks which they had just collected (Figure 2.28).

It was on Stone Mountain that several rocks with brown stains on their surfaces (the so-called 'rusty' rocks) were discovered, rocks which, when examined back here on Earth, were found to contain hydrated oxides of iron. Was this that long-sought lunar water? Or could it just be that some unknown Moon mineral becomes hydrated very rapidly on exposure to the moist terrestrial atmosphere? Isotopic studies on one rusty rock have suggested that the water does indeed have a terrestrial origin.

Another observation on Stone Mountain was that there seemed to be fewer craters there than on the supposedly younger plain below. But this could just be because mass wastage (and hence the rate of crater infilling) is more efficient on steeply sloping surfaces. This mass wastage process could also account for the brightness of the Descartes formation. Continuous removal of a radiation-darkened surface film would certainly keep the regolith looking bright.

We knew that the highland light plains were more heavily cratered than the oldest *mare* surfaces. But what did come as a surprise was quite how undulating the terrain was on the Cayley Plains. The astronauts had expected to find a relatively flat surface, albeit one pockmarked with craters. The broad scale undulations which they actually observed had not been apparent in pictures taken from lunar orbit.

*North Ray*

Stone Mountain was not, however, the prime sampling location for the Descartes formation. The third and final EVA was to take the astronauts up the lower slopes of Smoky Mountain towards the rim of North Ray Crater (Figures 2.27 and 2.29).

The approach to North Ray by Rover was rather more straightforward

Figure 2.28. *South Ray.* A long-distance view of the youngest crater at the site. Note the high abundance of huge boulders and also that the crater rim is elevated above the Cayley Plains. (NASA, Apollo.)

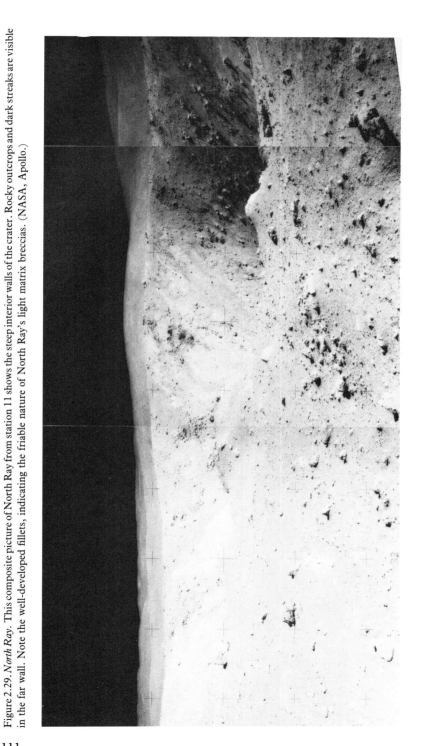

Figure 2.29. *North Ray*. This composite picture of North Ray from station 11 shows the steep interior walls of the crater. Rocky outcrops and dark streaks are visible in the far wall. Note the well-developed fillets, indicating the friable nature of North Ray's light matrix breccias. (NASA, Apollo.)

than had been the climb up to Cone Crater on Apollo 14. The terrain here was less heavily strewn with boulders, possibly because most of the rocks in the vicinity of North Ray were extremely friable in texture.

But there were a small number of more coherent boulders, and two of these were larger than any yet sampled by Apollo. The first one (called Shadow Rock because the regolith below an overhang was permanently shielded from direct sunlight) was some 800 yards from North Ray itself. The larger of the two boulders was very close to the rim and, because of its immense size (80 feet across), was promptly named House Rock (Figure 2.30).

The visit to North Ray presented our first opportunity for investigating a young highland crater at close quarters. We had missed seeing the interior of Cone by only a few yards. And what was immediately apparent to the astronauts here was that there were at least two distinct strata exposed in the walls of the crater (Figure 2.29). The uppermost layer was composed of friable breccias with light matrices. Deeper down were dark matrix breccias of which House Rock was a giant specimen. If

Figure 2.30. *House Rock.* This was the largest boulder sampled so far by Apollo, a block of dark matrix breccia exhumed from the floor of North Ray Crater and deposited on its rim. Described by the astronauts as being as big as a house. (NASA, Apollo.)

anything, the light matrix breccias seemed to be typical of the Descartes formation whereas the Cayley consisted of breccias with dark matrices. But, from a chemical point of view, there was not much difference between the two rock types. And there was still no sign of highland volcanism.

This then marked the end of man's first visit to the lunar highlands. We may not have found quite what we had expected, but Apollo 16 achieved just as much as its predecessor had done, if not more (Table 2.2). If we had known all the answers before going, there would have been little point in making the trip. But the question now was, where to send the last Apollo? We still had many questions about the Moon that needed answering and we now wanted to define a single mission to solve them all.

## 2.5    Taurus–Littrow

Before the flight to Hadley Rille, Alphonsus Crater had still been the prime landing site for Apollo 17. But Apollo 15 revealed another area which was even better (Figures 1.18 and 2.31). For not only did this new site also have dark halo craters, we could see ancient highlands, a landslide, a fault and what seemed to be a dark, and possibly very young, pyroclastic deposit. The Taurus–Littrow area on the southeastern border of Mare Serenitatis certainly had great potential and, for the very first time, one of the astronauts selected to take part in the actual landing was a geologist by profession, rather than a test pilot. Harrison Schmitt was going to add a new dimension to lunar exploration through his ability to make on-the-spot interpretations of geological structures. This certainly was going to be the grand finale, a mission designed to prove once and for all the power and flexibility of the Apollo concept.

### The South Massif

The Taurus–Littrow valley was created when blocks of crust were uplifted during the excavation of the Serenitatis Basin. Subsequent flooding of the valley with *mare* lavas has meant that the site is similar in many ways to Hadley–Apennine. But, whereas the landing approach on Apollo 15 could be made over level *mare*, the Taurus–Littrow valley is totally enclosed by mountains, some of which tower more than 7000 feet above the valley floor. The bold idea here was to land the LM in the middle of the valley so that the mountain massifs to the north and south would be readily accessible by Rover.

Once again the first EVA period of an Apollo mission was largely taken up with deploying the ALSEP, but there was just a little time left to make one short sampling traverse across the *mare* before returning the first bags of rock to the security of the LM (Figure 2.32). It was not until the second EVA, then, that the crew of Apollo 17 could make their first trip to the surrounding mountains, in this case to the foot of the South Massif.

Now the main problem with trying to sample a formation in the lunar highlands is finding a suitable exposure of bedrock. On previous occasions we had used craters as excavators, Cone at Fra Mauro and South Ray on the Cayley Plains. But the sides of some lunar mountains are so steep that, as on the banks of Hadley Rille, a thick regolith can never form there. No sooner does a thin layer of soil develop than it immediately gets shifted downslope. And this is precisely what happened (and on a very grand scale) on the flanks of the South Massif. High up on the mountainside were outcrops of bedrock, exposed some 100 million years ago by what can only be described as an avalanche, a landslide which partially filled a large crater at the foot of the hill and spread a light mantle of highland regolith out over the *mare* for several miles (Figure 2.32).

Exactly why this avalanche occurred on the South Massif is not known for certain, but the excavation of a cluster of (Tycho?) secondaries on the mountain top may well have triggered it off. One of Tycho's rays crosses the landing area and the age of the light mantle is certainly consistent with Tycho's youth.

Figure 2.31. *The Taurus Mountains.* This picture shows the location of the Apollo 17 landing site with respect to Mare Serenitatis, Mons Argaeus and the crater Littrow. Albedo differences are clearly defined in high Sun angle pictures such as this. The landing site is marked with a cross. (NASA, Apollo.)

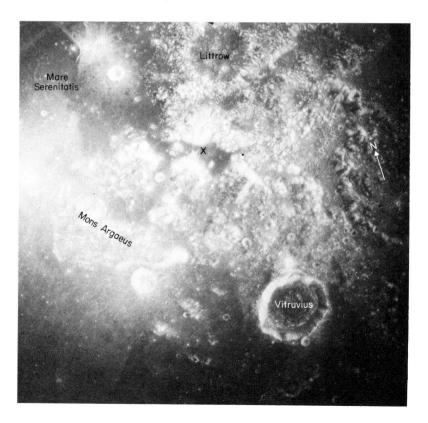

But, whatever the driving force behind it, this avalanche greatly facilitated our sampling of the South Massif. For, not only did the light mantle itself consist of regolith from high up on the mountainside, its removal also exposed outcrops of bedrock from which boulders later broke off and rolled down to the bottom of the hill, leaving clearly recognisable trails. Observations made at Station 2 seem to indicate that the greenish-grey boulders there were derived from higher levels in the Massif than the tan-grey ones. All of these rocks, however, were breccias of anorthositic gabbro, similar in many ways to those returned from the Cayley–Descartes site. A small fragment of a rock type called dunite was found in one, a sample which later turned out to be almost as old as the Moon itself.

### Shorty and Camelot

After leaving the foot of the South Massif, the astronauts took the Rover back across the Lee–Lincoln scarp and out towards the dark halo crater, Shorty, situated on one of the finger-like extensions of the light mantle (Figure 2.32). As they drove they found that the light mantle gradually became thinner and more contaminated with darker coloured regolith from the underlying *mare*.

And it was this contamination, in fact, which was actually responsible for the dark halo around the crater. Instead of being a volcanic fumarole, surrounded by black ash, Shorty was just an ordinary impact crater that just happened to have excavated dark *mare* materials from beneath a thin veneer of highland regolith (Figure 1.39).

But this was not immediately apparent during the mission, and there was one discovery which tended to point towards, rather than away from, a volcanic origin for Shorty. This discovery was, of course, that of the

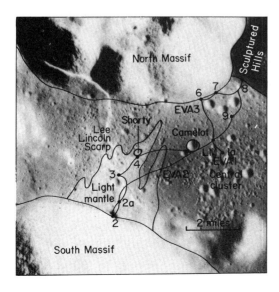

Figure 2.32. *The Apollo 17 landing site*. A close-up of the Taurus–Littrow valley, showing the major features and the three traverses. (NASA, Apollo.)

orange glass which caused so much excitement between the astronauts and mission control.

What happened was this. The crew were just about to sample one of the few large boulders on the rim of the crater when they noticed an orange-coloured patch in the regolith beside it. Now back on Earth an orange coloration such as this would normally indicate iron in its high oxidation state, and usually in a hydrated form. But most, if not all, of the iron found in the Moon so far had been highly reduced and waterfree. So if this orange soil at Shorty did indeed contain hydrated iron oxides, here would be compelling evidence for the occurrence of volcanism on the Moon during the last 100 million years.

What the astronauts did, then, was to take a core through the orange patch and dig a trench across it in order to recover a reasonably sized sample. As it turned out, however, this orange soil just consisted of anhydrous red, orange and yellow glass beads, similar in many ways to the green ones collected by Apollo 15 and chemically not dissimilar to the local *mare* basalts. Their orange coloration, far from being indicative of

Figure 2.33. *Boulder tracks.* The lower slopes of the North Massif were littered with large boulders; the tracks from two shown here can be traced upslope very easily. (NASA, Apollo.)

highly oxidised and hydrated iron, was actually due to a highly reduced ionic complex between iron and titanium. The search for water on the Moon would therefore have to continue elsewhere.

Like the green glass, the orange glass originally had a pyroclastic origin. But an exposure in the wall of Shorty confirms that the orange soil must have been redeposited on the rim of the crater during the Shorty impact event 30 million years ago. Orange (and devitrified black) glass was, in fact, widespread throughout the landing area. So the prediction that the dark mantle in the Taurus–Littrow area might consist of pyroclastics proved to be well founded. But the anticipated low age for the dark mantle was not borne out by analysis. The orange glass turned out to be as old, if not older, than the local *mare* basalts which, in turn, are among the oldest *mare* rocks on the Moon. The apparent youth of the valley floor does in fact still remain a mystery.

Leaving Shorty and the light mantle to head back towards the LM, the Apollo 17 crew just had enough time to make one final sampling stop, this time at the rim of a large crater on the valley floor known as Camelot. The large angular boulders of basalt on the rim of this crater must have been exhumed from depths as great as 300 feet. Their coarse textures suggest that the earliest basalt flows in the valley must have been thick and therefore able to cool relatively slowly. Basalts recovered from shallower craters elsewhere were finer grained, sometimes even glassy, in texture.

*The North Massif*
The final excursion on Apollo 17 was in the opposite direction to search for bedrock on the North Massif. Although there had been no landslide here, a number of even larger boulders had rolled down the mountainside, leaving tell-tale tracks to indicate their points of origin (Figure 2.33).

At the foot of the North Massif the astronauts concentrated on sampling two such boulders, the first of which was originally more than 50 feet long but had broken into five pieces. This fragmentation must have occurred at, or soon after, the time of the arrival of the boulder at its final resting place, after having rolled 1500 feet downslope from a rocky outcrop about one-third of the way to the summit of the mountain (Figure 2.34). This really was the nearest we ever got to sampling bedrock in the highlands.

What was so exciting about this boulder was that it contained a major geological contact. Half was bluish-grey in colour (with numerous exotic rock fragments embedded in it and abundant vesicles) while the other had a greenish tinge and had clearly been more strongly heated. The gas responsible for the vesiculation of the bluish half had been driven off during this heating event and the foreign fragments had been absorbed to generate a more homogeneous texture. Like the other Apollo 17 boulders, this one was sampled comprehensively and we in Sheffield had the privilege of working within an interdisciplinary consortium designed to unravel its complex history.

Driving eastwards along the foot of the North Massif towards the second boulder, the crew observed that the highland–*mare* contact was barely recognisable at close quarters. Millions of years of meteorite 'gardening' had diffused *mare* materials uphill and brought highland rocks down and out onto the *mare* floor.

Although smaller than the station 6 boulder, this one was intriguing in its complexity. Parts of it seemed to be suspiciously igneous in texture and there was a narrow vein of blue-grey material which cut through a large pale coloured clast (Figure 2.35). This boulder, like the first, was clearly going to keep a consortium of investigators busy for some years to come.

### Returning via the Sculptured Hills

Before returning to the LM there was one final highland rock unit left to be sampled, namely some elevated terrain adjacent to the North Massif, known during the mission as the Sculptured Hills. These hills were lower and more rolling than the stark massifs and, in places, seemed to be covered by that enigmatic dark mantle material. Unlike the massifs,

Figure 2.34. *The Station 6 boulder*. The largest Apollo 17 boulder, here on the North Massif, broke into 5 pieces. The South Massif can be seen on the other side of the valley. Note the development of an immature regolith on a ledge. (NASA, Apollo.)

which must be crustal blocks that were uplifted during the Serenitatis event, the Sculptured Hills are generally thought to be a Serenitatis ejecta deposit, similar in many ways to the knobbly mountains within the Orientale Basin (Figures 1.29 and 2.31).

Being rather less steep than the massifs, however, the Sculptured Hills were not so easy to sample unambiguously. For there were no large boulders lying around below them (Figure 2.32). Indeed, there were very few rocks of any size at all at Station 8. It is worth noting, however, that one of the few that were recovered (an anorthositic gabbro breccia) turned out to be at least 200 million years older than the Station 6 boulder at the base of the North Massif. If this rock was indeed typical of the Sculptured Hills, then the whole question of the origin of this rather enigmatic formation must be reexamined. Current theories make both the North Massif and the Sculptured Hills contemporaneous with the Serenitatis event.

Travelling back to the LM the astronauts made one final sampling stop, at a young crater in the valley known as Van Serg (Figure 2.32).

Figure 2.35. *The Station 7 boulder*. The complex lithology of highland breccia boulders is exemplified here by a vein of dark-coloured matrix which crosscuts a pale clast. The tongs provide the sense of scale. (NASA, Apollo.)

Here, a further collection of *mare* rocks was made and the unusual depth of the regolith in the valley was most clearly apparent. A thickness of nearly 100 feet, later confirmed by seismology, had been built up through extensive mass wastage from the surrounding massifs. Any such land-locked valley on the Moon will certainly accumulate more than the average thickness of dust cover.

The return of Apollo 17 (watched as it was from Earth through the eyes of the television camera on the abandoned Rover) marked the end of an era in lunar science. Some outstanding questions may well have been answered by an Apollo 18, but Apollo 18 was never going to fly. So we would now have to make do with what we had; namely thousands of pictures, hundreds of pounds of rock, and an endless stream of data from the surface experimental packages.

But before going on to discuss the tremendous significance of all this scientific material, it is worth while to consider what the forgotten astronauts had been doing. For Apollo was a three man mission and, while two of the crew were romping around in one-sixth *g*, the third was circling overhead in the command module, carrying out his own experiments and preparing for the return of his companions. These experiments, particularly on the later missions, were going to tell us a great deal about our planet from a global point of view.

## 2.6    Apollo's orbital tasks

One of the advantages of having an Apollo rendezvous in lunar orbit was that there was abundant time available to study the Moon from a distance. For, just as the Earth's weather can be monitored most efficiently from satellite, remote sensing of the Moon is best accomplished from lunar orbit. During their lone vigils, the Command Module Pilots therefore had to ensure that numerous experiments and photographic tasks were successfully completed, and they were certainly not the ones with the easy ride.

### Orbital photography

On the early Apollos, all orbital photography was carried out through the windows of the Command Module using handheld cameras. Although limited in quantity, pictures taken this way helped us to select, and then survey, future Apollo landing sites. The Cayley–Descartes area, for example, was surveyed by Apollo 14 and it was Worden's observations of dark halo craters in the Taurus–Littrow area which prompted the selection of a new site for Apollo 17 (Figure 2.31).

On the last three missions, however, orbital photography became much more sophisticated. Beginning with Apollo 15, two types of camera were mounted in the Scientific Instrument Module (SIM) bay of the Service Module (Figure 2.9). Now the functions of these two cameras were rather different, the metric camera being used primarily for mapping purposes, whereas the panoramic camera was designed to provide extremely high resolution pictures of surface features.

The resolution of the metric camera pictures was only 60 feet or so, but this was compensated for by the fact that the altitude and orientation of the CSM was simultaneously monitored using a laser altimeter and a stellar camera (see Chapter 5).

Both cameras produced sets of overlapping pictures which could be viewed stereoscopically, particularly useful when studying craters and for appreciating high Sun angle views. But the panoramic frames had a maximum resolution of a yard or so, making it possible to study in detail the after affects of an Apollo landing (Figure 2.36).

Mounting these two cameras in the SIM bay, however, had one major consequence. The Service Module was not destined to return to Earth and therefore, before reentry, the Command Module Pilot had to undertake a small spacewalk in order to recover the film packs.

But even on the later Apollos it was sometimes useful to have a human mind controlling the shutter. And, on each mission, the CMP had a number of specific photographic tasks to accomplish (Figure 2.37). Lighting was also important. Certain features are only readily observable at very low Sun angles and it was even desirable to take some views in earthshine. The earthshine photographs of the crater Aristarchus were particularly successful.

Interestingly enough, not all pictures taken by the CMPs were of the Moon itself because, when it is night above the far side, the surface of the Moon below is in total darkness. It would certainly not be a good place to get lost without a torch. But night time on the Moon's far side is an excellent environment from an astronomical point of view, particularly when it comes to studying dim objects. At these times, then, the CMP would turn his camera in the other direction to view such diffuse objects and phenomena as the Gum Nebula and the zodiacal light. In years to come the Moon's far side will doubtless become a major centre for observational astronomy.

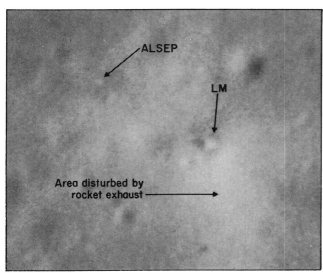

Figure 2.36. *Apollo 15 on the ground.* This highly enlarged panoramic frame shows the Apollo 15 LM and the LRV on the regolith at Hadley Apennine. Even the LRV tracks between them are just about visible. (NASA, Apollo.)

*Visual observations*

The study of the Moon from the Command Module was not, however, confined to photography. Apart from the passive operation of numerous other experiments in the SIM bay, there was also a requirement to make visual observations. In particular the astronauts had to keep their eyes open for Transient Lunar Phenomena (TLPs) and to examine the colours of certain lunar features.

As far as TLPs were concerned, none of the dust clouds and surface obscurations sometimes observed from Earth were noticed. The floor of Alphonsus, for example, remained cloudfree and there were no reddish glows in Aristarchus.

But light flashes were observed, only one of which was accounted for in terms of sunlight glinting on a spacecraft. A flash of light in Grimaldi, for example, may well have been associated with the gas emissions there for which we now have documentary evidence (see Chapter 5). But a second in Orientale may have been due to a small meteorite impact. One prob-

Figure 2.37. *King Crater*. This farside young crater was one of the many features selected for special study from lunar orbit. Note the smooth 'pool' of fine ejecta. (NASA, Apollo.)

lem with these observations, however, is that all astronauts reported seeing light flashes even with their eyes closed! The current theory is that these are caused by cosmic rays impinging on the retina. So it seems that the whole question of the nature of TLPs must remain open at this time.

As far as colour studies were concerned, the astronauts were rather more successful. The metric and panoramic cameras, of course, only used black and white film. But even the hand-held cameras were remarkably poor at capturing the Moon's true colour. The lunar surface was variously reproduced in shades of green, brown and blue. And parts that really were coloured (including patches of orange around Mare Serenitatis) generally did not show up in the final prints.

With the help of a pair of binoculars, a sketchbook, a tape-recorder and a colour wheel, however, the CMP could describe colour differences on the Moon quite accurately. And this was particularly valuable when it came to distinguishing lava flow units in the *maria*. Only in really extreme cases are these colour contrasts apparent in photographs (Figure 1.32) although they have been enhanced in Earth-based photographs using colour filters and superimposition (Figure 1.59).

The CMP, however, was not the only astronaut to be involved in detailed documentation. The collection of rocks on the lunar surface also had to be meticulous. And the processing of these samples back on Earth was by no means a straightforward operation. So here is a subject well worth some special mention.

## 2.7 Sample collection and documentation

Only by returning a comprehensive suite of rocks and soil from carefully selected locations on the Moon can we hope to build up a complete and accurate picture of the Moon's geological history. Only then can we possibly find out where the Moon came from and how it ended up orbiting the Earth in the way that it does. But geological sampling is not just grabbing those rocks which happen to be near the LM. Some sort of geological perspective is clearly needed here. For we need to know where a particular rock originated and how it ended up where it did.

### Documentation philosophy
On unmanned missions we cannot afford to be choosy, because any lunar sample is clearly better than none. But, having sent men to the Moon, we should at least try to sample the Moon as we sample the Earth. An igneous petrologist working in Cornwall, for example, rather than collecting granite pebbles from the beach, would search instead for an outcrop to photograph (using his geological hammer to provide a sense of scale) and then chip off the required sample. Its precise geological context would then have been recorded for posterity.

The shortage of bedrock outcrops on the Moon, however, means that we have to rely heavily on how a particular rock type was distributed at the landing site in order to establish its place and mode of origin. And this is not always easy. Nevertheless, we had to document the collection of rocks as comprehensively as possible for the sake of the future. For one thing, someone might one day invent a way to trace the arrival trajectory of a rock just from the way it happens to lie on the regolith. And this would certainly be out of the question for completely undocumented specimens.

### Soil collection

Among the least documented samples were the 'contingency fines', bags of dust collected during the first EVAs as safeguards against premature mission abortion. How and where these dust samples were collected was of no great consequence because all that really mattered here was that no mission should return from the Moon empty handed. No contingency sampling was scheduled on Apollo 17, so that more time could be made available for collecting better documented samples.

The collection of bulk soils was not as well controlled as that of individual rocks. We just needed to know from where they were scooped up and how deeply the scoop had penetrated on each occasion. Where the regolith was clearly stratified, the scoop depth was important because we wanted to collect uncontaminated samples from each level. A trench dug on the rim of Cone Crater, for example, yielded three distinctly different soils, confirming what the astronauts had observed at the sampling point. A similar exercise was undertaken at Shorty.

But there were also some very special soils to be returned. Permanently shadowed regolith, for example, was collected from below overhanging rocks and from beneath small boulders. And those of us who were searching for organic molecules on the Moon required samples from as far away from the LM (and its exhaust gases) as possible. These uncontaminated soils were returned in specially sealed containers. Another requirement was to lift a very thin surface film off the regolith using a velvet pad. And here it was absolutely crucial for the astronauts not to kick up dust over the sampling area. They therefore had to creep up to a sizeable boulder (the 'great Apollo 16 rock hunt'), lean over it and collect their soil from the other side.

### Core collection

In order to study the regolith in any detail, of course, we cannot just use a scoop. Some sort of coring device is clearly necessary if we are to preserve the regolith stratigraphy and, if possible, the orientations of individual grains. But, by employing such a technique, quantity is necessarily sacrificed for the sake of quality and the resulting cores should therefore be used sparingly and only for those studies which are directly concerned with regolith evolution, such as sedimentology and cosmic ray effects.

Techniques for lunar coring were greatly improved as the Apollo program progressed, in the light of experience gained on previous missions. The Apollo 11 cores, for example, were so disturbed that it looked as if the lunar regolith must be totally unstratified. But, by changing the drive-tube design, longer sections, with well-defined stratification, were recovered by Apollo 12. The lengths of the cores

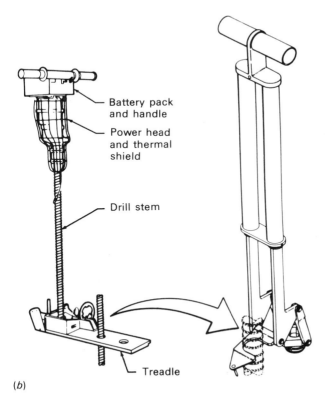

Battery pack
and handle

Power head
and thermal
shield

Drill stem

Treadle

(b)

Figure 2.38. *Drilling the regolith.* (*a*)
The drill core extractor is about to
be used here to remove a drill core at
the Apollo 17 landing site in preparation for the insertion of the
neutron probe. The electric drill
itself is stored on the rack to the
right of the picture. (NASA,
Apollo.) (*b*) a treadle was used
during drilling and a special device
was employed to extract the core,
operating like a jack. (NASA.)

collected then, however, were still considerably less than the drive-tube penetration depth, implying that some compaction had doubtless still occurred. But by making the tube wall thinner and enlarging the tube diameter, longer and still better preserved cores were returned by the last three missions.

In order to increase the coring depths still further, an electric drill had to be employed and, by joining several drill sections together, it was then possible to penetrate down to depths as great as 8 feet. This drilling operation, however, was found to be a particularly strenuous exercise for the astronauts in the Moon's low gravity field.

The internal diameter of a drill string was necessarily smaller than that of a typical drive tube. So drill sections are even more precious than cores. But the cost of their recovery could be offset to some extent by the fact that the holes left behind could be used to deploy heat-flow and neutron detector probes. This was a fine example of how closely the lunar surface activities were integrated, in order to make the most effective use of the limited EVA time available on each mission (Figure 2.38).

*Core preservation*

The lunar cores are so precious that, like some of the larger rocks, some are being preserved intact for posterity. For, just as an archaeologist should never excavate an entire site, because techniques could improve (or he might later want to look for objects which he is currently discarding), so we cannot know how methods for dissecting lunar cores could get better and we might one day need to carry out some vital experiment on a completely undisturbed section. Nevertheless, every single core was at

Figure 2.39. *Core dissection.* This photograph highlights the meticulous sectioning of a core in an atmosphere of dry nitrogen. (NASA, JSC–LRL.)

least X-rayed so that we could see what sort of stratification, if any, existed within it.

The dissection process itself was a truly painstaking one, designed not only to provide us with samples from within narrow depth ranges, but also to yield a permanent stratigraphic record.

Each drive tube liner (or, in the case of drill strings, each drill stem section) was laid on its side and cut in half longitudinally. The top half was then lifted off to reveal the upper outer surface of the enclosed cylinder of soil. The top two-thirds of this cylinder was then very carefully sectioned along its entire length, millimetre by millimetre. Whenever large pebbles were encountered they were picked out individually and great care was taken not to mix soils across obvious stratigraphic boundaries (Figure 2.39).

The upper surface of the core section exposed in this way was then impregnated with an inert resin, so that a permanent relief record of the core could be peeled off. This left behind an essentially undisturbed layer below, which could also be resin stabilised and then made into thin sections for microscopic analysis. No efforts were spared to make this core dissection process as perfect as was humanly possible.


### Peas and pebbles

The initial processing of bulk soils included the sieving out of all fragments larger than 1 millimetre in diameter, and the further separation of these small fragments into 1–2, 2–4 and 4–10 millimetre fractions. Not only was this sieving procedure a good way to obtain a statistically significant number of local rock fragments, it also greatly improved our chances of finding some really exotic rock types. For the tiniest particles are the ones which are most likely to have travelled the greatest distances, a good example of this being the anorthosites in the Apollo 11 soils, which must have originated in the highlands at least 30 miles away.

But the proportion of peas and pebbles in an average soil is rather low. So, in order to increase the numbers of large pebbles returned, a special rake was developed for Apollo 15 and this tool proved to be a very useful accessory (Figure 2.40). A large volume of soil could be sieved until sufficient pebbles had been collected. (This was indeed a laborious process near St George Crater because here rocks of any size at all were distinctly rare.) Using a rake also meant that we could collect pebbles that were large enough to be split and distributed among several investigators. This was not really practicable for sub-centimetre fragments.


### Collection of hand specimens

With larger rocks, of course, documentation was critical, because it was important here to know exactly where a particular sample was collected (for example how deeply it was buried) and how it was oriented. So each rock was photographed before collection, whenever possible, and the spot was then rephotographed with the sample removed. Included in each picture was a tripod device called a gnomon (the astronaut's equivalent of a geological hammer), which not only provided a sense of scale and colour, but also defined the vertical direction and cast a sharp shadow for orientation purposes (Figure 2.41).

The collection of large rocks was more or less systematic, the main idea here being to return as many different rock types as possible. Reading the astronaut's voice transcripts one can appreciate their excitement when an unexpected rock type turned up. Some samples were even given pet names by the astronauts, such as 'Big Bertha' (a huge Apollo 14 breccia (Figure 3.25) and 'Great Scott' (a *mare* basalt from Hadley Rille). Not all rocks had such good coverage, however, and some of the most exciting ones (such as rock 13) got no special mention whatsoever.

The collection of hand specimens became more scientific with Apollo 17, when the astronauts concentrated on the comprehensive sampling of boulders. With such complex rocks, it was clearly much better to collect several small samples from widely separated lithic units within a single boulder than to chip off just one large chunk. But individual rocks were by no means ignored on Apollo 17, because another innovation there was to collect samples between the major stops. The use of a long handled sampling tool meant that the passenger astronaut did not even have to leave his seat in the Rover to achieve this objective. And, as a result, the effective sampling area of the site was greatly increased.

Finally, of course, there were special requests for rocks as well as for soils. One such request was for a rock to be protected from the time of its

Figure 2.40. *Sampling tools.* (*a*) This tool was used to sieve pebbles out of a large volume of fine-grained regolith on the last three missions. (NASA, Apollo.) (*b*) Other equipment used for sampling. (NASA.)

(*a*)

collection by the use of a specially designed padded bag. This was because studies of surface erosion are greatly upset if the rock under study has been abraded in transit. But most investigators just waited patiently for the astronauts to return to Earth with their precious cargoes and, in the light of reports produced by the Preliminary Examination Team in the Lunar Receiving Laboratory (LRL), they then made their sample requests direct to the curator.

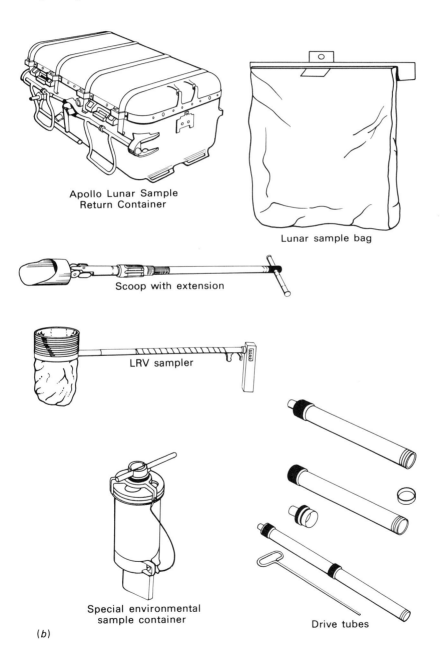

Apollo Lunar Sample
Return Container

Lunar sample bag

Scoop with extension

LRV sampler

Special environmental
sample container

Drive tubes

(b)

### Sample processing and distribution

The processing of the first lunar samples was initially hampered by the imposition of quarantine restrictions. These were absolutely necessary just in case the Moon happened to harbour some virulent organism that was capable of harming terrestrial life. Only when these quarantine

Figure 2.41. *Genesis Rock.* (*a*) Apollo 15 anorthosite on its 'pedestal' of anorthositic gabbro just prior to collection. Note the gnomon to the right of the picture which provides brightness, scale and orientation information. (NASA, Apollo.) (*b*) Details of the gnomon. (NASA.)

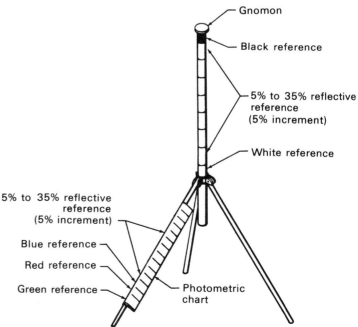

(*b*)

restrictions were lifted could sample handling continue in clean-air laboratories elsewhere, rather than under vacuum or in an atmosphere of dry nitrogen at LRL. Some experiments, however (such as the measurement of short lived radioactivities and the adsorption of gases), had to be carried out during this early period.

But most analyses were performed outside the Lunar Receiving Laboratory and this meant the beginning of a mammoth curatorial exercise. Because the distribution of lunar samples was not just a one way process. Each Principal Investigator (of which there were hundreds throughout the world) was responsible for keeping track of all the lunar samples in his possession (including their weights and states of preservation) and returning all residues when his work was complete. Wherever possible lunar samples returned by one investigator could then be reused by another.

But this cyclical distribution process meant, of course, that each primary sample had to have a unique identifier. This identifier took the form of a 5 digit catalogue number, the first one (or two) digits indicating the mission on which the sample was collected. Rock 13 was actually 12013, for example, whereas Great Scott was 15555. On the last two missions the second digit indicated the sampling stop and, as on Apollo 15, the last one indicated the sample type (e.g. coarse fines, large rock etc.). All hand specimens chipped from the Station 6 boulder, for example, have catalogue numbers of the form 76xx5 (Figure 2.42).

Figure 2.42. *Typical mugshot.* Each rock sample was photographed at LRL with its identification number on a revolving stage with an orientation cube alongside. This particular sample came from the vesiculated half of the Apollo 17, station 6 boulder. (NASA, JSC–LRL.)

Figure 2.43. *Sample division.* (*a*) Exploded views were produced during rock cutting to show the original locations of all subsamples. (NASA.)

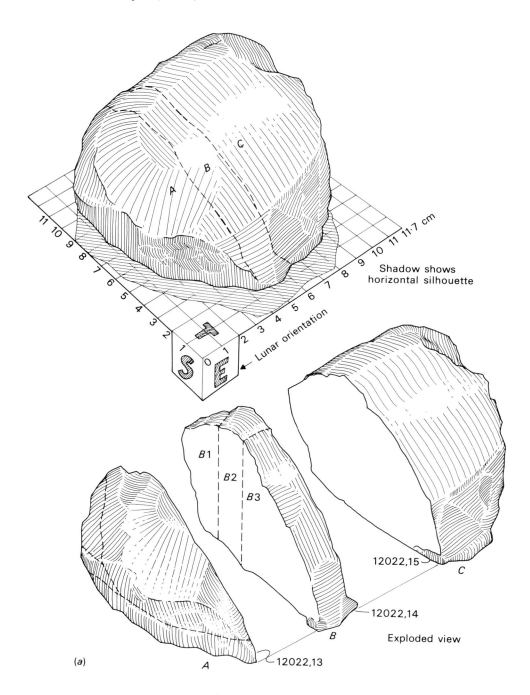

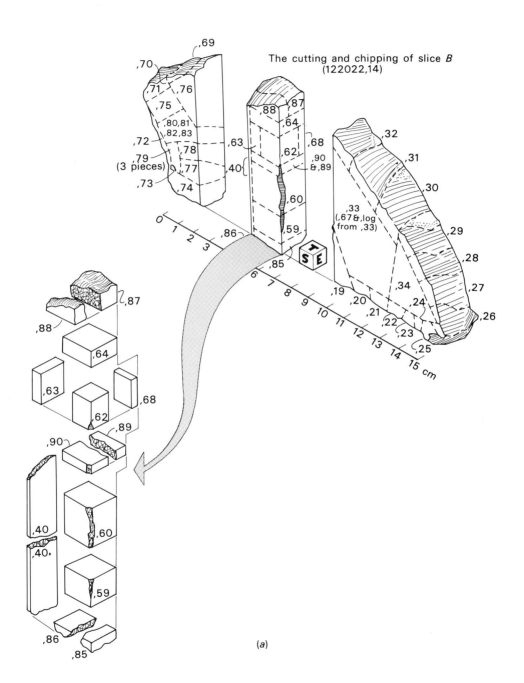

The cutting and chipping of slice *B*
(122022,14)

(a)

For some work it was important to know precisely where within a rock a particular subsample was derived. Therefore whenever a rock was cut up an exploded diagram was drawn to show the relationship between each numbered subsample and the complete rock in its lunar orientation (Figure 2.43a).

It need hardly be added of course that the Lunar Sample Curator's task is by no means an easy one. And it will not get any easier. Because although most lunar samples have now been examined more or less intensively, they still remain largely intact and this integrity must be preserved for as long as possible in order to benefit future generations of lunar scientists. For who knows how long it will be before we can get another opportunity to go back to the Moon for more? (Figure 2.43b).

But Apollo 17 was not in fact the last mission to return rocks from the Moon. This honour goes to Luna 24, which softlanded in Mare Crisium. So, before going on to discuss the scientific fruits of Apollo, let us briefly consider what has been gained from the Russian softlander program.

Figure 2.43. *Sample division.* (b) Histogram showing sample distribution. (Lunar & Planetary Institute.)

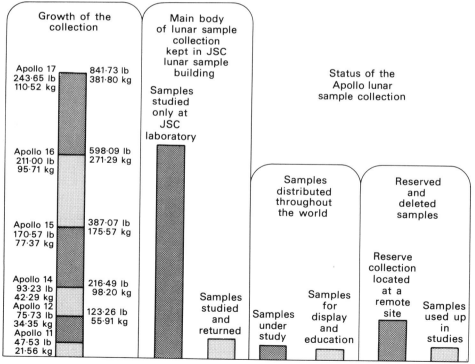

(b)

## 2.8 The Russian landers

The exploration of the Moon has by no means been just an American affair, but it is clear that the Russian effort was never on quite the same scale as the more comprehensive US program. During the early sixties, however, the Russians were certainly well ahead, at least as far as 'firsts' were concerned. Not only were they the first to reach the Moon and photograph its far side, they also put the first satellite into lunar orbit and were the first to make a soft landing.

### The first softlanders

After at least four failures, the Russians finally succeeded with Luna 9, beating Surveyor 1 to Oceanus Procellarum by only 4 months. The spherically shaped instrument capsule was separated from the main propulsion system shortly before touch down so that it could land in an undisturbed area. A shock absorbing system was used to cushion its 15 mph fall (Figure 1.18).

The outer parts of the capsule then unfolded to become leaf-shaped antennae for transmitting the first really closeup pictures of the Moon back to Earth (Figure 2.44). These radio signals were, in fact, intercepted by the radio telescope at Jodrell Bank, England and, as a result, the very first pictures of the lunar regolith were published in the West before their official release from Moscow. This caused a minor international incident, partly because, not knowing the scaling relationship between horizontal and vertical, the Jodrell Bank team had somewhat distorted the view (Figure 2.45).

Later that year, the same feat was repeated by Luna 13, the intervening craft in the Luna series having all been orbiters (Figure 1.18, Table 1.4). Once again the pictures were intercepted by Jodrell Bank, but this time publication here was withheld until the requested permission had been granted. Not only was the fine-grained nature of the lunar regolith confirmed, Luna 13 was also the first softlander to use a mechanical arm to investigate the Moon's physical properties. Furthermore, the relatively low level of natural radioactivity at each site was taken to indicate that the *mare* might have a basaltic composition.

### Sample return missions

After the successes of Lunas 9 and 13, the Russians concentrated on building unmanned spacecraft capable of returning lunar samples to Earth. This was to be achieved by combining the softlander technology developed for Lunas 9 and 13 with the return capability of the later Zond spacecraft.

If all had gone well (as far as the Russians were concerned that is), this approach could well have pre-empted Apollo by returning the first ever rocks from the Moon. For the flight of Luna 15 just happened (?) to coincide with that of Apollo 11. As fate would have it, however, Luna 15 crashed whereas Apollo succeeded.

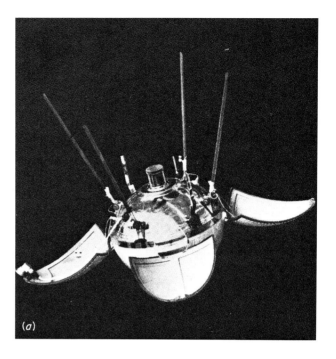

Figure 2.44. *Luna 9*. (*a*) landing and (*b*) inflight configurations. The instrument capsule was stored in the upper section and ejected shortly before landing.

The first unmanned sampling mission to succeed did not, in fact, return from the Moon until late 1970, by which time Apollo 12 had also come back safely. But Luna 16 went to a different area, landing in Mare Fecunditatis (or the Sea of Fertility) towards the Moon's eastern limb (Figure 2.46). And, although it only brought back about 4 ounces of soil (in the form of a core little more than a foot long), this tiny sample has greatly extended our understanding of the Moon, because a little can now

Figure 2.45. *The first close-up*. This undistorted first view of the lunar regolith returned by Luna 9 shows small craters and boulders. (The Associated Press Ltd.)

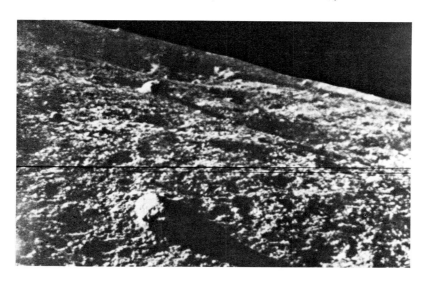

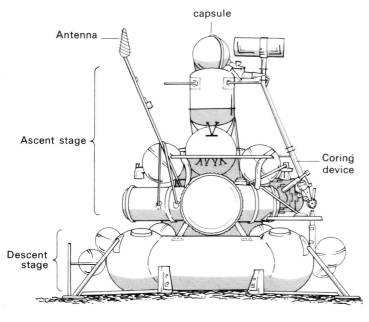

Figure 2.46. *Russian automated landers*. The three Russian automated lunar sampler spacecraft consisted of an instrument module, containing the spherical return capsule and the ascent engines, mounted on a larger descent stage. The coring device was mounted at an angle on the side of the spacecraft and was operated under the eye of a television camera. When the core had been withdrawn, it was coiled up and popped into the return capsule in preparation for return to Earth. The device pointing upwards to the left of the picture is a communications antenna.

Figure 2.47. *Luna 20 core*. This photograph shows a tray containing one ounce of highlands soil returned by Luna 20. Note that the stratigraphy could not be as well preserved as in the later Apollo cores. (NASA, JSC–LRL.)

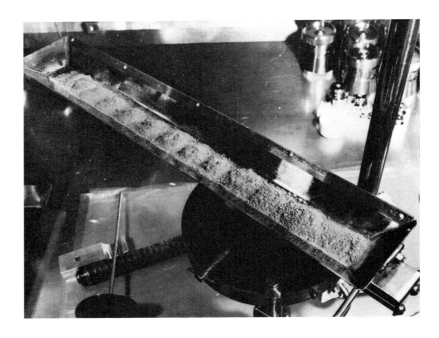

Figure 2.49. *Lunokhod panorama*. Part of a Lunokhod 2 panorama, taken in the vicinity of Le Monnier Crater in Mare Serenitatis, this photograph shows a sizeable crater and boulders. The final resting place of Lunokhod 2 is shown with a cross. (Courtesy C. P. Florensky, Vernadsky Institute, Moscow, USSR.)

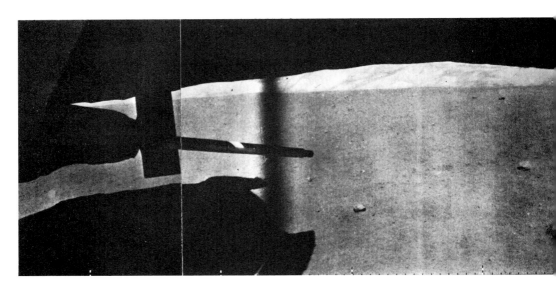

Figure 2.48. *Lunokhod.* Lunas 17 and 21 delivered Lunokhods 1 and 2 respectively to West Mare Imbrium and East Mare Serenitatis. This drawing shows some features of these roving vehicles, in particular the stereocameras, the communication system, the laser reflector, the magnetometer, solar panels, cosmic ray detector and survey camera.

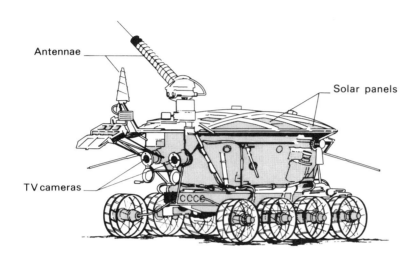

be made to go such a long way. We at Sheffield, for example, were able to obtain a precise crystallisation age on a basalt fragment weighing only 5 milligrams received through the Royal Society of London. Similarly, a detailed petrographic description and rubidium–strontium internal isochron age was obtained in the US for an exchanged 70 milligram boulder. The Luna specimens, at least as far as their scientific value is concerned, are therefore worth many times more, on a weight-for-weight basis, than the Apollo specimens.

But the Russian unmanned sample return program did not stop with Luna 16. Luna 20 later landed in the Apollonius highlands, to the north of Mare Fecunditatis, and more recently Luna 24 landed still farther to the north, in Mare Crisium (Figure 1.18).

The Luna 20 sample, weighing little more than an ounce, was particularly valuable because it came from a highland site very far removed from the influence of the Imbrium Basin. We can therefore have some confidence that the highland rocks found there are quite likely to have been rejuvenated by the Crisium (rather than the Imbrium) impact event (Figure 2.47). Indeed it is the age of Luna 20 fragments which is a major piece of evidence in favour of a cataclysmic bombardment of the Moon hundreds of millions of years after its formation (see Chapter 6).

As far as sample weight returned is concerned, Luna 24 must be considered the most successful of the series, returning 6 ounces. All these Russian cores had to be rolled up prior to their return to Earth, so they do not contain so much stratigraphic information as the Apollo ones. But they are certainly just as valuable. Quite how many more such samples we might expect the Russians to return, however, must remain pure conjecture at this time.

### The Lunokhods

The only other Lunas not mentioned so far are Lunas 17 and 21. These were very special missions, taking to the Moon the unmanned roving vehicles known as Lunokhods 1 and 2 (Figure 1.18). The Lunokhods, with their eight wheels (Figure 2.48), were remarkably successful, Lunokhod 1 operating for ten months in Mare Imbrium, whereas its successor roamed around on the edge of Mare Serenitatis for four months. They did not travel very far during these times (7 and 23 miles respectively), but numerous experiments were performed (including laser ranging, magnetometry and X-ray analysis) and photographic panoramas were also returned (Figure 2.49).

In years to come, vehicles not dissimilar to the Lunokhods may be used to explore the lunar poles or to investigate the surfaces of other planets. But the time has now come to return to Earth and see what has now been learnt about the most tangible products of lunar exploration, the rocks and the soils. It is here that we have learnt the most about our sister planet.

# 3. Moon rocks and minerals

## 3.1  An introduction to lunar petrology

Having heard how the first samples of lunar rock were brought back to Earth, let us now discuss what sorts of rocks they are, what minerals they contain and what their chemistries can tell us about lunar history. How do lunar rocks differ from terrestrial rocks and meteorites, for example, and are there any minerals which are unique to the Moon or which might have some commercial or practical value? But first we must define exactly what we mean by rocks and minerals and introduce the various methods which are used to investigate them.

*Basic rock types*

Any rock can usually be categorised as being igneous, sedimentary or metamorphic in origin. Igneous rocks are those formed by direct crystallisation from a silicate melt, whereas sedimentary rocks are the end result of the gradual accumulation of detrital material, usually in a marine or fluvial environment. Metamorphic rocks are simply igneous or sedimentary rocks which have been subjected to moderately high temperatures and pressures.

The various types of igneous rock are distinguishable from one another on the basis of their texture and mineralogy. Volcanic lavas (such as basalts) cool more rapidly than intrusive igneous rocks (such as granite) and they are therefore finer grained. For slow cooling helps to encourage existing crystals to grow larger by retarding the overall crystallisation rate.

But exactly which minerals form at any particular stage of the crystallisation sequence depends more on the initial chemistry of the silicate melt than on its rate of cooling. Granitic magmas, for example, are richer in silica than basalt lavas and therefore only they contain quartz as a major rock-forming mineral.

Natural silicate melts are, however, extremely complex systems and, from a chemical point of view, no two rocks can ever be exactly alike. Some will invariably be richer in trace elements than others. So, although all basalts contain the same two major minerals (plagioclase and pyroxene), the chemistries and relative proportions of each, and the nature of the minor phases present as well, may vary greatly from one basalt to another. There can even be chemical variations within a single mineral grain, variations which reflect the changing composition of the melt as crystallisation progressed and certain key elements were removed from it.

The absence of life or water on the Moon means that we are unlikely to find terrestrial type sedimentary rocks there. Chalk, for example, consists of the calcareous remains of tiny marine organisms, whereas

141

sandstone contains quartz grains (eroded out of granites) which have accumulated on river beds.

The lunar regolith, however, is certainly a sedimentary deposit, albeit a loosely consolidated one, having formed by the accumulation of debris from hills and craters nearby. Small meteorite impacts have the effect of shock welding this soil together to form what are known as 'instant' rocks, or regolith breccias.

When this welding occurs on a much grander scale, during major impact events, the interactions between adjacent mineral grains are more complex and the resulting rocks have metamorphic textures. In some cases the temperatures attained here were close to, or even above, that required for melting.

*Petrography*

In order to characterise lunar rocks we must first recognise the minerals present, by determining their chemical compositions and corresponding crystal structures. Given a certain chemical composition, only a very limited number of crystalline forms may be possible.

Now most minerals are oxides and, because oxygen atoms are so large, can just be thought of as closely packed arrangements of silica tetrahedra ($SiO_4$) with metal atoms fitting neatly inside the spaces in the lattice (Figure 3.1).

But some atoms are larger than others, and therefore require large holes to accommodate them. Similarly, other elements require more or less oxygen for electrical neutrality. So several different, more or less closely packed, arrangements of silica tetrahedra exist, and, with

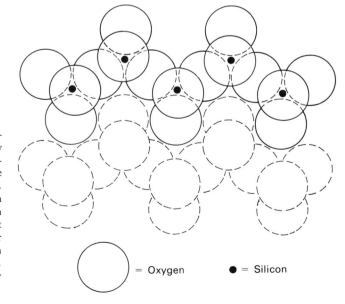

Figure 3.1. *Pyroxene structure.* Silicate minerals consist of closely packed arrangements of $SiO_4$ tetrahedra, with metal ions filling the interstices in the crystal lattice. Pyroxene is a chain silicate in which each $SiO_4$ unit shares an oxygen atom with each of two adjacent units. In quartz (Si:O = 1:2) all four oxygen atoms are shared whereas in olivine (Si:O = 1:4) none of them is. In aluminosilicates aluminium may replace silicon.

◯ = Oxygen     ● = Silicon

Table 3.1. *Common lunar minerals*

| Mineral Type | Mineral | Chemical Formula |
|---|---|---|
| Silicates | Olivine | $(Fe,Mg)SiO_4$ |
| | Pyroxene (ortho) | $(Fe,Mg)SiO_3$ |
| | Pyroxene (clino) | $(Fe,Mg,Ca)SiO_3$ |
| | Plagioclase | $CaAl_2Si_2O_8 - NaAlSi_3O_8$ |
| | Orthoclase | $KAlSi_3O_8$ |
| | Quartz | $SiO_2$ |
| | Cristobalite | $SiO_2$ |
| | Tridymite | $SiO_2$ |
| Metals, oxides, phosphates and sulphides | Ilmenite | $FeTiO_3$ |
| | Spinel | $(Mg,Fe)(Al,Cr,(Fe+Ti)/2)_2O_4$ |
| | Native iron | $Fe (+ Ni,Co)$ |
| | Troilite | $FeS$ |
| | Apatite | $Ca_5(F,Cl(PO_4)_3)$ |
| | Whitlockite | $Ca_3(PO_4)_2$ |

aluminium replacing silicon, and with several abundant metals on the Moon (Table 2.1), the possibilities for lunar mineralogy might seem to be endless. But, strangely enough, the total number of lunar minerals is really quite small (Tables 3.1 and 3.6), largely because of the absence of water there.

The amount of mineralogical information that can be extracted from a rock just by looking at its exterior is strictly limited, particularly if it has been weathered by micrometeorites. But, by examining a polished interior surface, the shapes and optical characteristics of minerals become clearly apparent. Indeed, some minerals are best studied by reflected light microscopy.

But, for silicate minerals, it is much better to make a very thin slice of rock and illuminate it from behind. And the reason for this is that silicates have a very useful property. Not only are they optically transparent, they can also rotate the plane of polarised light.

Now normal light is unpolarised. In other words its electromagnetic waves are randomly orientated about the direction of the light path. But this randomness can be changed by passing the light through a polarising filter. All waves then become aligned and cannot pass through a second polarising filter orientated at right angles to the first. The light will be totally extinguished under these conditions.

But, if a silicate mineral is placed between the two polarising filters, the extinction angle will no longer be 90°. And the exact value of this extinction angle can, in fact, help to identify the mineral. Similarly, thin sections viewed through crossed polarisers often exhibit brilliant interference colours, colours which may also be characteristic (Figure 3.2). So a polarising microscope in skilled hands is certainly a very powerful tool in mineralogical analysis.

*Chemical characterisation*

But it is one thing to recognise the minerals present in a rock and quite another to analyse them chemically. For the crystal structure of a mineral and its optical properties are only very crude guides to its detailed chemical composition.

The instrument that has done most to provide us with this detailed chemical information is the electron microprobe, now an essential part of any well-equipped mineralogy laboratory.

What happens in an electron microprobe is that the thin section to be studied is bombarded with electrons. These interact with atoms in the target mineral and X-rays are given off. Now some elements give off harder, or shorter wavelength, X-rays than others. So, by analysing the X-ray spectrum, we can obtain chemical maps of a mineral grain, each one showing the distribution of some common element (Figure 3.3).

The beauty of the electron microprobe is that it allows us to trace chemical variations across a single mineral grain. This chemical zoning can tell us much about the environment in which that particular crystal grew. Alternatively, the electron beam can be defocussed to yield an average chemical analysis. So the electron microprobe is clearly a very powerful analytical weapon.

But, for some elements, it simply is not sensitive enough. Only the commonest metals yield sufficient X-rays to be mapped in this way. For others we need to resort to such techniques as mass spectrometry (in which atoms are analysed according to their atomic mass), X-ray fluorescence spectrometry and neutron activation analysis.

Figure 3.2. *Thin section of Apollo 17 mare basalt.* Viewed through crossed polarisers the major minerals present show up as follows: pyroxene (grey), plagioclase (white) and ilmenite (black). (NASA, JSC–LRL.)

Neutron activation analysis is particularly appropriate for such rare elements as gold and platinum. Here the sample is bombarded with neutrons in a nuclear reactor and the unstable products of the interactions between these neutrons and the target atoms give rise to gamma rays of characteristic frequency. By analysing the sample's gamma ray spectrum, concentrations of a few parts per billion can be measured routinely.

Perhaps the ultimate analytical weapon, however, is the ion microprobe, an instrument which combines the high resolution of the electron microprobe with the extreme sensitivity of the mass spectrometer and its ability to separate different isotopes of the same element. As yet, however, few laboratories can afford such a powerful and sophisticated device.

*Simple silicates*

How then have these analytical methods helped us to analyse lunar minerals? What minerals exist on the Moon and why?

Well, after oxygen and silicon, two of the four next most common elements on the Moon are iron and magnesium (Table 2.1). And the simplest of all silicate minerals happens to be an iron magnesium silicate known as olivine, which can be best represented by the chemical formula $(Fe,Mg)_2SiO_4$. This formula just means that for every silicon atom present in olivine there are four atoms of oxygen and two atoms of iron or

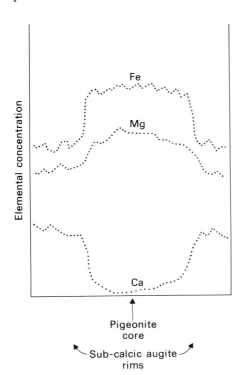

Figure 3.3. *Electron microprobe scans.* X-ray scans across a pyroxene grain, showing how the concentrations of iron, calcium and magnesium can vary from the rim of sub-calcic augite to the core of pigeonite.

magnesium. So there is no such thing as an olivine molecule, nor is olivine chemically pure. In other words it may have any composition lying between $Fe_2SiO_4$ (which is known as fayalite) and $Mg_2SiO_4$ (known as forsterite). Like many other minerals, then, olivine is best considered as a solid solution, rather than as a pure chemical compound.

In lunar rocks, olivines have variable chemical compositions, not only between one rock and another but also within a single olivine crystal. In general, magnesian olivines are more primitive (that is formed earlier) than iron-rich ones. So the iron content of the olivine in a rock may be a useful clue to its ultimate origin.

But some lunar rocks contain too much silica ($SiO_2$) to contain any free olivine whatsoever. And the excess silica in these rocks may have crystallised out in one of three forms: the two high temperature varieties, cristobalite and tridymite, or, as in rock 13, the somewhat lower temperature mineral, quartz. Like olivine, however, silica minerals are relatively minor components of lunar rocks (Figure 3.4).

The other common iron–magnesium silicate on the Moon, pyroxene, is rather more complex than olivine, because it may also contain calcium (Figure 3.5). All pyroxenes, in fact, can be considered as solid solutions of enstatite ($MgSiO_3$), ferrosilite ($FeSiO_3$), diopside ($CaMgSi_2O_6$) and hedenbergite ($CaFeSi_2O_6$). The ratio of silicon to oxygen in pyroxenes (1:3) is therefore lower than it is on olivines (1:4) but higher than in quartz (1:2).

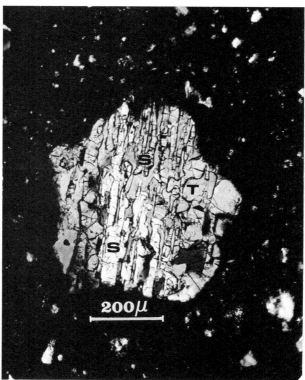

Figure 3.4. *Lunar 'granite'*. Rare fragments in the regolith and breccias consist of fine-grained intergrowths of quartz ($S$) and potassium feldspar ($T$). The scale bar represents a distance of one-fifth of a millimetre. (From Ryder, G., Stoeser, D. B., Marvin, U. B. and Bower, J. F. (1975). *Proceedings of the Sixth Lunar Science Conference*, pp. 435–49. New York: Pergamon.)

Figure 3.5. *Lunar pyroxenes.* The chemical composition of a lunar pyroxene can be visualised by plotting it on this ternary diagram. The names of some of the intermediate compositions, together with the end-members of the solid solution series, are given. Not all compositions are stable at all temperatures.

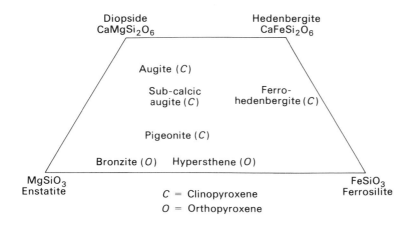

Figure 3.6. *Exsolution.* Lunar calcic pyroxenes may have decomposed on cooling into exsolution lamellae of sub-calcic augite and pigeonite. In this anorthosite sample the exsolution was on both a coarse and a fine scale. The crystal is about one-fifth of a millimetre across. The sample itself was a friable breccia from North Ray Crater and the matrix is just the same as the clasts but finely crushed. (From Brown, G. M., Peckett, A., Phillips, R. and Emeleus, C. H. (1973). *Proceedings of the Fourth Lunar Science Conference*, pp. 505–18. New York: Pergamon.)

But not all pyroxenes have the identical crystal structures. Those containing calcium have a monoclinic form, the calcium-rich ones being known as augites, whereas the calcium-poor ones are called pigeonites (Figure 3.5). When lunar clinopyroxenes form with intermediate compositions, they tend to decompose into alternating layers (or exsolution lamellae) of pigeonite and augite (Figure 3.6). The widths of these alternating bands are valuable clues, in fact, to the rate at which the rock must have cooled. The wider the lamellae, the slower the cooling rate must have been.

Calcium-free pyroxenes may have an orthorhombic structure instead of a monoclinic one and are given various names, according to their contents of magnesium. The so-called KREEP basalts, for example, contain the magnesium-rich pyroxenes known as bronzite and hypersthene. All in all, then, there is little doubt that pyroxene is one of the most informative of lunar minerals, particularly when it comes to unravelling the crystallisation history of lunar rocks. Petrologists can glean a great deal of information just from the chemical zoning across a single pyroxene crystal.

### Feldspars

But iron, calcium and magnesium are not the only common lunar metals. Another common metallic element is aluminium, which is only ever a minor component of lunar pyroxenes. Most aluminium on the Moon exists instead in the form of aluminosilicates, the commonest of which is a plagioclase feldspar called anorthite ($CaAl_2Si_2O_8$) (Figure 3.7).

This mineral is not, however, the only lunar plagioclase, nor is plagioclase the only lunar feldspar. Natural plagioclases are effectively mixtures of anorthite and the sodium-rich end-member known as albite ($NaAlSi_3O_8$).

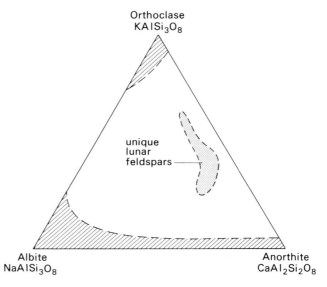

Figure 3.7. *Feldspars.* The three principal end-members of the feldspar series are orthoclase, anorthite and albite. Not all compositions are found, however, because of the different crystal structure of orthoclase.

We can see from these chemical formulae that albite has one more silicon atom and one fewer aluminium atom than anorthite. But this is simply to satisfy the different bonding requirements of calcium and sodium. Aluminium readily replaces silicon in the silicate lattice to satisfy such differences in oxygen requirement, so that anorthite and albite can have essentially identical crystal structures.

Plagioclase feldspars in lunar rocks tend to be chemically closer to anorthite than to albite. Indeed, the rock type anorthosite consists almost entirely of anorthite, because lunar anorthosites contain very little sodium. But not all lunar feldspars have the plagioclase crystal structure. This is because sodium cannot be replaced directly by potassium (a chemically similar alkali element) without some gross distortion of the crystal lattice. A much more open structure is necessary in order to accommodate the larger potassium ion.

Potassium may not be one of the commonest metals on the Moon, but there were situations during the formation of the crust when it became sufficiently concentrated for potassium feldspar to become a major mineral phase. Rock 13, for example, contains intimate intergrowths of potassium feldspar and quartz. Similar so-called 'granitic' fragments have been found elsewhere on the Moon but they are by no means abundant (Figure 3.4).

The novel crystal structure of potassium feldspar means that, unlike pyroxenes, feldspars do not form a continuous solid solution series between the three end members (Figure 3.7). The more open structure of potassium feldspar, however, does mean that it is capable of accommodating some rather more exotic elements. Indeed, this mineral is the major carrier of that alkaline earth element, barium.

### Metals and sulphides

Leaving aside for the moment the more exotic lunar silicates, the only other common minerals are opaque phases. And the most obvious of these is metallic iron, the sole carrier of lunar remanent magnetism. Magnets attached to the Surveyor footpads had already told us that magnetic phases were present in the lunar regolith. But it was not until the first rocks were returned by Apollo that we could confirm that it is indeed metallic iron, rather than magnetite, that is responsible for this magnetisation.

The very existence of free iron on the Moon indicates the highly reducing nature of the lunar crust. The existence of iron in its highest oxidation state (ferric) has yet to be proved conclusively, and most other metallic elements are also highly reduced.

Free iron is very rare on Earth but is common enough in meteorites. Indeed, much of the metallic iron in the regolith apparently originates from infalling meteorites. But there has also been reduction of iron in the soil by hydrogen atoms arriving at the surface of the Moon from the Sun in the form of the solar wind. These two factors account for the fact that the regolith tends to be more magnetic than the bedrock beneath.

Lunar metallic iron, like industrial steels, is by no means pure. It invariably contains appreciable quantities of nickel and cobalt, usually separating out into two phases, one rich in nickel (taenite) and one nickel-poor (kamacite). The cobalt and nickel contents of lunar metal fragments can, in fact, be used to decide whether a particular metal fragment has a lunar or meteoritic origin (Figure 3.8).

Closely associated with metallic iron in lunar rocks is a sulphide mineral known as troilite, also common in meteorites but very rare indeed on Earth. Although other sulphides do occur on the Moon (see Section 3.5), troilite is certainly the major carrier of lunar sulphur.

### Metal oxide minerals

The most abundant metal oxides on the Moon are ilmenites and spinels. Indeed, these two minerals are the major carriers of those two minor elements, titanium and chromium. In the *mare* basalts, in fact, titanium is so abundant that ilmenite can be considered as a major mineral. In highland rocks, however, titanium (and hence ilmenite) is almost totally absent. Although nominally having the chemical formula $FeTiO_3$, lunar ilmenites frequently contain appreciable amounts of magnesium.

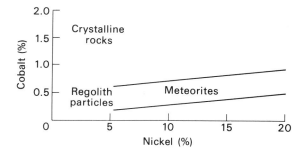

Figure 3.8. *Fe–Ni–Co.* The nickel and cobalt contents of lunar metal can be used to establish whether or not the metal is extralunar in origin. Meteorites are rich in nickel and poor in cobalt compared with lunar rocks.

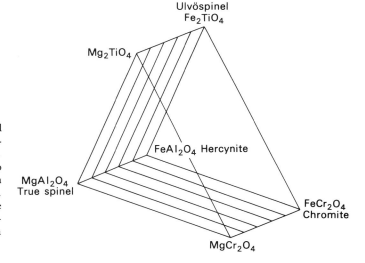

Figure 3.9. *Spinels.* The spinel diagram showing the major end-members of the solid solution series. Lunar spinels usually tend to be intermediate in composition between chromite and ulvöspinel. Like pyroxenes, spinels may be strongly zoned chemically, reflecting changes in melt composition during crystallisation.

Spinels are rather more complex from a chemical viewpoint. True spinel has the formula $MgAl_2O_4$ but magnesium can be replaced by iron and aluminium by chromium to yield chromite ($FeCr_2O_4$). Another end member of the solid solution series has the chemical formula $Fe_2TiO_4$ and is known as ulvöspinel (Figure 3.9). Like potassium feldspar, spinels may also be major carriers of some less common elements.

So much, then, for the minerals present in lunar rocks. Now we must turn to the rocks themselves. What can we learn from their textures and mineralogies?

## 3.2    The *mare* basalts

Long before Apollo it had become clear that the lunar *maria* must be volcanic plains rather than dried up seas or deep dust bowls as had once been believed. How else could we account for the obviously volcanic domes and wrinkle ridges (Figures 1.21 and 1.32)? And the chemical analyses performed by Surveyor indicated that the *mare* rocks themselves must be basaltic, just the sort of composition necessary to have resulted in thin lava sheets rather than towering volcanic piles.

Furthermore, we also knew that the Earth was not the only place in the Solar System where basaltic rocks occur. There are some meteorites (known as the basaltic achondrites) which also must have had a volcanic origin. So it was certainly no surprise when Apollo 11 returned nothing but basalts, rocks similar in many ways to the basaltic achondrites and to the rocks which comprise the Earth's oceanic crust.

### Regional occurrence

Since 1969, *mare* basalts have been recovered from several other locations on the Moon: Oceanus Procellarum, Mare Fecunditatis, Mare Crisium and the borders of Mare Imbrium and Mare Serenitatis. So basalt rocks, far from being local to Mare Tranquillitatis, are certainly widespread, at least on the Moon's near side (Figure 1.15).

But how do these rocks vary from one site to another, and in what ways do they differ from terrestrial and meteoritic basalts? Well the lunar *mare* basalts can best be grouped into three classes: the titanium-rich ones returned by Apollos 11 and 17, the titanium-poor ones returned by Apollos 12 and 15 and the aluminous ones typified by those brought back from Mare Fecunditatis by Luna 16 (Figure 3.10 and Table 3.2). Within each class there exist subtypes, such as the high and low potassium Apollo 11 basalts (distinguishable from one another on the basis of their potassium contents) and the Apollo 12 pigeonite and olivine basalts (which have different contents of silica).

### General petrology

When compared with terrestrial basalts, all three lunar *mare* types are strongly depleted in relatively volatile elements. Metals such as zinc and potassium, for example, are much less abundant in lunar basalts than in terrestrial ones. And there is as yet no sign of indigenous lunar water, much to the disappointment of those who were looking ahead to establishing the first lunar colony (Table 3.3).

One of the most interesting questions concerning lunar basalts, in fact, is how to explain the gross vesicularity of some of them (Figures 2.22 and 2.23). Just what was the gas phase responsible for those bubbles? Carbon monoxide is one possibility but the total amounts of carbon in lunar basalts are only a few tens of parts per million. It should, of course, be

Figure 3.10. *High titanium and low titanium basalts.* The amounts of ilmenite (black) in an Apollo 11 high titanium basalt (*a*) are clearly much larger than in a low titanium one from Apollo 12 (*b*). (Both specimens are 3.2 mm across.) (NASA, JSC–LRL.)

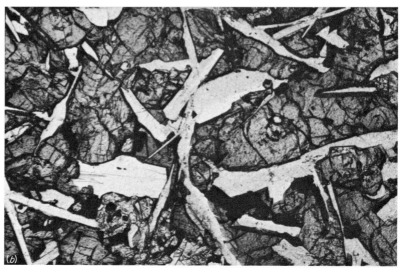

Table 3.2. *A summary of major element contents of typical mare rocks. Note in particular: (a) the high alumina content of the aluminous basalts; (b) the high titania contents of the Apollo 11 and 17 rocks; and (c) the high magnesia content of the green glass. Although typical, there exists a considerable spread in composition for each major basalt grouping*

| Elemental abundance (per cent by weight as oxide) | High titanium basalts | | | Low titanium basalts | | | Aluminous basalt |
|---|---|---|---|---|---|---|---|
| | Low K (A17) | High K (A11) | Orange glass | Low Si (A12) | High Si (A15) | Green glass | (A14) |
| $SiO_2$ | 39.5 | 40.4 | 38.9 | 43.6 | 48.9 | 45.4 | 46.3 |
| $Al_2O_3$ | 9.4 | 8.8 | 6.4 | 8.3 | 9.5 | 7.7 | 13.7 |
| CaO | 11.1 | 10.6 | 7.0 | 8.5 | 10.5 | 8.3 | 11.2 |
| FeO | 18.8 | 19.3 | 22.3 | 20.6 | 20.1 | 19.6 | 17.0 |
| MgO | 8.0 | 7.6 | 14.8 | 15.1 | 8.9 | 17.5 | 8.5 |
| $TiO_2$ | 11.8 | 11.8 | 9.0 | 2.6 | 1.8 | 0.4 | 2.8 |
| $Cr_2O_3$ | 0.4 | 0.4 | 0.7 | 0.7 | 0.3 | 0.4 | 0.4 |
| $Na_2O$ | 0.4 | 0.5 | 0.4 | 0.3 | 0.3 | 0.1 | 0.4 |
| $P_2O_5$ | 0.09 | 0.18 | 0.10 | 0.07 | 0.06 | 0.03 | 0.11 |
| $K_2O$ | 0.06 | 0.31 | 0.08 | 0.04 | 0.06 | 0.01 | 0.10 |

noted that the high vacuum of space would have greatly encouraged basalt lavas to froth on eruption. So the total amounts of gas necessary may not have been very large.

Another general chemical property of *mare* basalts (and of highland rocks as well) is their overall enrichment in refractory elements. When compared with the chondritic meteorites (which are generally taken to represent primordial Solar System material), lunar rocks tend to be enriched in such elements as zirconium, uranium, titanium and the rare earth elements (REE).

Within each broad basalt grouping, of course, there are extreme differences in chemistry, differences which result in a variety of mineral assemblages and textures. But temperature and pressure were also very important here. For it would have been very unusual for all parts of a lava flow to have crystallised out under exactly the same physical conditions to produce the same rock type throughout. For one thing, pressures would have been much higher at depth than at the surface of the flow. And, with lunar lavas starting to crystallise at temperatures above 1400 °C and not being completely solid even at 900 °C, a great deal could happen during crystallisation, exactly what depending on the initial chemistry of the lava and on how much time was available for differentiation.

*Basalt textures*

When crystallisation was very rapid (such as at the surface of a lava flow, where heat could be quickly radiated away) crystals had very little time to grow, making the rocks very fine-grained, or even glassy, in texture (Figure 3.11). But not all rocks quenched in this way are completely free of large crystals. Some magmas had partially crystallised before they

Table 3.3. *Concentrations in p.p.m. of a selection of minor and trace elements in lunar rocks. The seven most abundant are phosphorus, sulphur, manganese, zirconium, yttrium, barium and strontium*

| Element group | Element | | Anorthosite | KREEP | *Mare* basalt |
|---|---|---|---|---|---|
| Non-metals | Carbon | C | 10 | 20 | 20 |
| | Nitrogen | N | 5 | 10 | 10 |
| | Sulphur | S | 500 | 700 | 3000 |
| | Phosphorus | P | 70 | 7000 | 300 |
| | Chlorine | Cl | 200 | 50 | 20 |
| Transition | Manganese | Mn | 70 | 1000 | 2500 |
| metals | Zirconium | Zr | 20 | 1200 | 300 |
| | Yttrium | Y | 4 | 300 | 100 |
| | Vanadium | V | 5 | 50 | 100 |
| | Cobalt | Co | 0.8 | 40 | 40 |
| | Nickel | Ni | 30 | 30 | 40 |
| | Scandium | Sc | 0.8 | 20 | 70 |
| Alkalis and | Barium | Ba | 8 | 1000 | 50 |
| rare earths | Strontium | Sr | 200 | 150 | 100 |
| | Rubidium | Rb | 0.7 | 20 | 3 |
| | Lanthanum | La | 0.1 | 100 | 6 |
| | Cerium | Ce | 0.3 | 300 | 20 |
| | Samarium | Sm | 0.05 | 50 | 7 |
| | Europium | Eu | 0.8 | 3 | 2 |
| | Dysprosium | Dy | 0.3 | 60 | 10 |
| | Ytterbium | Yb | 0.02 | 40 | 6 |
| | Lutetium | Lu | 0.003 | 5 | 1 |
| Heavy metals | Hafnium | Hf | 0.01 | 30 | 5 |
| and | Indium | In | 0.2 | 3 | 0.4 |
| actinides | Gold | Au | 0.00001 | 0.0001 | 0.0002 |
| | Rhenium | Re | 0.000001 | 0.00002 | 0.00002 |
| | Uranium | U | 0.005 | 5 | 0.15 |
| | Thorium | Th | 0.015 | 20 | 0.5 |

reached the lunar surface. And when they cooled rapidly they formed what are known as vitrophyres, basalts with large, well-formed crystals (phenocrysts) set in a very fine-grained matrix.

All such fine-grained rocks are useful when it comes to deciding where basaltic magmas could have originated. This is because, with lavas cooling rapidly, little time was available for crystal fractionation. In other words, once formed, crystals could not migrate very far before the lava solidified completely. And this means, then, that the chemical compositions of fine-grained basalts may be quite close to those of their parent magmas, at least at their times of eruption if not at their points of origin deep inside the Moon.

It could be, however, that some fractional crystallisation did occur prior to (or during) eruption, or that some magmas became chemically contaminated with crustal material. In this case the determination of

Figure 3.11. *Quenched basalt.* This thin section of an Apollo 17 basalt shows very narrow crystals, indicating that this rock was cooled very rapidly after the parent lava was erupted. (Specimen is 1.4 mm across.) (NASA, JSC–LRL.)

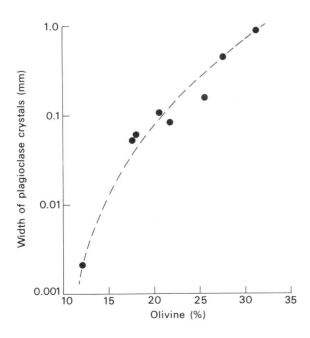

Figure 3.12. *Evidence for crystal settling.* Among Apollo 12 basalts there is a strong correlation between olivine content and the size of plagioclase crystals. This is expected because the slowest cooled lavas should contain the most evidence for olivine separation. (After Walker, D., Longhi, J., Kirkpatrick, R. J. and Hays, J. F. (1976). *Proceedings of the Seventh Lunar Science Conference,* pp. 1365–89. New York: Pergamon.)

primary magma compositions may not be just a question of analysing the finest-grained *mare* basalts.

But most lunar *mare* basalts are relatively coarse-grained, the crystals in some being more than one millimetre across. It could be argued that these should be called gabbros rather than basalts. But there is such a smooth continuum in grain sizes among them that they are invariably all called basalts. The term gabbro can then be reserved for the much coarser-grained (and chemically quite distinct) highland rocks.

### Basalt crystallisation

With the coarser-grained *mare* basalts, the case for fractional crystallisation may be much stronger. One rock, for example, contains more than 30% of olivine (normally a minor mineral compared with plagioclase and pyroxene) and this can best be accounted for by supposing that olivine (a dense mineral that crystallised early) settled out at the base of the lava pool.

One piece of evidence that strongly supports this olivine settling idea is that there is a strong correlation among Apollo 12 basalts between the grain size of plagioclase crystals and olivine content. This is what we would expect because olivine accumulation would have been most marked in lavas which cooled slowly and therefore resulted in crystals of the largest size (Figure 3.12).

The segregation of crystals must obviously affect the overall chemistry of the resulting rocks. Iron and magnesium are being removed from one part of the lava flow and are being added to another in a correlated manner. So the chemistry of a basalt depends not only on the initial chemistry of the magma but also on the crystallisation sequence.

Figure 3.13. *Mineralogical control.* Among the two groups of Apollo 15 basalts there is evidence for mineralogical control over their chemistry, more evidence for near-surface fractional crystallisation. (From Rhodes, J. M. & Hubbard, N. J., (1973), *Proceedings of the Fourth Lunar Science Conference*, pp. 1127–48. New York: Pergamon.)

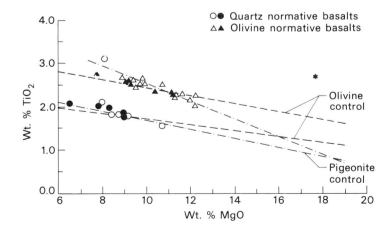

In the case of the Apollo 15 basalts the chemical effects of olivine (and pyroxene) control are quite marked. There are, for example, strong anticorrelations between magnesium and titanium and between iron and aluminium. Clearly, the more olivine and pyroxene a basalt contains, the less ilmenite and plagioclase it can have as well (Figure 3.13).

But the precise order in which the major *mare* basalt minerals crystallised is not always easy to establish. In some cases the textural relationships between the various minerals make their crystallisation sequence obvious. Crystals formed early, for example, tend to have better shapes than those formed later. In other cases, however, the relationships may be less clear cut.

One useful method here has involved the study of melt inclusions. When crystals formed they may have trapped small quantities of silicate melt. So analyses of these melt inclusions can tell us about the temperature and composition of the silicate melt at the time that the crystal formed (Figure 3.14).

The crystallisation of a particular mineral, however, was by no means instantaneous. Some minerals continued to crystallise as the temperature dropped by several hundred degrees. And even the sequence of crystallisation may have varied from one flow to another. In some lavas, for example, plagioclase may have formed before pyroxene, but in others this order of crystallisation may have been reversed.

But spinel, olivine and ilmenite are generally high temperature minerals, whereas phases such as potassium feldspar, quartz and phosphates were always the last to crystallise (Figure 3.15). The variable amounts of such accessory phases among *mare* basalts leads to a strong correlation between such geochemically dissimilar elements as potassium and lanthanum. Both elements were concentrated in residual liquids and then incorporated into accessory phases such as potassium feldspar and whitlockite (Figure 3.16). Some phases, such as metallic iron, may have crystallised both at high and low temperatures.

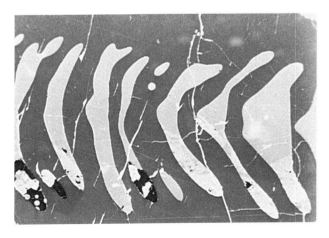

Figure 3.14. *Crystalline inclusions.* Inclusions in minerals such as ilmenite (shown here) can indicate the order of crystallisation in *mare* basalts. (From Haggerty, S. E., Boyd, F. K., Bell, P. M., Finger, L. W. & Bryan, W. B. (1970). *Proceedings of the Apollo 11 Lunar Science Conference*, pp. 1203–17. New York, Pergamon.)

Figure 3.15. *Crystallisation sequence.* (*a*) This is a typical sequence for mineral crystallisation in *mare* basalts.

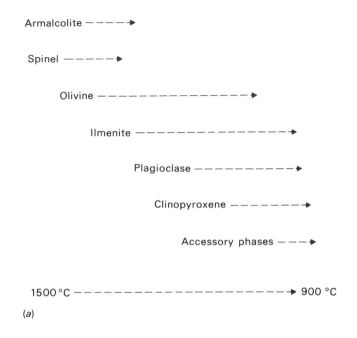

Armalcolite — — — — ➤

Spinel — — — — — ➤

Olivine — — — — — — — — — — ➤

Ilmenite — — — — — — — — — ➤

Plagioclase — — — — — — — ➤

Clinopyroxene — — — — — ➤

Accessory phases — — — ➤

1500 °C — — — — — — — — — — — — — ➤ 900 °C

(*a*)

(*b*) This thin section of an Apollo 12 *mare* basalt in reflected light shows the order of crystallisation very clearly. Chromite (Chr) followed by ulvospinel (Usp) followed by ilmenite (Ilm) and finally metallic iron (Fe). (Courtesy S. E. Haggerty.)

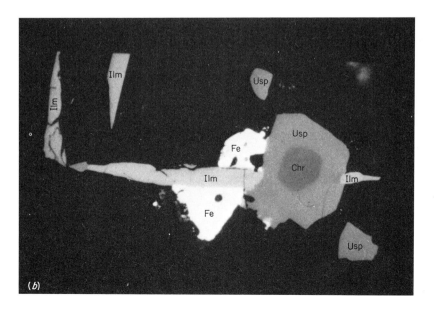

*Geochemical considerations*

Not all of the chemical variety among *mare* basalts can, of course, be due to fractional crystallisation. Most of it arises from the chemical differences which must have existed between the source regions for the various parent magmas. The titanium-rich Apollo 11 and 17 basalts, for example, cannot possibly have come from the same source as the Apollo

Figure 3.16. *Chemical correlations.* Some elements in lunar rocks are strongly correlated. In the case of manganese and iron this is because both metals tend to enter the same minerals. Lanthanum and potassium are geochemically dissimilar but both elements tend to become concentrated in residual phases rather than major minerals. The ratio of potassium to lanthanum is much lower than it is in Earth rocks, indicating that the Moon is depleted in volatile elements, of which potassium is one.

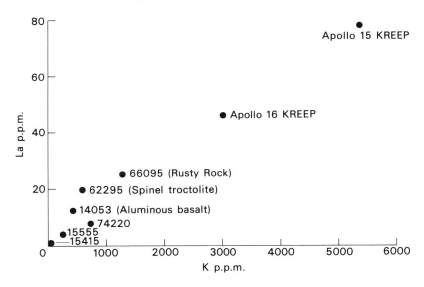

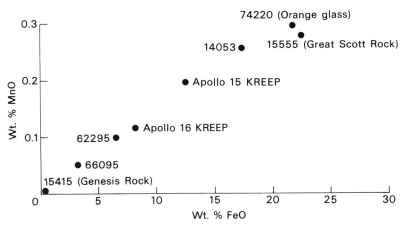

12 and 15 rocks. But the fact that certain key elemental ratios are relatively constant between basalt groups having different sources is really quite remarkable.

The low potassium–lanthanum ratio (compared with the chondritic value) in all *mare* basalts must mean, in fact, that the Moon (or at least that part of it from which *mare* basalts were derived) must be heavily depleted in relatively volatile elements, of which potassium is one. As such correlations also extend to highland rocks they will be discussed in a more general context later in this chapter.

Similarly, the *mare* basalts are all strongly depleted in those elements that readily dissolve in molten metallic iron. Gold and platinum, for example, are even rarer in lunar rocks (a few parts per billion at most) than they are in chrondritic meteorites. The most popular idea here is that these so-called siderophile (i.e. iron loving) elements were removed from the source regions of the *mare* basalts, or even from the entire Moon prior to its accumulation. Furthermore a corresponding depletion in chalcophile (i.e. sulphide loving) elements, such as lead, points to their extraction by molten iron sulphide.

These depletions clearly have a direct bearing on the origin of the Moon and its early evolution.

### The rare earths

The elements that can perhaps throw the most light on the origins of *mare* basalts are the rare earths, the commonest of which may be present in *mare* basalts at concentrations as high as 0.1%. They are clearly not so rare after all. In fact, of the minor elements only manganese, zirconium, barium, strontium and yttrium are more abundant than this (Tables 3.2 and 3.3).

The usefulness of the rare earths arises from their close chemical

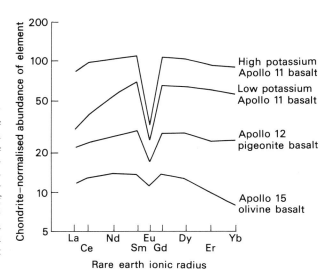

Figure 3.17. *Rare earths in mare basalts.* This plot shows the chondrite-normalised rare earth patterns for typical lunar mare basalts. Note the overall enrichments ($\times 10$ to $\times 100$) and the depletion of light rare earths in the low potassium Apollo 11 basalts and of the heavy ones in the Apollo 15 sample. Note also the correlation between the size of the europium anomaly (Eu) and the enrichment factor.

similarity. In other words they all tend to become segregated into the same minor minerals, such as the calcium phosphate phase whitlockite. But some are more readily incorporated into a particular mineral than others. So partial melting and fractional crystallisation have the effect of chemically fractionating the rare earths with respect to one another. The actual concentrations of rare earths are generally normalised to their abundances in the chrondritic meteorites in order to make such enrichments and fractionation patterns easier to appreciate (Figure 3.17).

One marked example of such chemical fractionation in *mare* basalts is apparent in one type of Apollo 11 basalt, namely the low potassium variety. Unlike other types, these rocks are relatively depleted in the light rare earths, such as lanthanum and cerium, a depletion that is clearly indicative of near surface fractional crystallisation. As we shall hear later, the formation of these basalts was also accompanied by fractionation of rubidium and strontium. The Apollo 15 olivine basalts, on the other hand, are relatively depleted in the heavy rare earths, such as ytterbium.

But clearly the most dramatic feature of *mare* basalt rare earth patterns is their depletion in europium, a depletion that is particularly marked in rocks which are strongly enriched in the rare earths as a whole. This depletion, instead of being indicative of near surface fractional crystallisation, arises from the fact that europium is the most easily reduced of all the rare earth elements. And, under the strongly reducing conditions which prevailed inside the Moon, some of it exists in its smaller, divalent (rather than trivalent) state. Being smaller, the divalent ions could then be accommodated into the major rock-forming mineral plagioclase, instead of being relegated to the accessory phases.

Where there is a negative europium anomaly, then, we can be sure that plagioclase (containing some of the europium) was removed from the *mare* basalt source region. And, sure enough, when we come to analyse the plagioclase in highland anorthosites we will discover exactly where this europium ended up. Some highland rocks have correspondingly positive europium anomalies. So it is to these highlands that we must now turn. How do highland rocks differ from the *mare* ones and how do we account for their existence?

## 3.3   The characterisation of highland rock types

A casual look at the Moon is all that is needed to see that the lunar highlands are quite different from the *maria*. Not only are they topographically higher and more rugged, they are also lighter in colour. And the most likely explanation for this albedo difference is that, unlike the *maria*, the highlands are not composed of iron-rich basalt (Figure 1.22). That this is indeed the case was first indicated by Surveyor 7, which revealed much higher concentrations of calcium and aluminium in the regolith near Tycho than its two immediate predecessors had found in the *maria* (Table 2.1). So how else do highland rocks differ chemically from *mare* ones, what sorts of rocks are they, when and where were they first discovered and how can we account for them?

## Early inferences

The first clues to the nature of the highland crust were provided by Apollo 11, even though Tranquillity Base was 30 miles or so from the nearest highlands. But Apollo 11 returned no large non-*mare* rocks at all. They were all *mare* basalts and regolith breccias.

Some particles in the soil, however, were found to be not just smaller examples of these basic rock types. Some were much more exotic, a few consisting almost entirely of white calcic plagioclase quite unlike the feldspar crystals in *mare* basalt. So could these be the rocks we were looking for? Their compositions were certainly consistent with the results from Surveyor 7.

Based just on these tiny Apollo 11 particles, then, the idea of a lunar crust consisting largely of anorthosite (a rock containing abundant calcic plagioclase) was born. This idea was to remain viable throughout Apollos 12, 14 and 15, and has now been shown to be essentially correct.

Anorthosite was not, however, the only non-*mare* rock type discovered among the Apollo 11 samples. A simple chemical analysis of the soil was enough to show that a third component must also be present, a magic ingredient rich in potassium (K), rare earth elements (REE) and phosphorus (P). The chemical composition of the Tranquillity Base regolith could not be explained simply by a two component mixture of *mare* basalt and anorthosite.

After careful searching, a few of the required fragments were found. Initially called Luny Rocks, these non-*mare* rocks had what was essentially a basaltic composition and soon became known as KREEP, an acronym which clearly highlights their anomalous chemistry.

## The occurrence of KREEP

The full chemical characterisation of KREEP came with the second Apollo mission, when breccias and soils consisting almost entirely of this component were returned from Oceanus Procellarum. Most of it was in the form of ropy brown glass, coated with a film of fine, pale-coloured dust. It was this coating which gave the KREEP-rich Apollo 12 soils (thought to be representative of Copernicus ray material) their pale appearance. Could it be, then, that lunar rays are due to the formation of such fine dust coatings?

The detailed chemical analysis of Apollo 12 KREEP showed it to be rich in many elements in addition to those implied by its acronym. Uranium, thorium and zirconium can be singled out for special mention here. But the names KREEPUTh and KREZP could not compete with KREEP, and so it is KREEP that has remained with us ever since, much to the annoyance of certain members of the lunar geological community (Table 3.3).

The reasons for the wide dislike of the term are (*a*) that it is not specific enough and (*b*) that it does not reflect the petrologic nature of the parent rock type. And even if we restrict ourselves to chemistry, there are still many different types of KREEP, some more KREEPy than others. The most KREEPy lunar rock of all, in fact, is rock 13, the KREEP parts of which contain several percent of potassium (Figure 3.18).

Rock 13 is a highland breccia, one of the very few returned by Apollo 12. And not only does it contain an extreme KREEP component, the rest of it has a granitic composition (already mentioned in the introduction to this chapter), consisting of potassium feldspar and quartz. So, prior to Apollo 14, we already knew of 3 highland rock types (anorthosite, KREEP and granite). And we had yet to sample the highlands directly (Table 3.4).

### The search for KREEP basalt

So far, all KREEP samples recovered from the Moon had been glasses or breccias. Somewhere there should be crystalline examples. And the chemistry of KREEP strongly suggested that its parent rock type should be a volcanic basalt.

Unfortunately, Apollo 14 did not help us very much here, even though most of the rocks returned from Fra Mauro did have a KREEP composition. Practically all were breccias (Figure 2.18).

Indeed, the only large crystalline example of KREEP that was found there (rock 14310) turned out to be an impact melt. But it did at least indicate the sort of mineralogy that we might expect to find, namely orthopyroxene, plagioclase and abundant accessory phases such as phosphates. Also found as clasts in Apollo 14 breccias were fragments containing abundant potassium feldspar, plagioclase, pyroxene and brown glass. Chemically related to the granites discovered by Apollo 12, these rare rocks are grouped together as monzonites/granites.

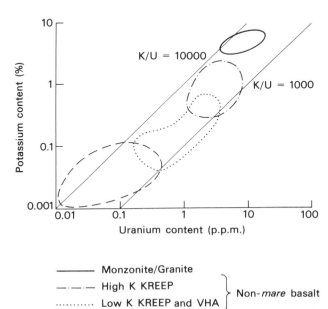

Figure 3.18. *Potassium and uranium in highland rocks.* Highland rocks contain very varied amounts of potassium, from less than 0.1% in the anorthosites, low K KREEP and VHA basalts up to more than 1% in the high K KREEP rocks and 5% in granites. Potassium–uranium ratios are relatively constant, despite the geochemical dissimilarity of the two elements, but the process that caused the enrichment of potassium in KREEP and granite (partial melting?) tended to increase the K/U ratio from as little as 1000 to as much as 10 000. (After Lovering, J. F. & Wark, D. A. (1975). *Proceedings of the Sixth Lunar Science Conference*, pp. 1203–17. New York: Pergamon.)

It was not until Apollo 15, then, that the very first crystalline samples of KREEP basalt were returned from the Moon, and even these were just small pebbles picked up at the Apennine front (Figure 3.19). Unlike 14310, the small Apollo 15 KREEP basalts were uncontaminated with meteorite debris. 14310 contained element excesses which could only have been added by an impacting projectile (see Chapter 4).

*Highland basalt/anorthositic gabbro*

During the first three missions, a feeling was steadily growing that there must be at least one major highland rock type still to be found on the Moon. The lunar highlands could not just consist of anorthosite, KREEP and a little bit of granite.

The reason for this confidence lay in the fact that chemical analyses of the ubiquitous glass beads from the regolith tended to form a number of well-defined clusters, one of which did not correspond to a known rock type.

When we analyse rocks we can have a sampling problem if the rock we are analysing is not much larger than the minerals contained within it. If we are not very careful we could end up with an analysis of a mineral rather than that of a rock.

But glasses are different. Not only are they homogeneous within themselves, they were also formed during major impact (or volcanic) events during which large volumes of parent rock (or magma) must have been homogenised. The grain size problem therefore does not exist with glass analyses.

But how can we tell that a particular glass composition corresponds to a primary rock type rather than to just a mixture of, say, anorthosite and KREEP? Well the answer here is that, strictly speaking, we cannot. And the fact that some mixing does occur is clearly apparent. KREEP glass

Table 3.4. *A summary of major element contents of the principal highland rock types. Actual compositions within each group may vary considerably from these typical values. Note in particular: (a) the high silica and potassium contents of the monzonites and granites; (b) the high alumina and low iron in the anorthositic rocks; and (c) the generally low titanium and iron contents*

| Element abundance (as oxide) | Anorthosite | Anorthositic gabbro | Dunite | Troctolite | Norite | KREEP | Monzonite | Granite |
|---|---|---|---|---|---|---|---|---|
| $SiO_2$ | 44.1 | 45.3 | 40.7 | 42.9 | 49.8 | 48.0 | 59.6 | 73.1 |
| $Al_2O_3$ | 35.5 | 28.7 | 1.3 | 20.7 | 18.4 | 14.9 | 20.6 | 12.4 |
| $CaO$ | 19.7 | 16.2 | 1.1 | 11.4 | 10.5 | 7.4 | 7.9 | 1.3 |
| $FeO$ | 0.2 | 4.1 | 11.9 | 5.0 | 6.0 | 9.2 | 3.3 | 3.5 |
| $MgO$ | 0.1 | 4.4 | 45.4 | 19.1 | 14.5 | 7.4 | 3.3 | 0.1 |
| $TiO_2$ | 0.02 | 0.3 | 0.03 | 0.05 | 0.08 | 2.2 | 0.6 | 0.5 |
| $Cr_2O_3$ | 0.01 | 0.10 | 0.11 | 0.11 | 0.3 | 0.3 | 0.3 | 0.4 |
| $Na_2O$ | 0.3 | 0.5 | 0.01 | 0.2 | 0.3 | 0.9 | 0.9 | 0.6 |
| $K_2O$ | 0.02 | 0.09 | 0.002 | 0.03 | 0.05 | 0.6 | 4.7 | 6.0 |

compositions, for example, are invariably more dispersed than analyses of larger KREEP rocks.

Extreme glass compositions (such as KREEP, granite and anorthosite) are clearly easier to characterise than the more nondescript intermediate rock types. But one glass composition (first referred to as highland basalt) was clearly not a simple mixture. The proportions of KREEP and anorthosite required to account for its aluminium content, for example, were not sufficient to explain its high content of iron. We may not have been able to characterise this new component very precisely but we could at least be certain that it did exist. Soon it was to become more widely referred to as anorthositic gabbro.

### The search for highland crustal material

The chemical relationships and differences between the various types of highland (and *mare*) rocks are perhaps best represented on a plot of magnesium against aluminium (Figure 3.20). KREEP rocks have between 13 and 24 % of $Al_2O_3$ and 6–12 % FeO, whereas anorthosites have as much as 35 % $Al_2O_3$ and virtually no iron at all. Only the most aluminous *mare* basalts (such as one that turned up as a clast in an Apollo 14 breccia) have alumina contents as high as KREEP basalt and these are much richer in iron. The rock composition known as highland basalt (or anorthositic gabbro) contains between 26 and 32 % of $Al_2O_3$ and up to 7 %

Figure 3.19. *Crystalline KREEP.* This thin section is of a sample of KREEP basalt, showing narrow laths of plagioclase and broader crystals of orthopyroxene set in a matrix of brown glass. (NASA, JSC–LRL.) (Courtesy A. J. Irving.)

FeO. And it is this composition, in fact, which is closest to the average composition of the lunar crust. But this we could not know at the time of Apollo 15.

One of the primary objectives of Apollo 15, then, was to establish what the lunar highlands really did consist of. Was it anorthosite or anorthositic gabbro? The idea here was that the Imbrium impact should have uplifted highland crust (in the form of the Apennines) which could be sampled at the foot of Hadley Delta (Figure 1.25).

Interestingly enough, neither rock type was found to be characteristic of the Apennine Front. The underlying rocks clearly had a KREEPy, rather than anorthositic, composition. In the author's view this is strong evidence in favour of the entire area affected by the Imbrium Basin having once been the site of extensive flooding by KREEP lavas. The original anorthositic crust could well have been stripped away prior to this early volcanism by the excavation of an even larger impact crater, namely the Gargantuan Basin (see chapter 5).

Figure 3.20. *Highland rock types*. The principal highland rock types are well separated from one another (and from the *mare* rocks) on this plot of magnesium oxide against alumina. The increasing alumina content of high K KREEP, low K KREEP, VHA basalt, anorthositic gabbro, gabbroic anorthosite and anorthosite is paralleled by their decreasing potassium contents. Note that both types of glass are more magnesium-rich than the associated *mare* basalts. The sample of dunite consists almost entirely of olivine.

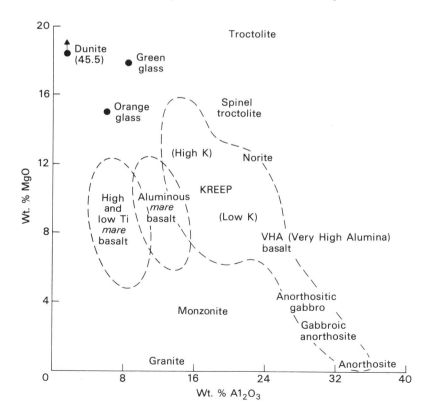

But whatever the true story may be, the fact is that the Apennine Front did at least yield the first large crystalline samples of anorthosite (Genesis Rock) and the first anorthositic gabbro. Clearly, anorthositic crust of some sort was not that far away (Figure 2.41).

Most of our typically highland samples, then, were collected during the last two missions, particularly at Cayley–Descartes. Highland rocks were also returned by Luna 20, the only one of the Russian sample return missions to visit a highland area (Figure 2.47).

*Chemical characterisation of the highland crust*

Despite the fact that we now have large rocks (rather than just glass beads) that can be considered as representative of the principal types of highland rocks, strangely enough this still does not tell us exactly what the primary rock types are in the lunar highlands. For, as with KREEP at Fra Mauro, nearly all of the samples are fragmental rocks, or breccias. In most cases original textures have been erased and in others there is strong evidence for impact melting.

There now seems to be no doubt that anorthositic gabbro must be a primary rock type. But the hiatus between KREEP and anorthositic gabbro has now been filled to some extent by rocks known as 'very high alumina basalts'. Similarly, there are rocks called gabbroic anorthosites which fill the chemical gap between anorthositic gabbros and the true anorthosites. The terminology for highland rocks certainly starts to become rather confusing at this point.

It would of course be very strange indeed if there was no mixing of rock types at all during major impacts in the highlands. But how much of this mixing has been responsible for amalgamating the various chemically defined groups and how much have mixtures reseparated? Bearing in mind the saturation of the highlands with large craters we might have expected to find just one average highland rock type rather than a wide spread in compositions.

Interestingly enough, highland soil compositions show a much smaller range in chemistry and do not exhibit the clustering apparent from large rock analyses. Could this mean that even large impacts were inefficient at homogenising the lunar crust? Or might it be more likely that the energy involved was large enough to bring about partial melting and subsequent reseparation of refractory anorthosite from easily melted KREEP? It is an intriguing possibility, but one that is still difficult to resolve. It is hard to believe that reseparation could have been more effective than mixing.

Returning to the chemical characterisation of highland rock types, the KREEP rocks are clearly the most enriched in rare earths (with enrichment factors over chondrites of several hundred) and have strong negative europium anomalies. Also the light rare earths are enriched with respect to the heavier ones, in contrast to the *mare* basalts (Figures 3.21 and 3.17).

The anorthosites on the other hand have very low rare earth contents and positive europium anomalies. It is here then that the plagioclase removed from the KREEP (and *mare*?) basalt source regions must have ended up. The anorthositic gabbros are only slightly enriched in rare earths, exhibiting small negative europium anomalies. The various highland rock types are also well separated on the basis of their potassium and uranium contents. And once again there are direct correlations here between geochemically dissimilar elements (Figure 3.18).

But perhaps the most significant characteristic of highland rocks from a chemical point of view is their enrichment in magnesium at the expense of iron. Iron–magnesium ratios are distinctly lower (often less than 1:1) in almost all highland rocks than in even the most aluminous of *mare* basalts (mostly greater than 2:1). Magnesium–chromium ratios are also invariably higher in highland rocks (Figure 3.22). So one of the crucial questions to be answered is how did the highland crust come to be relatively enriched in magnesium rather than iron?

The chemistries of highland rocks clearly cannot be discussed in isolation. It is of course important to relate chemistry to mineralogy and texture. But the chemical characterisation of highland rock types came first. And their textures tend to be related more to the physical conditions prevailing during their formation than to their initial chemistry. So let us now discuss the textures of highland rocks and introduce some more exotic types such as dunites and troctolites.

Figure 3.21. *Rare earths in highland rocks.* Unlike *mare* basalts, highland rocks show an enrichment in light rare earths compared with the heavier ones. In this diagram the positive europium anomalies of the more anorthositic rocks are apparent. The pattern labelled 'average highland composition' is simply a mixture of 3 parts highland basalt (or anorthositic gabbro) and 1 part low K Fra Mauro (KREEP) basalt. Note that this composition is slightly enriched in REE. (From Taylor, S. R. & Jakes, P., (1974), *Proceedings of the Fifth Lunar Science Conference*, pp. 1287–305. New York: Pergamon.)

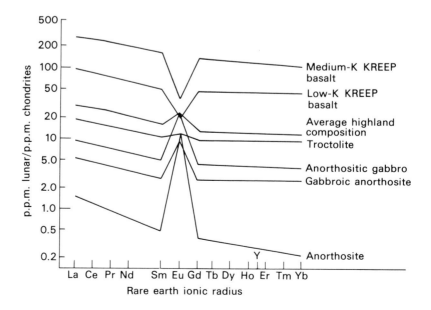

## 3.4    Highland rock textures

Having characterised the principal highland rock types chemically, we must now decide what the textures of highland rocks can tell us, not only about the genesis of the rocks themselves but also about the formation and evolution of the lunar crust, and indeed of the Moon as a whole.

One way to categorise highland rocks texturally is to divide them into 3 broad groups: (*a*) coarse-grained rocks containing crystals larger than a millimetre or so across; (*b*) finer-grained rocks which at least appear to have primary igneous textures; and (*c*) metamorphic breccias of various grades and lithologies.

The last group is by far the largest of the three and may itself be broken down into two subgroups: monomict breccias arising from the impact metamorphism of a single rock type from group (*a*) or (*b*), and polymict breccias which were ultimately derived from at least two different sources.

### Coarse-grained highland rocks

Prominent among the coarse-grained highland rocks are the anorthosites. But there are some rather more exotic specimens too, such as dunites, norites, troctolites and pyroxenites. These contain olivine or orthopyroxene as well as (or instead of) plagioclase.

These rocks, although by no means numerous among the returned specimens, are particularly valuable to us because they may be the sole surviving remnants of the primitive lunar crust. All other highland rocks were clearly formed more recently, either as a result of igneous activity

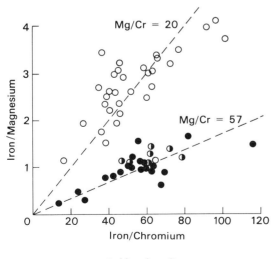

○ *Mare* basalts

◑ KREEP basalts

● Anorthositic rocks

Figure 3.22. *Magnesium, iron and chromium.* Almost all highland rocks have lower iron–magnesium ratios than the *mare* basalts. This plot shows that the dispersion of magnesium–iron ratios in each group is closely linked to chromium content. The highland rocks with the lowest iron–magnesium ratios must contain a component rich in magnesium and chromium. And the formation of the lunar crust must have preferentially removed magnesium from the source regions for the *mare* basalts, because their magnesium–chromium ratios are 3 times lower than in highland rocks. (After Wanke, H., Palme, H., Baddenhausen, H., Dreibus, G., Jagoutz, E., Kruse, H., Spettel, B., Teschke, F. & Thacker, R. (1974). *Proceedings of the Fifth Lunar Science Conference*, pp. 1307–35. New York: Pergamon.)

within the Moon, or as a consequence of impact metamorphism at its surface.

So what makes us think that these coarse-grained rocks are indeed primitive cumulates? Well, for one thing, their coarse textures are indicative of slow cooling rates. And this in turn implies formation at some depth rather than near surface crystallisation of, say, an impact melt.

But texture is not everything here. The plagioclase in these rocks is very calcic (>98%) and the mafic minerals (olivine and pyroxene) are very magnesian. Both of these chemical characteristics are usually taken to be primitive. The lack of the chemical zoning which is usually observed in the crystals in more rapidly cooled rocks is also significant here. All in all, then, the textures and chemistries of these coarse-grained rocks are highly indicative of their deep-seated origins.

### Petrologic classification

This first group of rocks contains the ANT suite, here the acronym ANT stands for Anorthosite–Norite–Troctolite and implies a close genetic relationship between these three rock types.

Anorthosites are those rocks which contain more than 90% plagioclase, small amounts of pyroxene (mainly orthopyroxene), spinel and ilmenite and no olivine whatsoever. Norites contain approximately equal amounts of orthopyroxene and plagioclase (more than 40% of each) with only minor amounts of other minerals, such as troilite and whitlockite.

Figure 3.23. *Recrystallised troctolite.* A thin section of troctolite 76535 shows the 120° triple junction, mosaic texture that is characteristic of recrystallised highland rocks. The minerals present are olivine (lower right and left) and plagioclase. (From Dymek, R. F., Albee, A. L. & Chodos, A. A. (1975). *Proceedings of the Sixth Lunar Science Conference*, pp. 301–41. New York: Pergamon.)

Troctolites, on the other hand, contain olivine as the major ferro-magnesian mineral (about 60%), together with about 35% plagioclase and only minor amounts of orthopyroxene and other minerals (Figures 3.23 and 3.24).

Although not included within the ANT suite, the rare lunar dunites and pyroxenites are closely related, a dunite being a cumulate rock consisting almost entirely (greater than 90%) of olivine, with only minor amounts of plagioclase, pyroxene and spinel.

The distinction between dunites and troctolites, and that between pyroxenites and norites, is really only one of degree, with pyroxenites just containing more pyroxene than norites. As always, then, when it comes to giving specific names to individual rock types, problems can arise if intermediate rock types turn up later. And this could well happen as we explore the Moon still further.

*Shock features*

Although these coarse-grained highland rocks may exhibit primitive textural (as well as chemical) features, they invariably show evidence of shock metamorphism.

The most obvious shock effect is the crushing of crystals to form what are known as cataclastic textures. Such cataclastic rocks contain large rock and mineral fragments embedded in a matrix of finely crushed fragments of identical lithology. These rocks (which can best be

Figure 3.24. *Troctolite mugshot.* This photograph of lunar troctolite 76535 shows how coarse grained some lunar rocks can be. (NASA, JSC–LRL.)

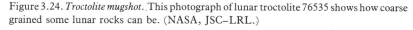

described as monomict breccias) were found to be common in the vicinity of North Ray Crater, but they also turned up elsewhere, particularly as clasts in Apollo 17 boulders (Figure 3.26).

Intense shock did, of course, have profound effects on crystal structures. Plagioclase, for example, may not only have been intensely fractured, it may also have been converted in the solid state into a glassy form known as maskelynite (otherwise referred to as diaplectic or thetomorphic glass) with the passage of the shock wave through the crystal. Indeed, studies of such crystal defects and glass can tell us much about the peak shock intensities (several tens of kilobars) to which rocks were subjected during brecciation (Table 3.5).

### Classification of highland breccias

But impact metamorphism on the Moon did not just give rise to cataclastic textures. In some cases, the temperatures attained were sufficiently high for extensive recrystallisation to occur (Figures 3.23 and 3.24). The resulting textures are then known as granulitic. Mineral grains are packed closely together in the form of a mosaic, with no intervening fine-grained matrix and with characteristic 120° triple junctions. The Apollo 15 Genesis Rock and the largest of the Apollo 17 troctolites are typical examples of such recrystallised breccias.

Less severe brecciation may not have completely destroyed preexisting textures and, where multiple source rocks were involved, the resulting breccia may contain abundant evidence of its polymict nature. Parts of the rock may have very different lithologies from other parts.

Table 3.5. *Stages in the development of lunar metamorphic breccias, with temperature and degree of shock increasing downwards*

| Effects due to temperature | Effects due to shock |
|---|---|
| | Fracturing |
| | Variations in extinction |
| | Decrease in porosity |
| Plastic deformation of crystals | Crystal dislocations |
| | Formation of diaplectic (or thetomorphic) glass or maskelynite |
| Partial melting of fine-grained matrices | |
| | Welding of adjacent grains |
| Formation of vitric breccias | |
| Annealing of dislocations and cosmic ray tracks | |
| Recrystallisation of matrices – increase in mean grain size | |
| Release of volatiles – formation of vugs – deposition from vapour phase | |
| Partial melting of clasts and their absorption | |
| Total melting | |
| Vaporisation | |

Such complex textures were clearly discernible in photographs taken on the lunar surface, particularly at the Apollo 14 and 17 sites (Figures 2.18 and 2.35). But several generations of brecciation may in some cases be discernible even within the rock specimens that were returned to Earth. The huge Apollo 14 breccia known as Big Bertha, for example, contains a clast within a clast within a clast (Figure 3.25). Clearly, for the long history of this particular rock to have been preserved so well throughout multiple periods of brecciation, it cannot have been subjected to such extreme temperatures as those experienced by Genesis Rock.

Whether or not the earlier generations of brecciation in Big Bertha might even predate the excavation of the Imbrium Basin is, however, an arguable point. For it could be that a single major impact could have led to successive waves of brecciation, as the ejecta from the basin swept outwards from the crater, incorporating crustal materials on its way. So we are clearly still a long way from associating particular breccia structures (even within the large Apollo 17 boulders) with specific lunar events. This must await much more detailed sampling of the Moon.

*Intermediate stages of metamorphism.*
Cataclastic and recrystallised textures clearly represent just two extremes in a continuum of metamorphic grades. Between them are a sequence of breccia types distinguishable from one another by the peak temperatures to which they were subjected.

Figure 3.25. *Big Bertha*. This large Apollo 14 breccia shows evidence for its complex history. With the help of thin sections, as many as four generations of clasts are visible. (NASA, JSC–LRL.)

When lunar rocks were heated, it was invariably the finest-grained regions which responded first to the changing environment. For the smaller the mineral grains, the more physical contact they had with adjacent grains and the more rapidly they could react with one another. This was particularly true for rocks formed from the lunar regolith, which is just a physical mixture of minerals that were never in chemical equilibrium with one another.

The first stage of thermal metamorphism occurred at temperatures as low as 950 °C, with the plastic deformation of crystals and the mobilisation of late stage (i.e. low melting point) minerals. As the temperature of metamorphism rose, increasing amounts of glass were formed incorporating small mineral fragments, with this glass being moulded around the larger clasts (Table 3.5).

In such low grade breccias there may be little in the way of chemical differences between clasts and matrix. But visually the two parts may be very different. The presence of glass in the matrix (together with its fine grain size) makes it much darker in colour and may give the breccias a characteristic light–dark structure. Perhaps the best-known example of this type of rock was the aptly named Black and White Rock from the Apennine Front. Dark matrix breccias were also characteristic of the Cayley Plains.

As the peak temperature experienced during brecciation rose further, decreasing amounts of glass resulted because there was a greater tendency for recrystallisation to occur. Recrystallisation tends to produce rather more equidimensional shaped crystals (Figure 3.23) and to restrict the grain size distribution. In low grade breccias, grain sizes may range from micron-sized fragments in the matrix all the way up to coarse-grained clasts of cumulate origin. Recrystallisation tends to encourage the growth of intermediate sized crystals at the expense of the smaller ones.

Gradually, the outlines of clasts became more and more indistinct and a number of chemical changes may have occurred. Volatiles were expelled, for example, creating vugs within the breccia, inside which exotic minerals may have crystallised directly from the vapour phase. Some truly perfect crystal forms have been discovered in this particularly favourable environment. Here, in the thick ejecta blankets of large basins, temperatures as high as 1000 °C may have been maintained for several hours, if not days. One of the most reliable methods for classifying lunar basin breccias, in fact, has been on the basis of their volatile element content.

*Impact melts*

As peak temperatures rose still further, breccias became less and less like breccias and more and more like fine-grained igneous rocks. Indeed, the extensively recrystallised matrices of some highland breccias look suspiciously like the KREEP and Very High Alumina basalts which were otherwise thought to be primary (that is internal) in origin.

Unless the rock has been completely melted, however, we do have

Figure 3.26. *Interrelationships between rock types.* (*a*) This diagram shows how the various types of highland rock may be related, through a cycle of impact metamorphism and melting. Shock was responsible for the formation of cataclastic anorthosites (*b*), whereas large scale impact melting could have given rise to melt rocks and feldspathic residues. (Specimen 2.7 mm across.) (NASA, JSC–LRL.)

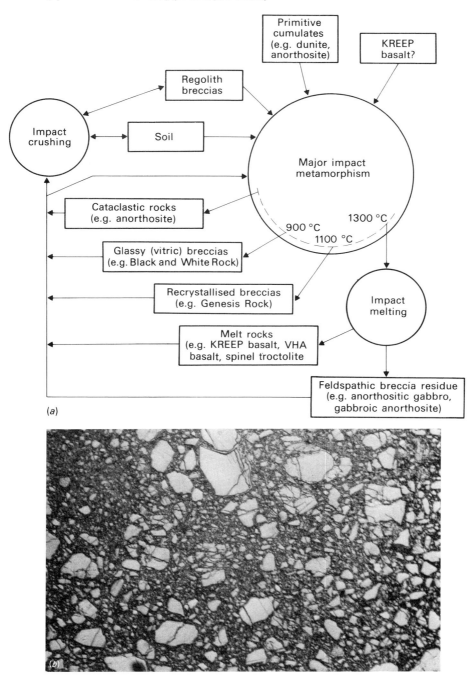

some clues to its mode of origin. Some rocks, for example, contain small crystals (known as chadacrysts) enclosed inside larger ones (known as oikocrysts). This so-called poikilitic texture is generally considered to indicate that the rock crystallised from an impact melt.

The origin of such rock types as KREEP, VHA basalt and spinel troctolite (a highland basalt type containing an abundance of spinel) is clearly very strongly linked to the question of impact melting. Could these fine-grained highland rocks, for example, just be those parts of the crust that were partially remelted? The chemistries of these rocks are certainly consistent with various degrees of partial melting of a mixture of anorthositic gabbro and a little troctolite or dunite, the more extensive the melting, the more feldspathic the end-product.

A number of schemes have been proposed to describe the genetic relationships between highland melt rocks and the more primitive components (Figure 3.26). So perhaps the greatest constraint here is that some of them contain no excesses of certain elements (such as gold) which are usually taken to indicate the presence of meteoritic material. Before moving on to discuss the origins of lunar rocks in general, however, let us make a minor digression to review some of the more exotic lunar minerals and their occurrence in samples returned from the Moon.

## 3.5  Exotic lunar minerals

So far we have only considered the most widespread of lunar minerals, the common silicates and some less abundant phases, such as ilmenite, spinel and metallic iron. But what about the rarer and more exotic minerals? Are there any that are unique to the Moon or that might have some practical or intrinsic value when it comes to exploiting the Moon's resources? Have we, for example, found any minerals (such as diamond) which are worth bringing back to Earth for other than scientific reasons?

*Hydrated minerals*
Perhaps the most important question to ask is whether or not there is any water on the Moon. For, unless water can be extracted (or manufactured) from lunar rocks, the prospects for establishing a permanent lunar base must remain poor. This is because the cost of importing all the required water from Earth might be prohibitive, however efficiently it could be recycled.

At one time it was thought that running water was responsible for the formation of sinuous rilles on the Moon. But, realising that they must have a volcanic origin, we must now ask ourselves if there is any evidence at all for the existence (or even former existence) of water on the Moon.

Well, the answer to this intriguing and all-important question is still elusive. Lunar igneous rocks are certainly very dry, with the only hydrated mineral discovered so far being what characterised the so-called 'rusty' rocks that were collected during the Apollo 16 mission. But even here, in fact, water may be terrestrial rather than lunar in origin.

This particular hydrated mineral was an oxide of iron which was

Table 3.6*a*. *Less common silicates, oxides and metals*

| Mineral type | Mineral | Chemical formula |
|---|---|---|
| Silicates | Zircon | $Zr_2SiO_4$ |
| | Pyroxferroite | $(Fe,Ca)SiO_3$ |
| | Nepheline | $KNa_3(AlSiO_4)_4$ |
| | Tranquillityite | $Fe_8(Zr,Y)_2Ti_3Si_3O_{24}$ |
| | Maskelynite | $CaAl_2Si_2O_8 - NaAlSi_3O_8$ |
| | Amphibole? | $(Ca,Na,K)_{2-3}(Mg,Fe_3,Al)_5$ |
| | | $(F_2(Si,Al)_2Si_6O_{22})$ |
| Metal oxides | Haematite? | $Fe_2O_3$ |
| | Rutile | $TiO_2$ |
| | Corundum | $Al_2O_3$ |
| | Perovskite | $CaTiO_3$ |
| | Baddeleyite | $ZrO_2$ |
| | Armalcolite | $FeMgTi_4O_{10}$ |
| | Zirkelite | $CaZrTiO_5$ |
| | Zirconolite | $CaZrTi_2O_7$ |
| Metals | Nickel | $Ni$ |
| | Copper | $Cu$ |

Table 3.6*b*. *Less common minerals, including sulphides*

| Mineral type | Mineral | Chemical formula |
|---|---|---|
| Carbonaceous minerals | Graphite | $C$ |
| | Cohenite | $(Fe,Ni)_3C$ |
| | Moissanite? | $SiC$ |
| | unnamed mineral? | $Al_4C_3$ |
| | Aragonite? | $CaCO_3$ |
| Phosphorus minerals | Schreibersite | $(Fe,Ni)_3P$ |
| | Monazite | $Ce(PO_4)$ |
| Sulphides | Mackinawite | $FeS$ |
| | Pentlandite | $(Fe,Ni)S$ |
| | Chalcopyrite | $FeCuS_2$ |
| | Chalcopyrrhotite | $(Fe,Ni,Cu)S$ |
| | Chalcocite | $Cu_2S$ |
| | Sphalerite | $(Fe,Zn)S$ |
| | Molybdenite? | $MoS$ |
| | Niningerite | $(Mg,Mn,Fe)S$ |
| Chlorides and hydroxide | Lawrencite? | $FeCl_2$ |
| | Sylvite | $KCl$ |
| | Akaganéite? | $FeO(OH)$ |

actually first discovered in an Apollo 14 breccia. Here it was described as goethite, a mineral with the chemical formula FeO(OH). A more detailed study using X-ray crystallography, however, proved that this mineral was really akaganéite, which has basically the same chemical formula as goethite, but may contain as much as 5% chlorine and has a slightly different crystal structure (Table 3.6).

Hydrogen isotope analyses of the water in lunar akaganéite, however, imply that it is probably the hydration product of an unstable iron chloride mineral (known as lawrencite or $FeCl_2$) which is frequently found in meteorites. In other words it is lawrencite, rather than akaganéite, which exists on the Moon. This lawrencite must react very rapidly on exposure to our moist atmosphere to make the rocks rusty. So it now looks as though we must search elsewhere for genuine lunar water. And proving that such a thing does not exist could be just as difficult as proving that it does.

*Metals, sulphides, carbides and a phosphide*
The occurrence of metallic iron in lunar rocks has already been attributed to the Moon's highly reducing conditions. It is therefore not surprising to find that no highly oxidised lunar minerals have yet been discovered. The few that have been reported (such as aragonite, haematite and mica) must surely be terrestrial contaminants.

But iron (albeit containing cobalt and nickel) is not the only metal that occurs as such in lunar rocks. Occasionally small fragments of native copper turn up and one particle of pure nickel has been reported. Fragments of soft indium metal and what was effectively stainless steel, in the soil were of course contaminants. Indium is incredibly rare on the Moon but was widely used for sealing gaskets on Apollo equipment.

Associated with the metallic iron phase in lunar rocks are one or two other minerals which are also characteristic of iron meteorites. One of these is a carbide known as cohenite, a mineral which has the chemical formula $(Fe,Ni)_3C$ and is structurally, if not chemically, identical to the cementite phase in industrial steels.

Another phase common in iron meteorites is a phosphide known as schreibersite, $(Fe,Ni)_3P$. But on the Moon this mineral is only a minor carrier of phosphorus. Most of the phosphorus in KREEP, for example, is in the form of the calcium phosphates whitlockite (a major carrier of rare earths) and apatite (a major carrier of chlorine and fluorine).

Sadly we have yet to discover any diamonds on the Moon, which has been somewhat of a disappointment because it had been thought that the high pressures reached during meteorite impacts might have been sufficient to convert meteoritic carbon into this precious mineral. The fact is, of course, that high temperatures as well as high pressures were involved. So most meteoritic carbon must just vaporise on impact, although one or two fragments of graphite have been reported.

The only other carbonaceous minerals reported so far have been moissanite (with the formula SiC and possibly also a terrestrial contaminant) and an enigmatic aluminium carbide phase, which, if it really exists, is certainly unique to the Moon.

As far as sulphides (other than troilite) are concerned, there is another iron sulphide (mackinawite), a nickel-bearing one (pentlandite), three copper-bearing ones (chalcopyrite, chalcopyrrhotite and chalcocite) and an iron–zinc sulphide known as sphalerite.

These rare sulphides often condensed from the vapour phase inside vugs (holes) in lunar breccias, where their tiny well-formed crystals are recognised. If other sulphides are going to turn up in future, then it is in this environment that they are most likely to be found.

But one sulphide which is almost certainly not lunar is molybdenite (MoS). This was commonly used on Apollo as a lubricant. The magnesium sulphide niningerite is also a contaminant in the regolith but a meteoritic (rather than a terrestrial) one. Niningerite is a characteristic mineral component of enstatite chondrites.

*Metal oxides*

The low oxidation state of the Moon and the very rapid cooling of lunar lavas has meant that a number of unusual metal oxide minerals exists in lunar rocks. One of these is unique to the Moon and was therefore given a new name: armalcolite (after the three Apollo 11 astronauts Armstrong, Aldrin and Collins), a name that certainly rolls off the tongue better than one alternative that was at first suggested, namely magnesian ferropseudobrookite.

Having the chemical formula $FeMgTi_4O_{10}$, armalcolite is closely related, both chemically and crystallographically, to the iron-free end-member karrooite $MgTi_2O_5$, which does occur on Earth.

Now the reason that armalcolite is present at all in lunar rocks is that there were some titanium-rich lavas on the Moon which cooled very rapidly indeed. For armalcolite is only a stable mineral at temperatures above 1010 °C and will readily decompose by reacting again with the melt phase. In the high titanium Apollo 11 and 17 basalts, armalcolite can be seen to have crystallised even before spinel, but here (Figure 3.27) it is usually seen to be mantled by ilmenite with exsolved lamellae of rutile ($TiO_2$). The extent of this decomposition is clearly a valuable clue to the cooling rate of the lavas.

Figure 3.27. *Armalcolite*. Certain lunar rocks contain an iron/titanium oxide mineral known as armalcolite, after the names of the three Apollo 11 astronauts. Its presence is taken to indicate rapid cooling rates. The enclosing mineral is ilmenite, with exsolved rutile. (Courtesy S. E. Haggerty.)

Armalcolite may be the only metal oxide that is unique to the Moon but it is certainly not the only exotic one. In particular, there are a number of oxides of zirconium, one of the commonest of the Moon's minor metals (Table 3.5).

One of the most widespread of these oxides is baddeleyite, having the basic chemical formula of $ZrO_2$. With its open crystal structure and its capacity for incorporating atoms with high bonding requirements, baddeleyite is a major carrier of rare earth elements, niobium and hafnium. Other zirconium-bearing oxides include zirconolite and zirkelite. The only other metal oxides which have been reported in lunar rocks are perovskite (which although having the basic chemical formula $CaTiO_3$ may also incorporate zirconium) and corundum, or $Al_2O_3$.

*Exotic lunar silicates*

As far as silicates are concerned, basalts which are highly deficient in silica may crystallise small amounts of the ultrabasic mineral, nepheline. Zirconium may occur most frequently in metal oxide minerals, but it is not totally excluded from silicate phases. One lunar zirconium silicate, Zircon ($Zr_2SiO_4$), for example, is well known on Earth as a carrier of uranium as well as zirconium. Another, however, is completely novel, not only in chemical compositions but also in crystal structure. It occurs as bundles of foxy red crystals in the fine-grained interstitial material (sometimes called the mesostasis) in *mare* basalts and has been aptly named tranquillityite, after the site of its first discovery, Tranquillity Base (Figure 3.28).

Tranquillityite is basically an iron–titanium silicate, but zirconium (or another large metal ion such as yttrium) may occupy one of the sites in the crystal lattice in accordance with its basic chemical formula of

Figure 3.28. *Tranquillityite*. This crystal in the Apollo 14 melt rock 14310 is of tranquillityite, a uniquely lunar silicate, rich in zirconium and yttrium. The adjacent crystal is one of whitlockite, a phosphate mineral rich in rare earth elements. Both minerals occur in the fine-grained mesostasis, or residuum. (Courtesy G. M. Brown, Institute of Geological Sciences, London.)

$Fe_8(Zr+Y)_2Ti_3Si_3O_{24}$. It can be seen from this formula that tranquillityite is by no means a simple silicate. But it is a significant one. For, of all lunar minerals, tranquillityite is perhaps the most important carrier of the naturally radiogenic elements, uranium and thorium, despite the fact that it represents at most 0.1% of rocks on a volume for volume basis.

Tranquillityite may have been the first uniquely lunar silicate to be discovered, but it has not been the only one. Another turned up in an Apollo 11 basalt and belongs to the group of minerals known as the pyroxenoids. The pyroxenoids are chemically similar to pyroxenes and are also chain silicates. The arrangements of these chains in the two mineral groups, however, are somewhat different.

This new lunar mineral is structurally similar to the terrestrial pyroxenoid known as pyroxmangite. But instead of manganese, it contains iron and has therefore been given the name pyroxferroite. Its basic chemical composition can best be expressed by the formula $(Fe,Ca)SiO_3$.

Like armalcolite, pyroxferroite is unstable and its presence is indicative of rapid cooling (Figure 3.27). Like tranquillityite, it tends to be one of the last phases to crystallise out. Because it is here in the mesostasis that the chemistry of the melt phase is most exotic. Before the very last drops of silicate melt crystallised out, in fact, two immiscible silicate liquids may have separated from one another, one rich in potassium and silica and the other rich in iron. It is the iron-rich phase which yielded pyroxferroite and tranquillityite. The silica-rich phase crystallised to produce such minerals as quartz and potassium feldspar, the latter in some cases with extraordinary compositions (Figure 3.7).

### The future for lunar mineralogy

Despite the fact that Apollo and Luna brought back samples from only a handful of sites on the Moon, it is unlikely that there are many lunar minerals remaining to be discovered, at least with high abundance. It is of course possible that small amounts of several exotic phases have so far escaped detection in volatile deposits (e.g. sulphates and chlorides like sylvite (KCl) discovered already) or unusual rock types. But as far as hopes for the discovery of abundant concentrated ores are concerned, we can now be reasonably sure that no such deposits exist on the Moon. Small amounts of such ores would surely have found their way to at least one of the landing sites. And the total absence of water on the Moon has meant that all elements have remained finely disseminated throughout the Moon, by terrestrial standards. Any concentration of lunar elements was on a global scale and intimately associated with the origins of lunar rocks. The formation of the calcium and aluminium-rich lunar crusts and the segregation of metallic iron and iron sulphide deep in its interior, for example, are among the chemical fractionation processes that have occurred on a global scale. So it is to the discussion of these global processes that we should now turn.

### 3.6   The origins of lunar rocks

Establishing where, when and how lunar rocks originated inside the Moon has already involved many years of patient analysis, thought and experiment in laboratories all over the world. But, even today, there are still major disagreements over the most fundamental of issues, such as when the segregation of metallic iron may have occurred, and to what extent primary magma compositions can be determined directly from rock analyses. So the best that can be done here is just to review the most important opinions, arguments and constraints and then see how they can best be fitted together to form an acceptable and self-consistent model for the Moon's evolution.

We have been trying to do just this for our own planet for many years and it is only recently that, through plate tectonics, we are beginning to understand clearly the origins of terrestrial igneous rocks. But, even so, we are still very much in the dark when it comes to unravelling the first thousand million years of Earth history. And here on Earth we can travel far and wide, to collect suitable samples and study volcanoes and sea floor spreading directly.

*The problem*

Our sampling of the Moon, on the other hand, has been restricted to just a few hours of study at a handful of landing sites and to the collection of just a few hundred pounds of rocks and soil. Furthermore, lunar volcanoes have been extinct for thousands of millions of years. So how can we possibly hope to understand the Moon's history of volcanism and early bombardment on the basis of such limited information?

But the restriction of volcanism to the earliest lunar epoch does at least mean that lunar geology may be relatively simple. As there was relatively little time available on the Moon for igneous activity to develop, there has never been any continent building there, at least in the terrestrial sense. And, together with the absence of water on the Moon, this means that lunar rocks tend to be less varied than terrestrial ones. So, as one part of the lunar highlands may be very much like another, our current sampling of the Moon may, in fact, be more than adequate.

What we have to do here is build a four-dimensional model of the Moon, working with what is effectively two-dimensional material; in other words, rocks collected from its very surface. What we really want to know is how the three-dimensional Moon (in other words its interior as well as its crust) has evolved in the fourth dimension, that of time.

Fortunately, most lunar rocks are very old indeed and have spent many millions of years exactly where they were collected. Only the very first few hundred million years of lunar history, in fact, are poorly represented in the form of the rock specimens returned by Apollo.

So, of the geophysical constraints on the origins of lunar rock types, their crystallisation ages are obviously crucial. Similarly, we must take into account structural details and temperatures in the lunar interior, in so far as these can be inferred from seismology and heatflow studies. But these are all subjects to be covered in the last two chapters. So let us for the moment confine ourselves to the geochemical constraints and see what these alone can tell us about the evolution of the Moon and its rocks.

*The question of volatiles*

Perhaps the most powerful geochemical constraint on the origin of lunar rocks is their almost universal depletion in volatile elements. Their potassium–lanthanum ratios, for example, are relatively constant and are distinctly lower than in terrestrial rocks and meteorites (Figure 3.15). And the fact that this same ratio is found in all types of lunar rock means that the Moon (or at least that part from which all the rocks at the surface were derived) must be strongly depleted in elements as volatile as potassium. And the general consensus of opinion here, in fact, is that this depletion must predate the formation of the Moon as a planet. The constancy of ratios such as this shows the close genetic relationship which must exist between highland and *mare* rocks.

But this is not to say that there are no local concentrations of volatiles on the Moon. A few samples (such as the Apollo 16 rusty rocks and the orange and green glasses) contain unusually large amounts of certain easily vaporised elements, such as zinc, cadmium, bismuth, chlorine and lead. Electron microscope studies of the green glass beads, for example, revealed cubic crystals of sylvite on their surfaces and it is likely that metals other than potassium are also present as their chlorides in such environments, because, in many cases, metals are more volatile as the chlorides.

It has been suggested that such volatile elements may, in fact, be extralunar in origin, perhaps added to the lunar crust during cometary impacts. But the isotopic composition of the lead in these deposits is unequivocally lunar and we must therefore conclude that there are pockets of volatiles in the crust which have occasionally been mobilised, either during igneous fumarolic events or as a direct consequence of meteorite impact. This is an intriguing conclusion, and one which we will be returning to when we discuss extinct radionuclides in the final chapter.

*Is the Moon enriched in refractories?*

Not only is the Moon as a whole depleted in volatile elements, it may also be enriched in refractory elements, elements such as uranium, titanium and the rare earths. If this is indeed the case then there are important consequences here for the origin of the Moon as a whole.

Some rocks, of course, are highly enriched in certain refractory elements. KREEP basalt, for example, contains high concentrations of rare earths. But these are just geochemical enrichments and other samples, such as the anorthosites, may be correspondingly depleted. There is some evidence, however, to suggest that the Moon as a whole may have several times the rare earth content of the primitive chondritic meteorites.

An enrichment factor of 5, for example, is that required for a mixture of anorthositic gabbro and KREEP to have no net europium anomaly. A similar factor is found in the Apollo 15 green glass beads, generally accepted to be the most primitive of *mare* materials. The green glass has essentially no europium anomaly, implying that its source region inside the Moon can never have had crystals of plagioclase removed from it.

A similar (or perhaps slightly smaller) enrichment factor is required to account for the rather large amounts of heat which are currently flowing out through the Moon's crust, as measured by the Apollo 15 and 17 heatflow probes. The heatflow results will of course be discussed in some detail later, but suffice it to say here that the Moon certainly has to be slightly enriched (×3?) in the refractory heat-producing elements uranium (i.e. 30–40 parts per billion) and thorium in order to account for these results.

### The removal of siderophiles and chalcophiles

If the question of the Moon's refractories is controversial, then the mechanisms by which lunar rocks came to be depleted in siderophile and chalcophile elements are even more so.

That lunar rocks are indeed depleted in elements that readily dissolve in a molten metal or sulphide phase is not in question. The amounts of gold and rhenium in *mare* basalts, for example, are incredibly low (Table 3.4). No, what is arguable is precisely when this extraction occurred. Some would say that it must have taken place even before the Moon became a planet. Others have suggested that it segregated soon after the Moon formed, sinking to form layers (or pods) several hundred miles below the lunar crust, or even a small core at its centre. Such suggestions are of course tightly constrained by certain geophysical measurements, such as the Moon's mass distribution and its electrical conductivity. These measurements suggest that the Moon is depleted in iron by a factor of about 3 relative to the chondritic meteorites.

Figure 3.29. *Tungsten–lanthanum correlation.* Tungsten–lanthanum ratios in lunar rocks are remarkably constant, and lower than in chondritic meteorites. Not only does this demonstrate that the Moon is strongly depleted in siderophile elements, it also shows that this depletion must have occurred before the formation of all lunar rocks analysed to date. (After Delano, J. W. & Ringwood, A. E. (1978). *Proceedings of the Ninth Lunar Science Conference.* New York: Pergamon.)

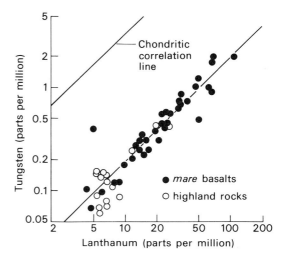

A final approach here has been to suggest that the removal of siderophiles, from highland rocks at least, may have been local rather than global in nature. In other words the sinking of a metallic iron phase (and the consequent extraction of siderophiles) could have occurred within large impact melts (and perhaps even in lava lakes).

Perhaps the most powerful argument in favour of the Moon having lost its siderophile (and chalcophile) elements prior to its formation lies in the remarkable constancy of tungsten–lanthanum ratio in lunar rocks of all types (Figure 3.29). Like potassium and lanthanum, tungsten is only readily incorporated into late stage minerals. But tungsten is a siderophile element whereas lanthanum is not. So if one of these rocks had ever been in equilibrium with large amounts of molten metal, this correlation should not exist. Tungsten should always be heavily depleted in rocks from which metallic iron has been removed. A similar close relationship holds between lanthanum and phosphorus, another siderophile element.

### The formation of cumulates and the early crust
So much, then, for the general geochemical constraints on the origins of lunar rocks. We must now turn to specific types and the best place to start

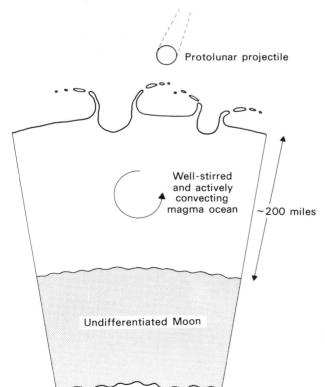

Figure 3.30. *Formation of the magma ocean.* The intense bombardment of the Moon during the final stages of its accretion would have melted its surface down to a depth of 200 miles or so.

is with the Moon's feldspathic crust. How did the surface of the Moon come to be so rich in calcium and aluminium?

Well seismology tells us that the lunar crust must be about 40 miles in thickness assuming a global aluminium content of 8%. And to form a layer of anorthosite as deep as this requires the flotation of plagioclase in a magma 'ocean' extending more than 200 miles down inside the Moon. In other words, in order to produce such a thick feldspathic cumulate, almost half of the Moon, if not more, must at some stage have been completely molten.

But plagioclase is not generally the first to crystallise from a silicate melt (Figure 3.14). Olivine usually precedes it and olivine, being dense, would inevitably have sunk to the bottom of the actively convecting global magma chamber to produce a thick zone rich in magnesium and poor in silica. This olivine cumulate may well have trapped some silicate liquid, but this liquid would not have had a europium anomaly because plagioclase had yet to crystallise out of it (Figure 3.32).

The evidence for cumulates in the early Moon has led to the popular idea that the energy responsible for the early heating of the Moon was provided by gravitation, as incoming matter came together during the final stages of lunar formation and its kinetic energy was converted into heat (Figure 3.30).

Figure 3.31. *The chilled crust.* After the intense bombardment phase, a chilled crust would have formed, having the composition of the liquid beneath it, in other words rich in magnesium and chromium. The absence of nickel in the highlands today implies that siderophile elements must have been removed from the Moon prior to the formation of this crust.

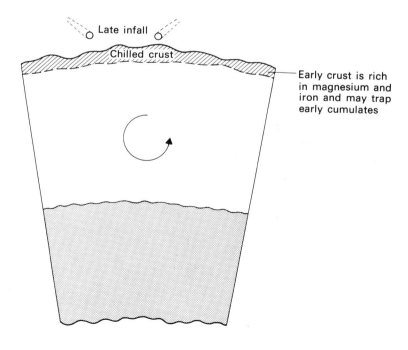

But at some point this kinetic energy input would have fallen well below that necessary to prevent a 'mushy skin' forming on top of the melt. So the formation of such a chilled surface layer would have been an inevitable consequence of the efficient loss of heat from the surface by radiation into space. And this zone (which would have been augmented by, say, the last 1% of infalling protolunar matter) would have constituted the earliest lunar crust.

It would not, however, have been an anorthositic one. Instead, it would have had a composition similar to that of the molten Moon beneath it, in other words rich in magnesium, chromium and iron. This component is at least chemically recognisable in highland rocks and its low nickel content is powerful evidence in favour of the Moon's global depletion in siderophile elements. The formation of this early crust would have prevented the sinking of all olivine cumulates and therefore goes some way towards accounting for the ancient dunites and troctolites (Figure 3.31).

### Later cumulates and differentiates

As the crystallisation of the magma ocean progressed still further, olivine crystals would have become increasingly rich in iron and chromium, and the melt must have gradually become saturated with silica. Consequently, orthopyroxene would have started to crystallise out and sink alongside the olivine (Figure 3.33).

As the temperature continued to drop, calcium-rich plagioclase would also have started to crystallise but, being less dense than the melt, this

Figure 3.32. *Early olivine cumulate*. One of the first minerals to crystallise would have been olivine, which would have sunk to produce a magnesium-rich cumulate without a europium anomaly, but with small amounts of trapped liquid.

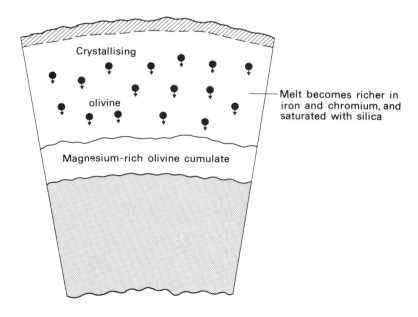

plagioclase would not have sunk. Instead it would have risen to form a thick feldspathic cumulate just beneath the primitive crust, in places trapping small amounts of pyroxene with it to form norites and pyroxenites. At this point, the major separation of europium (i.e. out of the melt and into the crust) would have occurred (Figure 3.33).

The continuing bombardment of the Moon by large projectiles would have ensured that the primitive magnesium-rich crust would not have retained its identity as a discrete rock unit. Instead, it must have become thoroughly dispersed within the feldspathic crust, where it is now only recognisable by virtue of its chemical signature.

But the flotation of plagioclase did not mark the end of the early differentiation phase. The remaining liquids would have continued to crystallise and segregate, forming olivine, clinopyroxene and ilmenite cumulates, and finally leaving a residuum, rich in the incompatible elements which are characteristic of KREEP, with its highly fractionated rare earth pattern. The global differentiation of the Moon would therefore have led to an overall upward segregation of potassium, thorium and uranium, the principal producers of radioactive heat in the Moon (Figure 3.34).

Although grossly over simplified, this picture does at least give some idea of how the lunar magma ocean must have differentiated, and certainly reflects the current consensus of opinion in this field.

Figure 3.33. *Subsequent cumulates*. As the silica content of the melt increased, orthopyroxene would have crystallised with the olivine, and calcic plagioclase would have risen to form a feldspathic cumulate crust. The ferromagnesian minerals became gradually more iron-rich.

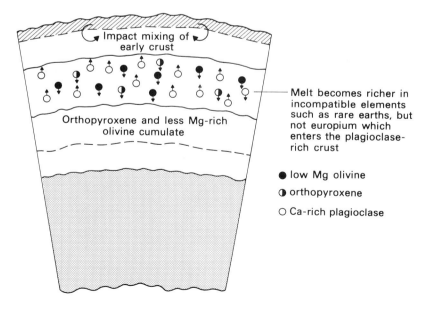

## Highland rock origins

So far we have just considered how cumulate rocks (such as dunite, troctolite and anorthosite) could have built up in the lunar highland crust. But what of those rocks which are more typical of highland terrain, namely the anorthositic gabbros and feldspathic basalts? To what extent are these primary rock types?

Well the question of the origin of KREEP is certainly a controversial one. Some believe that certain parts of the lunar surface were flooded by genuine KREEP lavas, in other words partial melts from the lunar interior. Others, however, feel that KREEP is just the result of impact melting and the subsequent chemical differentiation of the melt. The remarkably similar rare earth patterns in KREEP rocks with different enrichment factors certainly favours this mixing model. A third possibility is that the final residuum from the Moon's early global differentiation was brought to the surface during major impact events. This would certainly help to explain the existence of such extreme rock types as the granite and high potassium KREEP found in rock 13.

One piece of evidence which tends to favour genuine KREEP lavas on the Moon is the existence of small basalt fragments which are completely uncontaminated by meteoritic debris. Impact melts, for example, would be expected to contain excesses of certain siderophile elements from the

Figure 3.34. *Late cumulates and KREEP*. The melt became increasingly rich in iron, titanium and incompatible elements. Clinopyroxene and ilmenite would have sunk and the plagioclase would have become progressively rich in sodium. The melt would have become extremely rich in uranium, thorium, zirconium and rare earths, particularly the light ones.

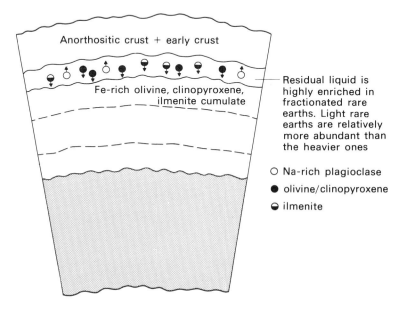

Anorthositic crust + early crust

Fe-rich olivine, clinopyroxene,
ilmenite cumulate

Residual liquid is highly enriched in fractionated rare earths. Light rare earths are relatively more abundant than the heavier ones

○ Na-rich plagioclase
● olivine/clinopyroxene
◖ ilmenite

projectile responsible. But even here it could be argued that the segregation of metallic iron within the impact melt could have rapidly removed those all important siderophiles. So we are clearly still some way from reaching a firm conclusion about the origins of KREEP and related basalts.

The situation with anorthositic gabbro and gabbroic anorthosite is even less clear cut. Because most of these rocks are breccias, which may or may not have polymict origins, having been subjected to severe shock and extensive thermal metamorphism. Whether anorthositic gabbro really was once a primary igneous rock (rather than just a mixture of anorthosite, KREEP and the primitive magnesium-rich component) and whether such rocks as KREEP and VHA basalts are the products of partial melting of anorthositic gabbro are still very much open-ended questions at this time.

### Mare *basalts by partial melting*

The situation with the *mare* basalts is rather different because these are rocks which quite clearly were derived from considerable depths within the Moon. How can we be so sure of this?

Well our most important clues come from experimental petrology. This is a field of study in which silicate melts are investigated under laboratory conditions to find out how certain silicate compositions will behave under various conditions of temperature and pressure.

Now there is one fact that we can be reasonably sure about from such studies and that is that the source regions for the *mare* basalt did not themselves have the composition of *mare* basalt. This is because *mare* basalt deep inside the Moon would exist as a rock type known as eclogite, the density of which exceeds that of the Moon as a whole. Furthermore, *mare* basalt is not the sort of composition we would expect the lunar interior to have. Based on terrestrial and meteoritic studies we would have anticipated a composition closer to that of pyroxenite or dunite.

The generation of *mare* basalt magmas, then, was through a process of partial (rather than total) melting. And this is where experimental petrology really comes into its own. Taking a likely source rock composition (in other words a pyroxenite), we can establish at what pressure and temperature a liquid can be obtained from it that matches, say, the composition of a fine-grained low titanium *mare* basalt. (We should remember here that coarse-grained basalts could have suffered some degree of fractional crystallisation and may therefore not be representative of the original parent magma.)

Well, in the case of the *mare* basalts, they can indeed be produced by partial (a few per cent) melting of a pyroxenite source rock, but only at pressures of several tens of kilobars. And these pressures are only attained at considerable depths within the Moon. Only a small degree of partial melting would have been required here, however, because, on a volume basis, the *mare* basalts form very small proportion (say 1%) of the lunar crust and only 0.1% of the entire Moon.

Such studies, for example, suggest that the sources zone for the Apollo 15 green glasses must be some 200 miles below the lunar surface. This happens to be exactly where those olivine cumulates should have ended up. And the negligible europium anomaly in the green glass is certainly consistent with the postulated absence of such an anomaly in this olivine cumulate (Figure 3.35).

The high pressures necessary to produce *mare* basalt magmas should be compared with the low pressures required for the production of KREEP (Figure 3.36). The composition of KREEP corresponds to that of what is called a low pressure cotectic. In other words, KREEP can be generated by minor partial melting of the highland crust very close to (or even on) the lunar surface. And it is this low pressure origin for KREEP that makes it difficult to decide whether or not it really is a genuine igneous rock. Only by thorough equilibration with crustal materials could KREEP have originated from any appreciable depth.

### A sequence of basaltic lavas

If KREEP lavas were perhaps just impact melts then at least the aluminous basalts (the first *mare* rocks), were produced by partial melting of the lunar interior. It seems, in fact, that the source zone for the aluminous

Figure 3.35. *Mare basalts by partial melting of cumulates*. The compositions of *mare* basalts can best be explained if they were generated by partial melting of cumulates as a zone of melting gradually moved in towards the centre of the Moon. The close relationship between KREEP and the high titanium basalts is apparent from their complementary rare earth patterns. The absence of a europium anomaly in the green glass is consistent with an origin from the early olivine cumulate.

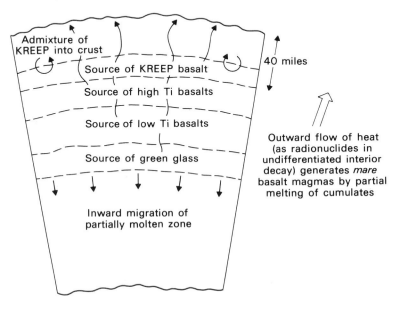

basalts must have contained plagioclase, and, because this mineral is not stable at high pressures, aluminous basalt magmas cannot have come from very deep down inside the Moon, 60 miles at the very most.

The next *mare* basalts to be erupted on the Moon were the high titanium basalts characteristic of the eastern *maria*. These must have come from a 100 mile deep titanium-rich and iron-rich clinopyroxene, olivine and ilmenite cumulate (relatively depleted in light rare earths compared to the heavy ones because of the removal of KREEP). Finally, the low titanium lavas must have originated from an earlier pyroxenite layer (with its high zirconium–niobium ratio), just above the still more primitive olivine cumulates, the source of the green glass (Figure 3.35).

The heat necessary for basalt magma generation must have come from the naturally radioactive elements uranium and thorium that were still trapped inside the Moon's undifferentiated interior. For, with these elements in the original magma ocean having been segregated upwards during the Moon's early differentiation, the *mare* basalt source zones would not have been able to melt themselves. In other words, if they had already contained sufficient radioactivity to melt them they would never have solidified in the first place.

Figure 3.36. *KREEP – a low pressure cotectic*. The compositions of feldspathic highland basalts fall close to the cotectic line (A–B–C) on a low pressure phase diagram. This implies that KREEP must have been generated close to the lunar surface, or even on it as an impact melt. (From Hays, J. F. & Walker, D. (1973). *Earth Planet. Sci. Lett.*, **42**, 897–961.)

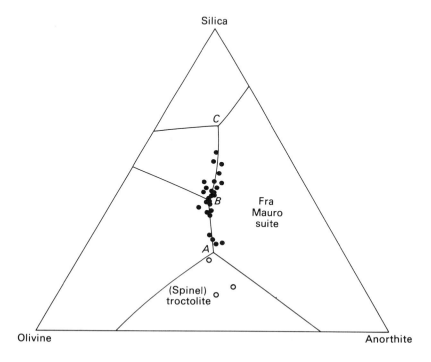

This picture of a zone of melting which gradually migrated towards the Moon's centre, generating basaltic magmas on the way, must of course be a gross oversimplification of reality. For one thing, how did the magma manage to reach the lunar surface at all? To what extent does convection affect the model and how were magmas modified during their upward migrations? Some investigators believe, for example, that the aluminous basalts may have been the product of low titanium magmas which incorporated plagioclase as they penetrated through the lunar crust. This would certainly account for their relatively late eruption at the Luna 16 and Apollo 12 landing sites.

The whole question of how and why basaltic magmas were erupted from depth is clearly a crucial one and we will be returning to it when we move on to discuss gravitational anomalies, the shape of the Moon and its detailed thermal history. Why, for example, were *mare* basalt lavas largely confined to the Earth-facing hemisphere?

But it is now time to put aside global problems for a while and focus our attention down to a smaller scale. In other words, we must consider the Moon's thinnest and outermost layer, the lunar regolith, the boundary between the Moon's crust and the interplanetary environment.

# 4. The regolith

## 4.1 The fabric of the regolith

In the preceding chapter our attention was focussed on the larger samples
returned from the Moon; namely, the crystalline rocks and breccias.
Only when describing some of the less common components (such as the
Apollo 11 Luny Rock and the Apollo 15 green glass), in fact, did we make
any mention at all of the smaller particles. But the Moon's surface is
everywhere covered with a thick layer of dust known as the regolith. And
the time has now come to investigate this regolith in rather more detail.
Among the crucial questions to be answered are: (*a*) what does it consist
of, apart from being just a mechanical mixture of fragments chipped off
larger rocks; (*b*) how did it develop; and (*c*) how does it interact with its
interplanetary environment? We will begin by investigating the fabric of
the regolith and then move on to study the destructive and chemical
effects of the solar wind, cosmic rays and micrometeorites.

### Physical and mechanical properties

Before the days of Surveyor, one of our chief fears was that the Moon
(and the *maria* in particular) might be covered with a thick layer of
unconsolidated dust, capable of having a quicksand effect on any spacecraft which attempted a landing. The Luna and Surveyor probes succeeded in dispelling our worst fears (Figure 2.1) but before Apollo we

Figure 4.1. *Man's first footprint on the Moon.* This bootprint made by Neil Armstrong on
Apollo 11 conveys a good visual impression of the fine grain size, bearing strength and
cohesiveness of the lunar soil. Erosion processes on the Moon are so slow that it will remain
recognisable for millions of years. (NASA, Apollo.)

still did not know quite how fine the soil really was, nor how far down we would have to drill before encountering bedrock. One of the prime objectives of the early Apollos then was soil mechanics; experiments had to be performed to determine the porosity, cohesiveness and bearing strength of the regolith.

These experiments showed that the porosity of the uppermost few inches may be as high as 45%, but this value gradually decreases with depth, particularly where the soil has been compacted by impact (Figure 4.1). The bulk densities of lunar soils may therefore be little more than half that of their parent rocks. And it is this porosity, coupled with the complete absence of water in the regolith, which is responsible for some strange seismic effects, as we shall discover in the next chapter.

But what about the depth of the regolith? Well our earliest clues here came from cratering statistics. A marked decrease in the ratio of crater depth to crater diameter must indicate the crater depth at which a coherent substratum is encountered during impact. And, as expected, only those craters with diameters larger than the critical value are found to have blocky rims.

These studies were complemented by active seismometry on the later Apollo and indicate regolith depths of 30–50 feet in the highlands (and on the landlocked and ancient *mare* plain of Taurus Littrow), less than 15 feet on the young western *maria* and only a few inches on the lip of Hadley Rille (Figure 2.25) and on the rims of young fresh craters such as North Ray (Figure 2.29). The thickness of the regolith at any particular point is therefore related to local mass wastage processes as well as to the antiquity of the surface itself.

*Regolith growth*

The regolith must grow in thickness, then, not only by the impact excavation of underlying bedrock, but also by horizontal mass transport, as at the foot of the North Massif. What has to be established now is the relative importance of these two processes. In other words, does the regolith grow mostly by soil deposition or by vertical turnover? To answer this question we must first turn our attention to the core samples and discover what stratification, if any, exists in the lunar regolith. Examining the Apollo 11 cores it seemed at first that the lunar regolith must be gardened very effectively. It soon transpired, however, that these samples had been grossly disturbed during collection. Well-defined strata, some much less than an inch thick, were clearly apparent in some of the drive tubes and drill cores returned by subsequent Apollos (Figure 4.2). And such stratigraphy can only be preserved if the regolith grows by the gentle deposition of successive layers of dust. It would certainly not survive for very long if the regolith is regularly turned over by meteorite impacts to depths of several feet.

But where does this debris come from? If the regolith is becoming deeper at this point, then somewhere else it must presumably be thinning out. How can the regolith grow at all? Well the solution to this apparent paradox is really not very hard to find. The debris responsible must arrive from local impact craters, inside which the regolith remains much thinner than elsewhere until they are themselves filled up. All core samples were taken from carefully selected areas between craters because no one would have expected to find much in the way of soil stratigraphy in the floor of a fresh crater.

Meteorites clearly do excavate the soil very effectively, then, but most of the excavated soil is deposited at some distance from the impact point, otherwise there would be no crater to be seen. And the only way that the mean depth of the regolith can continue to grow at all, other than by downhill transport of soil from the flanks of nearby hills and the abrasion of larger rocks to form soil fillets (Figure 4.3), is for relatively large impacts to excavate underlying bedrock. Without such larger impacts, soil will simply be transported back and forth without becoming any deeper.

As large impacts must always have been rarer than the smaller ones, the rate of regolith growth in any particular area must, in general, decrease with time. The deeper the regolith becomes, the fewer will be the impacts that are sufficiently large to penetrate right through it.

And it now seems, in fact, that the regolith is only turned over regularly down to a depth of an inch or so. The last time that soil at the base of the Apollo 15 drill string was gardened in this way, in fact, was some 500 million years ago, when the Palus Putredinis regolith was at least six feet thinner than it is today. Ever since then this dust layer has been protected from all but the most penetrating extralunar influences (i.e. the galactic cosmic rays) by overlying soil strata. The deep drill cores are therefore not only sections through space, but also sections through time.

Figure 4.2. *Coarse stratum*. This coarse-grained layer in the Apollo 12 core tube is strong evidence against regular turnover of the regolith to depths of more than an inch or so. (NASA, JSC–LRL.)

*Regolith maturity*

But what is actually responsible for the stratification that is observed in lunar cores? In other words what enables us to distinguish one stratum from another? Well firstly there are sometimes obvious differences in colour to be seen. The Apollo 12 double core, for example, contained a zone of light coloured KREEP glass, whereas the Apollo 17 section through the rim of Shorty Crater included grey and black layers.

Usually, however, it is grain size differences which are most striking, with some layers being distinctly coarser than others, despite their being compositionally similar to one another (Figure 4.2). So how do soils differ in their grain size distributions, how are these differences related to soil maturity, and what sorts of particles characterise soils of different maturities?

Well the most immature soils simply consist of coarse-grained fragments which have been spalled off larger rocks. This is the sort of soil that we would expect to find developing on a recently exposed rock surface

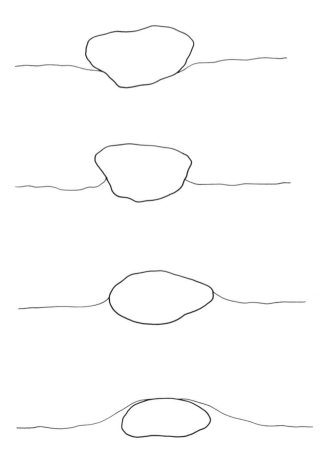

Figure 4.3. *Fillets*. The abrasion and erosion of rock surfaces results in the formation of banks of soil, or fillets.

(Figure 2.34) or in the ejecta blanket of a crater, such as North Ray, which was large enough to have penetrated right through to bedrock (Figure 2.29). The inch-thick coarse layer found in the Apollo 12 core must have originated in this way, as it consisted solely of basalt fragments, with no glasses, breccias or exotic particles in it. But most soils have been more or less reworked. And this impact reworking process gradually decreases the mean grain size, by creating finer and finer dust. Usually, however, some sort of equilibrium is established such that, instead of the mean particle size decreasing for ever, the finest particles start to aggregate together to form regolith breccias, or melted by impact to produce glassy aggregates, both of which tend to be more abundant as coarse particles (Figure 4.4). The net result of this is that most lunar soils

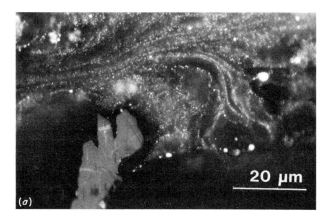

Figure 4.4. *Glassy aggregates.* (*a*) Glassy aggregates contain tiny droplets of metallic iron and have complex flow structures. (*b*) These dark, magnetic and irregularly shaped glassy particles are produced during minor meteorite impacts and are indicative of soil maturity. (From Pillinger, C. T. (1979). *Rep. Prog. Phys.*, **42**, 897–961.)

have a grain size distribution that can best be described as between fine sand and coarse silt, with a median grain size of 80 microns. (Figure 4.5).

Among the parameters which can provide an estimate of the relative maturity (or age) of a sample of soil then, are its mean grain size and the proportion of impact glass within it. The oldest soils, for example, are usually very fine grained and contain the highest proportions of glassy aggregates. So, as these glassy aggregates are primarily responsible for making soils so dark in colour, low albedo may also be indicative of soil maturity. Unlike other glasses (such as brown KREEP and green spheres) glassy aggregates were produced by very minor impacts into the uppermost inch of regolith and contain abundant finely divided metallic iron. This makes them strongly magnetic, as well as dark in colour (Figure 4.4).

### Soil transport

We can clearly never speak of the age of the regolith as a whole, only of the ages of its individual strata. The light-coloured soil at the Apollo 12 site, for example, may well have originated as a ray deposit from Copernicus Crater. But, in places, it is overlain by much more mature *mare* soil and here it has suffered very little from exposure to the interplanetary environment.

One of the most important consequences of impact comminution and soil transport is that the finest fragments have, on average, travelled the greatest distances. In other words, the process which was responsible for

Figure 4.5. *Grain size distribution.* The distribution of grain sizes in an Apollo 12 soil sample is represented here in two ways. Lunar soils are best described as fine sand to coarse silt. (After Heywood, H. (1971). *Proceedings of the Second Lunar Science Conference*, pp. 1989–2001. Cambridge, Mass.: MIT.)

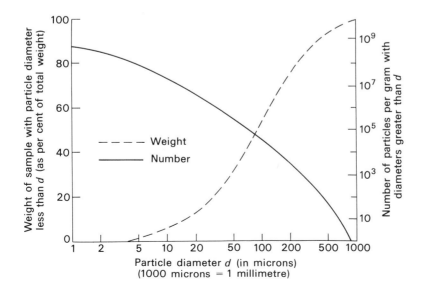

making them smaller is the same as that which dispersed them, the more hops, the farther the distance travelled.

It also explains why anorthosites and KREEP particles were found in the Apollo 11 soil and not among the larger specimens. This is why in most cases the proportion of a particular lithic component in a soil is inversely related to the distance of the sampling area from its point of origin (Figure 4.6). Similarly, when the soils are sieved, KREEP tends to be concentrated in the finest fractions. This obviously makes the dating of individual soils rather meaningless, as different size fractions may have different ages, and also suggests that another soil transport mechanism might be in operation, one which preferentially moves the tiniest particles and leaves the larger ones behind. One such process has now actually been demonstrated in the laboratory, an electrostatic effect which is induced by solar electrons charging up the lunar surface. What happens is that strong electric fields may be generated which levitate the tiniest grains to build 'fairy castles' which promptly fall down. Just how significant such a process is on the Moon, however, still remains to be proved conclusively.

Figure 4.6. *Soil types at Taurus–Littrow.* This map shows the distribution of major soil components over the Apollo 17 landing area. *Mare* basalt fragments increase in abundance from the foot of the South Massif out across the light mantle to Station 4. Note the ubiquitous occurrence of orange glass at the site. (From Rhodes, J. M., Rodgers, K. V., Shih, C., Bansal, B. M., Nyquist, L. E., Weismann, H. & Hubbard, N. J. (1974). *Proceedings of the Fifth Lunar Science Conference*, pp. 1097–117. New York: Pergamon.)

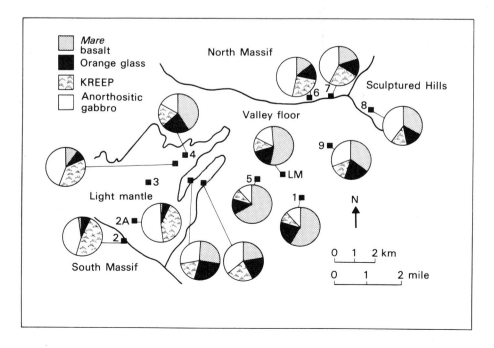

*Regular glasses*

There is one type of soil particle which we should discuss in a little more detail here, namely the glass sphere. Although the chemistries and origins of the orange and green glasses have already been reviewed (and other types were shown to be crucial for the chemical characterisation of highland rock types) we have not yet considered how they ended up in such perfect forms (Figure 2.24). Nor have we explained why glass spheres tend to be coated with volatile deposits.

Although most examples of this type of lunar glass are spherical, there are a small proportion of beautiful dumbbell forms. Such shapes are strongly suggestive of formation in a jet of silicate liquid which broke up in mid-flight. A dumbbell was presumably a section of such a jet which solidified just as it was about to divide into separate droplets.

But under what circumstances could these jets of molten silicate have been generated in the first place? Were they volcanic or impact generated? Might they even be both? In one hypothesis the glass beads were produced when meteorites splashed into already molten pools of volcanic lava.

Perhaps the greatest objection to the impact origin is that, unlike glassy aggregates, glass spheres do not contain occluded rock fragments. They are usually perfectly transparent and homogeneous. And some (like the Apollo 15 green glass) are chemically very uniform, with compositions which can only really be accounted for if they originated from the Moon's deep interior. So, in these cases at least, a volcanic origin (most probably a lava 'fountain' powered by the explosive release of volatiles from the erupting lava) seems to be the most likely explanation. And the volatiles responsible for the lava fountain must then have condensed back on to the surfaces of the spheres to produce their chemically unusual coatings. The short lifetime of molten lavas at the lunar surface makes the likelihood of simultaneous meteorite impact too remote to be acceptable.

But not all glass spheres could have been produced in this way. For some have anorthositic compositions and there have certainly never been any anorthositic volcanoes on the Moon. So for these glasses (and for the irregular glasses as well) an impact origin (by much larger projectiles than those responsible for the glassy aggregates) seems inescapable.

That the chemistries of the principal highland and *mare* rock types are preserved so well in these glasses just confirms that they cannot just be samples of melted regolith. Even the Copernicus cratering impact, large as it was, only produced irregular shaped glasses. It could well be, in fact, that the glassy spheres with highland rock affinities may be associated only with basin-forming events. They are certainly to be found within the Imbrium ejecta blanket, although here they may have more or less recrystallised to form what are morphologically very similar to meteoritic chondrules (Figure 4.7).

One final intriguing observation here is that the aluminous *mare* basalt composition is much more abundant among glasses in the regolith than it is among large hand specimens. Quite why this is so is still a mystery.

But what happens below the regolith? Are there undisturbed layers of solid rock waiting to be sampled? This now seems unlikely. There is probably a gradual transition from a fine grained regolith into a coarse rubble layer which in turn overlies the highly fractured upper lunar crust. So let us turn our attention instead to the even smaller scale and investigate what happens when lunar soil grains are exposed to inter-planetary space. How are they altered by long term exposure to the solar wind, cosmic ray particles and micrometeorites?

## 4.2   Solar wind chemistry

The greatest barrier to astronomical research today is without doubt our atmosphere. It may seem to be transparent enough, but atmospheric turbulence means that we have to site our largest optical telescopes on the summits of high mountains if we are really going to benefit from them.

And the very transparency of the Earth's atmosphere is really just an illusion anyway, one that arises from the fact that only millions of years of evolution have made our eyes sensitive in one of the few atmospheric windows in the electromagnetic spectrum through which radiation may pass. No creature is going to evolve X-ray eyes if there are no X-rays for it to detect at the Earth's surface.

Earth-based astronomy is therefore restricted to viewing the Universe from the near infrared to the near ultraviolet, and also in certain restricted parts of the radio spectrum. So, in order to study cosmic

Figure 4.7. *Lunar chondrule.* Glass spherules may have recrystallised, like this plagioclase one from the Apollo 16 core tubes. They then resemble meteoritic chondrules. (From Meyer, H. O. A., McCallister, R. H. & Tsai, H–M. (1975). *Proceedings of the Sixth Lunar Science Conference*, pp. 595–614. New York: Pergamon.)

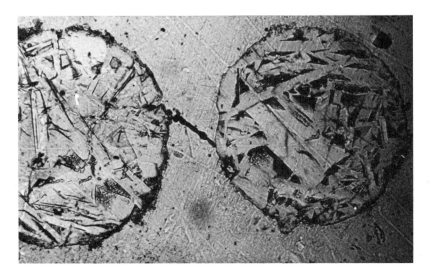

infrared and ultraviolet radiation (as well as gamma-rays and X-rays), we have to resort to launching our telescopes on high altitude balloons, rockets or satellites. In the past few years these new fields of astronomy have taught us a great deal about such exciting objects as neutron stars, black holes and stars which are just being born. So space research has opened up whole new vistas in astronomy, subjects which were certainly undreamt of thirty years ago.

Astronomy is, of course, mainly concerned with interpreting extraterrestrial electromagnetic radiation, as this represents by far our largest source of information about the Universe. But the Earth also collects solid matter from elsewhere in the Solar System (in the form of meteorites and cosmic dust) and also atomic and subatomic particles from distant stars and galaxies as well as from the Sun. These materials bring with them vital clues about chemical and physical conditions elsewhere in the Universe.

The problem is that little of this cosmic debris ever reaches the surface of the Earth in a recognisable form. Only the most energetic cosmic rays, the tiniest micrometeorites and those objects which are large enough (but not too large) to be collected as meteorites on the ground, manage to pass through the Earth's atmosphere largely unscathed. The remainder are (a) deflected by the Earth's magnetic field (b) trapped or burnt up in the atmosphere or (c) totally vaporised on impact. So space exploration provides us with excellent opportunities for studying interplanetary debris and exotic radiations well away from the terrestrial atmosphere. And one phenomenon that we are particularly interested in is the solar wind.

*The nature of the solar wind*

Our first inklings about the very existence of the solar wind came from studies of comets, those frozen wanderers from the outer reaches of the Solar System that assume such splendid proportions as they approach the Sun. What produces those magnificent tails, some of which may be millions of miles long?

Well the most obvious driving force is solar radiation pressure. But some comets have two tails, one composed of dust and the other made up of excited molecules of gas. And, while the dust tail invariably points in an antisolar direction, and so must indeed be produced by radiation pressure, the gas tail is slightly curved and must consist instead of gases (released from the comet's nucleus as it heats up) which are being blown outwards by a wind of electrically charged particles from the Sun, the solar wind.

Satellite studies have now taught us that the solar wind is composed mainly of hydrogen and helium nuclei. These particles follow curved paths away from the Sun because of their strong interaction with the Sun's rotating magnetic field. We also know that solar wind ions travel at a few hundred miles per second (the exact velocity depending on the activity of the Sun at the time), whereas the electrons which ensure that the solar wind as a whole is electrically neutral move much more slowly.

We conclude that, unlike sunlight, which originates from the solar photosphere, the solar wind must come from the outermost part of the Sun's atmosphere, the corona, where gas temperatures are as high as several million degrees. For only at this high temperature are atoms so highly ionised and fast moving. The solar wind can best be viewed, then, as an extension of the Sun's outer atmosphere, which reaches out beyond the orbit of Mars.

But the solar wind is, of course, a very diffuse atmosphere. In the immediate vicinity of the Earth–Moon system, in fact, the number of solar wind ions that would pass through a one square inch aperture in one second is only about 600 million, which amounts to an interplanetary gas pressure of only one ten-thousand-million-millionth of a terrestrial atmosphere, or as near to a perfect vacuum as one could ever hope to achieve in the laboratory. So, although the solar wind has an effective temperature of several million degrees, there is so little of it that its heating effect is minimal.

### Solar wind composition

The fact that the solar wind is really just an extension of the Sun's outer atmosphere means that we could, through the solar wind, perform a chemical and isotopic analysis of the Sun from a distance. Spectroscopic analyses certainly enable us to detect at least some components of the solar photosphere. But a few elements (such as the rare gases) never show up in spectra and isotopic studies are impractical for all but the lightest elements.

Before Apollo, some success had already been achieved in this direction, using satellite-borne solar wind spectrometers. Atoms such as carbon, nitrogen and iron, for example, were among the first elements to be detected in this way. But spectrometers can only measure mass–charge ratios and this can lead to a certain amount of ambiguity (and loss of resolution), because solar wind ions can exist in a large number of ionisation states.

It would be much more satisfactory if we could collect the ions together and analyse them as atoms. And it was this prospect which led to our first attempts to sample the Sun directly, in the form of the Apollo Solar Wind Composition Experiments, flown on Apollos 11, 12 and 14.

The principle behind the Solar Wind Composition Experiment is delightfully simple. A sheet of aluminium foil is exposed to the Sun for as long a period as possible and it is then returned to Earth to see what atoms have become trapped inside it (Figure 4.8 (a)). The high velocities of the solar particles ensure that they will be embedded to depths of at least several hundred ångströms.

The major limiting factor for this experiment, however, was time, not an abundant commodity on any of the Apollo moonshots. Only those elements which are relatively abundant in the solar wind, in fact, could possibly have been trapped in measurable quantities in the limited time that was available. But there were also serious problems of impurities in the aluminium foil to contend with. And atmospheric contamination had also to be considered. There would have been more carbon, nitrogen and oxygen adsorbed onto the foil, for example, than could possibly have been embedded into it from the solar wind.

The net result of all this was that the only elements which could be measured successfully were the light rare gases: helium, neon and argon. These could be extracted from the foil by melting it, and analysed with a mass spectrometer. But these three elements are particularly interesting because they are the ones that are difficult, if not impossible, to detect spectroscopically. And we very much wanted to know what their isotopic compositions are in the solar wind in order to compare with terrestrial, lunar and meteoritic values. Because rare gases are totally immune from the chemical fractionation processes which so radically affect the abundances of chemically active elements, they are useful when it comes to investigating the physical processes involved in planet formation.

What then did the Apollo Solar Wind Composition Experiments tell us about the Sun? Well for one thing they confirmed that helium ions are striking the lunar surface at a rate of several million per square inch every

Figure 4.8. *Apollo Solar Wind Composition Experiment.* (*a*) Sheets of aluminium foil were used to collect solar wind atoms. (NASA, Apollo.)

second. And this is just what we would expect if the Sun contains about 10% of helium in its atmosphere.

As it turned out, however, not all experiments gave quite the same number here. The flux of helium in the solar wind (and the small but significant proportion of the rare isotope, helium-3) seemed to vary in a systematic way with the solar activity level during the mission (Figure 4.8(b)).

Obviously the isotopic composition of helium in the Sun cannot change appreciably over such short time periods, so helium must be fractionated isotopically by the interplanetary magnetic and electric fields. This is not very surprising (in view of the very different mass–charge ratios of the two isotopes) because the force on a charged particle depends on that charge, whereas its subsequent acceleration will depend on its mass.

### Neon and argon

It was also feasible to measure neon in the foils, not only the two common isotopes (neon-20 and neon-22) but also the much rarer one, neon-21. The ratio of helium to neon was found to be about 600:1.

This solar neon turned out to be the most unfractionated neon ever to be found in the Solar System. In other words the ratio of neon-20 to neon-22 in the foil (13.6:1) was even higher than that in the most primitive of meteorites (12.5:1).

Any degree of physical fractionation tends to deplete the lighter isotope. So the formation of the earliest meteorites must have been accompanied by neon loss and the preferential loss of neon-20. The neon-20–neon-22 ratio in the Earth's atmosphere is even lower (9.8:1), so neon losses from the Earth must have been ever more extreme. This would certainly account for its scarcity on Earth.

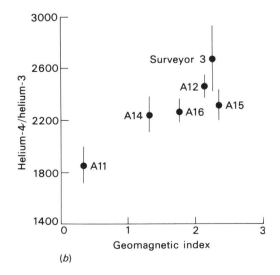

Figure 4.8. *Apollo Solar Wind Composition Experiment*. (b) The ratio of helium-4 to helium-3 in the solar wind varies with the activity of the Sun's surface. (After Geiss, J., Buchler, F., Ceruth, H., Eberhardt, P. & Filleux, C. (1972). NASA SP 315, 14.1–14.10.)

Argon also has three stable isotopes but, unlike neon, argon (particularly argon-40) is a major component of the terrestrial atmosphere. But theoretical nuclear physics tells us that argon-40 should not be present in the Sun anyway. This is because this isotope is only produced by the radioactive decay of potassium-40, an isotope that is very much rarer in the Sun than it is on Earth. We can confidently say, then, that all of the argon-40 extracted from the foils must have been terrestrial argon.

This is very useful because, based on the amounts of argon-40 in a foil analysis (and on the atmospheric ratios of the other two isotopes, i.e. argon-36 to argon-40 and argon-38 to argon-40) we can then make corrections for the light isotopes. What we are then left with is pure solar wind argon.

This argon was, in fact, found to be some 40 times less abundant still than the neon. But amounts of each element were quite consistent with theory and with the known fact that barely 1 % of the Sun's mass consists of elements heavier than helium. A total of less than 5 million such atoms are striking every exposed square inch of the Moon every second.

### Lunar soils as solar wind collectors

Important as these experiments were, they were not really part of lunar science as such. They simply used the surface of the Moon to collect solar wind atoms. And the return and subsequent analyses of aluminium struts from the Surveyor 3 spacecraft (which had been exposed to the solar wind for 30 months) successfully achieved much the same objective. In addition, these results provided us with time-averaged fluxes for the three rare gases in the solar wind.

But the lunar regolith has been collecting solar wind atoms for not just months but millions of years. And so we might expect to find some really novel solar wind chemistry in the soil samples. Here, we should be able to detect not only the heavier rare gases, krypton and xenon, but also light and chemically active species, such as hydrogen, carbon and nitrogen.

Lunar soils, particularly the more mature ones, are extraordinarily rich in hydrogen and helium, both derived from the solar wind. It is not unusual for a gram of soil to contain one cubic centimetre of gas. Indeed the outermost layers of soil grains must contain hydrogen gas at pressures of several tens of atmospheres. But a whole gram of lunar soil would of course never be profligated in this way. Mass spectrometers are now so sensitive that they can detect the helium in a grain of soil weighing less than a microgram. We will certainly never run short of helium on the Moon, then, and, what is perhaps more important, we might be able to use trapped solar wind hydrogen to generate water, a precious commodity indeed in such a dry environment, by reacting solar wind hydrogen with lunar rock.

It was, in fact, quite feasible to detect all stable isotopes of krypton and xenon (as well as neon and argon) in lunar soil. And a strong correlation was found between all rare gas isotope abundances and grain size (Figure 4.9). The finest particles invariably contain the highest concentrations of rare gas isotopes, as we would expect for gases which are located in the very outermost layers of each grain. For a given mass, the tiniest grains have the largest surface area on which to trap rare gas atoms, and this grain size correlation is therefore strong evidence (if we really needed such evidence) for their solar origin.

The study of solar wind components in lunar soils did not of course stop here. We wanted to know exactly how the atoms are accumulated, what physical processes might occur after they have become trapped and what their ultimate fate might be. Presumably the soil cannot continue to accumulate solar wind ions forever. Ion implantation is a rapidly growing research field (particularly in the manufacture of semiconductor materials) and the Moon is an ideal natural laboratory in which such extreme radiation effects can readily be studied.

Well, first of all it appears that solar wind ions are indeed trapped in a surface layer only a few hundred ångströms thick. But some are then knocked further in (Figure 4.10), while others diffuse outwards and are released when the surface is eroded away by ion sputtering to contribute to the lunar atmosphere (see Chapter 5). The net result is that not only is there an overall migration of gases towards the centre of the grain, but the gases also become more and more fractionated. The more mature a soil is, for example, the lower its ratio of helium to neon will be. Helium,

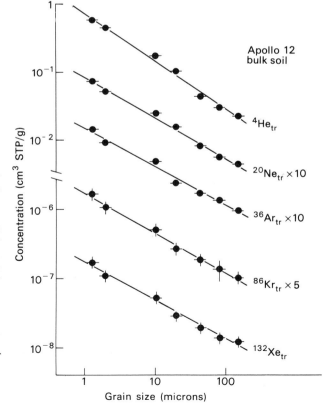

Figure 4.9. *Rare gas grain size correlation.* The concentrations of solar wind rare gases in lunar soils are inversely correlated with grain size, indicating their surface location. This correlation breaks down for the largest sizes because fines may become aggregated to form regolith breccias without appreciable gas loss. Note the break in scale for krypton and xenon. (tr = trapped.) (From Eberhardt, P., Geiss, J., Graf, H., Grögler, N., Mendia, M. D., Mörgeli, M., Schwaller, H., Stettler, A., Krahenbühl, U. & von Gunten, H. R. (1972). *Proceedings of the Third Lunar Science Conference*, pp. 1821–56. Cambridge, Mass.: MIT.)

being less massive than neon, is more easily lost from the grains (Figure 4.14).

This fractionation process is not just limited to elemental ratios. Older soils have lower helium-3–helium-4 ratios (because helium-3 is lost even more easily than helium-4) and proportionally less neon-20. There are a number of parameters, then, based on solar wind rare gases, which relate to soil maturity. But what might such a maturity age mean? How might it be related to ages of the individual rock particles, or to the proportion of glassy aggregates in the soil?

Well, what we are in effect measuring here is the time that the soil constituents have spent on the *very surface* of the regolith. For, while glassy aggregates may form at depths of an inch or so, solar wind rare gases can only be accumulated on sunlit surfaces. But any soil is a constantly reworked mixture of particles. And this ensures that there will always be a good correlation between glassy aggregate content and solar wind gas concentration. And this is exactly what we find (Figure 4.11). But it would be a brave scientist indeed who would attempt to estimate absolute exposure ages based on rare gas contents of lunar soils. The gas

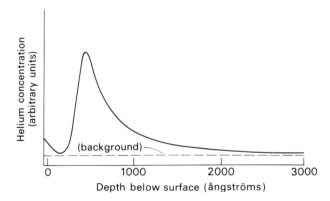

Figure 4.10. *Helium profile*. The experimentally determined depth profile for solar wind helium-4 in a sample of lunar glass shows that solar ions are embedded to depths of several hundred Ångström units, from which they may migrate towards the centre of the particle. (After Muller, H. W., Jordan, J., Kalbitzer, S., Kiko, J. & Kirsten, T. (1976). *Proceedings of the Seventh Lunar Science Conference*, pp. 937–51. New York: Pergamon.)

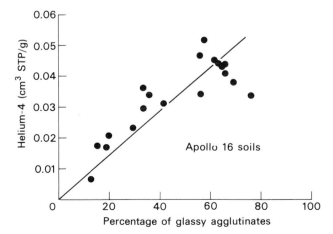

Figure 4.11. *Correlation with glassy aggregate content*. There is a strong positive correlation between the amounts of solar wind helium in a soil sample and the content of glassy aggregates. Both are indicative of soil maturity. (After Charette, M. P. & Adams, J. B. (1975). *Proceedings of the Sixth Lunar Science Conference*, pp. 2281–9. New York: Pergamon.)

recycling processes in the regolith are just too complex to allow us to do this as yet. We can at least say, however, that certain soils (such as the Apollo 15 green glass and Apollo 12 KREEP, both of which contain very small amounts of solar rare gases) must have been exposed to the solar wind for very much shorter periods of time than more mature *mare* soils.

### Carbon and nitrogen

So much then for hydrogen and rare gases. But what about other elements in the solar wind? Well, the three next most common elements in the Sun after hydrogen and helium are oxygen, carbon and nitrogen. Oxygen is, of course, far too abundant in lunar rocks for its solar contribution to be recognisable. Carbon and nitrogen from the Sun have now been detected, but the characterisation of both was by no means straightforward.

The search for carbon was, of course, initially tied up with the search for lunar life. Historically, what happened was this. First of all, it was found that methane (or $CH_4$) is invariably released from lunar soils whenever they are crushed, dissolved or heated. The question was: could this methane be a reaction product of solar wind carbon and solar wind hydrogen? Dissolving the soil in acid always released the most methane, and this tended to suggest that the carbon might be in a combined form (such as carbide), which released methane as a reaction product. As much as one-fifth of the total carbon in lunar soils (up to 200 parts per million) may be released in this way.

We in Bristol then undertook dissolution studies using acids which were 100% isotopically labelled with heavy hydrogen, or deuterium (D). And what we found was that, while some of the methane was indeed $CH_4$, the rest (by far the majority) was pure $CD_4$. In other words it was a reaction product.

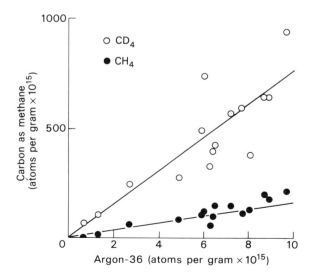

Figure 4.12. *Solar carbon*. Solar wind carbon is released as methane from lunar soils by acid dissolution. Some of this methane is present as such, while some is a reaction product. Both species, like helium, correlate well with glassy aggregate content.

The strange thing was that there was a very good correlation between the amounts of $CH_4$ and $CD_4$ that were released, and that both methane species correlated very well with solar wind rare gas content. So each type of methane seems to be a good indicator of exposure age (Figure 4.12).

It now looks then as if the carbon responsible for both types of methane must ultimately be of solar origin. What presumably happens is that, while some solar carbon is reduced by hydrogen, part of the remainder dissolves in metallic iron, produced during the formation of the glassy aggregates, from which a fraction may be extracted as methane (Figure 4.13). These glassy aggregates certainly have higher carbon contents than any other type of lunar particle (as much as several hundred parts per million) and yield the most methane. Some lunar carbon must come from meteorites (and some may even come from the Moon itself), but the majority of it must originate from the Sun.

The solar nitrogen story is very similar to the carbon one, with some of the nitrogen in lunar soils being in the form of ammonia, whereas much of the rest is associated with metallic iron. The most heavily exposed soils have the highest nitrogen contents and, as with carbon and rare gases, concentrations are highest in the finest grain size fractions and in those soils which have the highest contents of glassy aggregates.

Figure 4.13. *Formation of a glassy aggregate.* A schematic representation of the regolith surface during the impact of a micrometeorite. The outer surfaces of exposed soil grains become saturated with solar wind atoms (dark boundaries). Tiny impacts produce glass-lined craters and generate metallic iron by hydrogen reduction. (From Housley, R. M., Cirlin, E. H., Paton, N. E. & Goldberg, I. B. (1974). *Proceedings of the Fifth Lunar Science Conference*, pp. 2623–42. New York: Pergamon.)

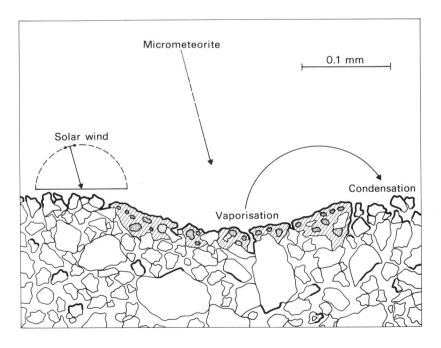

When it comes to numbers no one would pretend to be able to calculate the flux of carbon or nitrogen in the solar wind from the abundances of these elements in the lunar soil. But the amounts present are certainly consistent with fluxes of the order of 600 000 atoms per square inch per second.

Maybe one day we will be able to measure these (and other) elements more directly by extending the aluminium foil experiments. But, for the present, we must be satisfied with the knowledge that we have at least succeeded in proving the existence of elements in the solar wind other than hydrogen and rare gas isotopes. It is really just the rarity of indigenous lunar carbon and nitrogen on the Moon (because they form volatile compounds) that makes a solar origin for what does exist easily recognisable.

### Rare isotopes

But it is also important to be able to measure the isotopic composition of these two elemental species in the solar wind. Ratios of the heavy stable isotope of carbon (carbon-13) to the lighter one (carbon-12) are in fact found to be quite variable in lunar soils, and are higher than in all types of

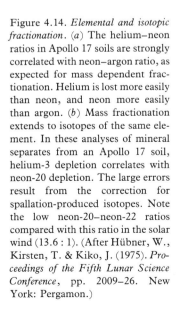

Figure 4.14. *Elemental and isotopic fractionation*. (*a*) The helium–neon ratios in Apollo 17 soils are strongly correlated with neon–argon ratio, as expected for mass dependent fractionation. Helium is lost more easily than neon, and neon more easily than argon. (*b*) Mass fractionation extends to isotopes of the same element. In these analyses of mineral separates from an Apollo 17 soil, helium-3 depletion correlates with neon-20 depletion. The large errors result from the correction for spallation-produced isotopes. Note the low neon-20–neon-22 ratios compared with this ratio in the solar wind (13.6 : 1). (After Hübner, W., Kirsten, T. & Kiko, J. (1975). *Proceedings of the Fifth Lunar Science Conference*, pp. 2009–26. New York: Pergamon.)

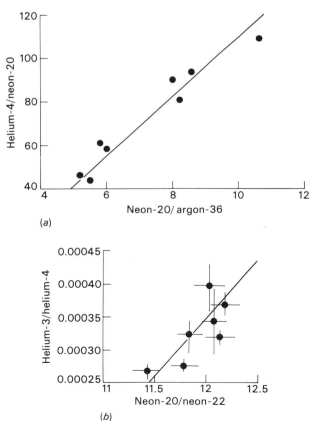

terrestrial carbon, with the exception of certain marine carbonates. So could it be that this high isotopic ratio is characteristic of the Sun? Or does it just reflect the physical fractionation processes which take place in the regolith?

One possible explanation for this carbon-13 enrichment is what is known as 'hydrogen stripping'. What happens is this. Solar wind protons react with already implanted carbon to produce methane which, being a gas, will gradually escape by diffusion. But heavy methane (i.e. methane containing carbon-13) will diffuse more slowly than light methane. And there will therefore be a gradual enrichment in the heavier isotope. Much the same process has been proposed to account for corresponding, but smaller, enrichments in nitrogen-15.

This hydrogen stripping process should not, of course, operate only on previously implanted solar wind atoms. So we should, and indeed do, find enrichments of oxygen-18 and silicon-30 in the outer surfaces of lunar rocks. Here then is a mechanism for the production of lunar water from solar hydrogen and lunar oxygen.

But there are, of course, not just two isotopes of carbon. There is a third one (the radioactive isotope carbon-14), which occurs naturally in our atmosphere because of cosmic rays. Indeed, if cosmic rays did not react with nitrogen in our atmosphere to produce carbon-14, we would not be able to date long-dead organic material by the C-14 method.

The half life of carbon-14 is so short that the Moon's original carbon-14 could not possibly have survived to the present day. But just as carbon-14 is produced in our own atmosphere, so it should also be present in lunar rocks which are directly exposed to cosmic rays. But what we find is that the outermost surfaces of lunar rocks contain marked excesses of carbon-14, excesses which can only be accounted for if the solar wind contributes about 100 atoms of carbon-14 per square foot of the Moon every second. So the solar wind must in fact contribute three carbon isotopes to the Moon.

Similarly, the solar wind must add about 500 atoms per square foot per second of radioactive tritium (or hydrogen-3). As far as the other stable hydrogen isotope is concerned, however (deuterium or hydrogen-2), we would not expect to find very much in the Sun, as it is unstable at solar temperatures. And sure enough the deuterium–hydrogen ratios in lunar soils are very low indeed.

This then concludes our short account of the solar wind and its effects on the chemistry of the regolith. So far we have only detected isotopes of helium, neon, argon, krypton, xenon, carbon, hydrogen and nitrogen, but in the future we will no doubt discover others and also learn much more about the complex chemical and physical processes that these atoms undergo. But it is now time to move up to higher energies. For not all solar particles move at speeds of a few hundred miles per second. Some are ejected from solar flares 100 times faster than this, while ions moving close to the velocity of light are impacting the Moon from distant parts of the Galaxy. Clearly such energetic particles must have significant effects on the chemistry and structure of lunar samples.

### 4.3 Cosmic ray chemistry

The amounts of rare gas collected by the Solar Wind Composition Experiments were of course extremely small. But suitably sensitive mass spectrometers had already been developed by those scientists who were thoroughly experienced at analysing the minute traces of rare gases trapped in meteorites. Solar physics is certainly one field in which the meteoriticists were better equipped than most to cope with the first samples to be returned from the Moon. They were also well practised at working with small and highly precious rock fragments, and already very familiar with Solar System problems and cosmic rocks: exotic minerals, strange chemical and isotopic anomalies and the great antiquity of the rocks themselves.

In addition, they fully appreciated that man was going to succeed in returning rocks from the Moon. For the rest of us it was still just a pipe dream to be believed when, and only when, Moon rocks were back here on Earth ready to be examined.

But by then, of course, the farseeing meteoriticists were ready to start their first analyses. So the period between 1969 and 1972 became a meagre one indeed for meteorite research itself. Only after Apollo 17 had borne fruit did meteoritics once again return to a high level of activity, inspired and rejuvenated by a new understanding of lunar science.

*Cosmic rays*

But meteoriticists were rather overwhelmed by the sheer amounts of rare gas that they did extract from lunar soil. And there did not seem to be any obvious way to relate solar wind rare gas abundances to the ages of specific lunar features. There is little point in developing a dating method which cannot be applied to some significant aspect of lunar history. And the best we can possibly hope for here is to establish how long a particular fragment has spent completely unshielded on the lunar surface, not an age with any Moon-shattering significance.

For the total unshielded exposure time of such a particle may not even have been continuous. The particle could, for example, have spent a much larger period buried at a depth of, say, an inch in the regolith, during which time no solar wind atoms would have been accumulated.

But the particle would not have escaped completely from ionic interactions under these conditions. Far from it in fact. For some ions can penetrate not just inches, but several feet into the hardest lunar rock. Solar particles somewhat more energetic than solar wind protons are known as suprathermal ions, whereas more energetic ones still are emitted during solar storms and are known as solar flare ions or solar cosmic rays. The most penetrating ions of all originate from well beyond the Solar System and these, travelling at speeds close to the velocity of light, are known as galactic cosmic rays.

And it is these cosmic ray ions (rather than the much more abundant solar wind ions) which really can help us to understand how the lunar regolith develops, by enabling us to date such events as the formation of small impact craters. So what chemical traces have cosmic rays left behind in lunar rocks and how can these traces now help us to unravel lunar history?

Well, first of all it is not the cosmic ray atoms themselves which are readily detectable in lunar samples, but rather the products of the nuclear reactions between cosmic ray protons (and helium nuclei) with atoms in lunar rocks. The rare gases found in the lunar soil must certainly include a tiny direct contribution from cosmic rays, but this will invariably be overwhelmed by the much more abundant solar wind component. For the number of solar ions of a particular isotope decreases very rapidly indeed with increasing energy (Figure 4.15).

So how do fast moving protons react with atoms in lunar rocks and what isotopic products might we expect to find? Well there are basically two types of nuclear reaction involved here. The first is called spallation, whereas the second involves reactions initiated by secondary neutrons. The two reaction types will be considered separately, as they provide independent and complementary clues to regolith evolution.

*Spallation*
Cosmic ray spallation can be likened to the first 'break' in a game of snooker. The pack of red balls represents some target atom in a lunar rock, whereas the white cue ball corresponds to an energetic cosmic ray proton. When the cue ball strikes the pack of red balls, some are split off and go their separate ways. In much the same manner, cosmic rays can strip protons, neutrons and larger fragments off any nuclei that they happen to encounter (Figure 4.16).

The more energetic the cosmic ray, the more protons and neutrons it might remove, and the greater may be the atomic weight difference between the original target atom and the spallation product. So the commonest spallation products in lunar rocks should be those initiated

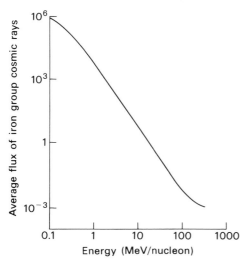

Figure 4.15. *Cosmic ray spectrum.* This plot shows the abundance of iron group cosmic ray nuclei at the lunar surface as a function of energy. The vertical axis is in units of ions per square centimetre per second per unit solid angle with energies greater than that shown by the horizontal axis. Note the strong energy dependency among solar cosmic rays, and the flattening off at higher energies, where galactic cosmic rays start to contribute to the flux.

by the more abundant, lower energy, cosmic rays and should therefore correspond to the removal of just one or two protons (or neutrons) from the most abundant of lunar isotopes. As this process happens very quickly, the nature of the isotopes produced by spallation is not dictated at all by nuclear forces and they may therefore be quite exotic.

Many will in fact be radioactive, while others, although stable, may be isotopes which are normally rare in nature. The stable isotopic ratios of certain elements may therefore be drastically modified. The neon-21–neon-20 ratio in a lunar rock which has been exposed to cosmic rays, for example, may be much higher than this ratio in the solar wind. Both neon isotopes are generated by cosmic rays (from sodium and magnesium) and in approximately equal amounts, but, as we heard earlier, neon-21 is a relatively minor component of solar wind neon.

### Spallation radioisotopes

In order to use spallation reaction to calculate the exposure age of a rock, we must first measure its content of the product isotope and then find out how efficiently this isotope is generated by cosmic rays. It is here that radioactive spallation isotopes are so valuable. For, not only does radioactivity allow us to measure minute quantities of rare isotopes, it also provides a clock with which to calculate production rates.

This can best be illuminated by means of examples. There are many spallation radioisotopes present in lunar rocks, but two of the most useful ones (at least as far as regolith studies are concerned) are sodium-22 and

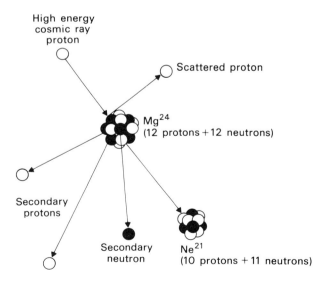

Figure 4.16. *Cosmic ray spallation.* High energy cosmic ray protons cause nuclear transmutations in the lunar regolith by stripping neutrons and protons off target nuclei. Here, neon-21 is being produced from magnesium-24.

aluminium-26. What makes these two particularly valuable is the great difference in their half-lives, the half-life of a radioisotope being the time required for it to decay away to half of its initial abundance.

Now the half life of sodium-22 is only 2.6 years, whereas that of aluminium-26 is 700 000 years. And what this means is that while sodium-22 effectively monitors solar flare activity over the current solar cycle, aluminium-26 provides a measure of cosmic ray intensity during the last 5 million years or so. For, while these isotopes are being created by cosmic rays, they are also decaying away, and so most of the radioactive atoms which are present in lunar rocks today must have been produced within, say, the last five half lives of the appropriate isotope. Manganese-53 ($t_{\frac{1}{2}} = 3.9$ million years), argon-39 ($t_{\frac{1}{2}} = 270$ years) and argon-37 ($t_{\frac{1}{2}} = 35$ days) provide useful complementary information over longer, intermediate and shorter time scales respectively.

An example of what we find is that Apollo 12 rocks tended to have more sodium-22 (relative to aluminium-26) than those returned by, say, Apollo 11. This is because there was an intense solar flare just before Apollo 12 and cosmic rays from this flare boosted concentrations of sodium-22 without markedly affecting the amounts of aluminium-26.

Normally, of course, the concentration of a short-lived isotope such as sodium-22 will have reached a steady state, at which point it will be decaying just as rapidly as it is being formed. But short term variations in solar activity are capable of upsetting this fine balance. And studies of solar activity over long periods of time could eventually help us to understand some knotty problems in terrestrial meteorology and climatology, such as when we might have another ice age. There are so many subtle ways in which the Sun affects our everyday lives that we should try and do all we can to understand this complex interaction. And the surface of the Moon is ideally suitable as a solar laboratory.

In the case of the longer lived aluminium-26 and manganese-53 a steady state concentration may never have been attained in the first place.

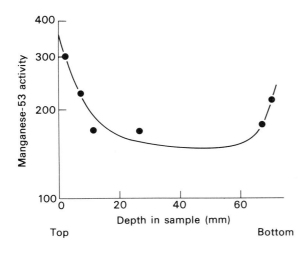

Figure 4.17. *Depth dependency in a lunar rock.* This plot shows the depth dependency of spallation produced radioisotope manganese-53 (half-life = 3.9 million years) in a slice taken through an Apollo 14 breccia. The high levels of activity at the bottom of the rock indicate that it must have rolled over at least once since its ejection from Cone Crater 25 million years ago. (After Herpers, U., Herr, W., Kulus, H., Michel, R., Thiel, K. & Woelfe, R. (1973). *Proceedings of the Fourth Lunar Science Conference*, pp. 2157–69. New York: Pergamon.)

In other words, these isotopes may still be building up to the point at which their decay rates equal their production rates. And under these non-equilibrium conditions we can say something definite about the exposure history of the sample (Figure 4.17).

In the Apollo 12 core tube, for example, sodium-22 concentrations

Figure 4.18. *Depth dependencies in the regolith.* This plot shows the depth dependencies of two cosmogenic radionuclides in the Apollo 12 core. Whereas levels of sodium-22 are consistent with theory, those of aluminium-26 do not reflect the penetrative capabilities of cosmic ray protons, and indicate that vertical mixing in the regolith must be effective down to a depth of several inches. The depth to which cosmic rays penetrate depends on the density of the target, and the dimensions of depth/density are g/cm². (From Rancitelli, L., Perkins, R. W., Felix, W. D. & Wogman, N. A. (1971). *Proceedings of the Second Lunar Science Conference*, pp. 1757–72. Cambridge, Mass.: MIT.)

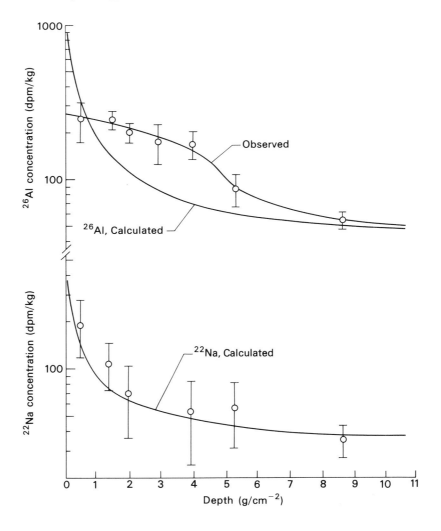

were found to decrease sharply with depth as expected. But concentrations of aluminium-26 in the uppermost inch or so of the core were relatively constant (Figure 4.18). Only below this depth did the amounts begin to decrease with depth. We conclude, then, that while there may be very little movement in the regolith over periods of a few years (as indicated by the sodium-22 concentrations) the uppermost inch or so of the regolith is actively gardened on a million year time scale. If this was not the case, the surface layer would contain a much higher (i.e. steady state) concentration of aluminium-26. Active turnover clearly prevents such a steady state situation from being attained.

*Spallation exposure ages*

Once the concentration of a radioisotope has reached a steady state it can, of course, no longer be used to calculate an exposure age. For however much longer that rock continues to be exposed to cosmic rays, the radioisotope will no longer accumulate further.

But, knowing the half-life of the isotope, what we can do is calculate its production rate from its steady state concentration. And we can then assume a similar production rate for a stable isotope which has similar numbers of protons and neutrons in its nucleus (because spallation is a purely physical process and treats protons and neutrons equally) and so calculate an exposure age for the rock.

Radioactive sodium-22, for example, can be used to estimate the production rate for neon-21, because both isotopes are generated from magnesium and aluminium. More useful still is a radioactive isotope of krypton, krypton-81. Because, using krypton-81, we can than calculate an exposure age simply by measuring a single rare gas isotope ratio. Indeed, krypton-81–krypton-83 spallation ages are considered to be the most reliable of all exposure ages, because krypton-81 has such a long half-life (210 000 years).

There is one major uncertainty left, however, and that has to do with shielding. Was the spallation isotope in question produced when the rock was fully exposed for a short period, or when it was partially (or completely) buried for a much longer time? More than one spallation method may be useful here, because different spallation reactions require protons having different energies and penetrative capabilities. Higher energies are required, for example, to produce argon-38 from iron-56 than are necessary to produce the same isotope from calcium-40.

The exposure ages calculated may then range from a few million years (for rocks which originate from young craters such as South Ray) all the way up to more than a thousand million years for some mature soils, such as those returned by Luna 20. When we have discussed the various other methods for estimating exposure ages, based on secondary neutrons, radiation damage and micrometeorite craters, we will see how it is possible to reconstruct certain aspects of recent lunar history from these results.

## Secondary products

So much then for the major spallation products, but what is the fate of the minor nuclear debris which is also produced? Well, the protons are slowed down very efficiently, because of their positive charge. Whenever they approach an atomic nucleus, mutual repulsion results in some of the momentum of the proton being transferred to the crystal lattice.

Indeed, it is only the fastest cosmic rays which can initiate spallation reactions in the first place. For only they have sufficient energy to penetrate all the way to the nucleus of the target atom. The slower cosmic ray protons (like those in the solar wind and those chipped off target nuclei by spallation) will simply pick up spare electrons and remain trapped in the crystal lattice as hydrogen atoms.

In much the same way deuterium, tritium and helium-3 also occur as spallation by products. Indeed, essentially all of the deuterium present in lunar samples must be there as a byproduct of spallation, as this isotope is very rare in the Sun.

But we would also expect large numbers of secondary neutrons to be produced. And these neutrons will react further with the target, and much more readily than protons because, being electrically neutral, they are not repelled by an atomic nucleus.

Presumably the Sun is itself emitting a constant stream of neutrons (as well as protons and electrons) in the solar wind. But, outside atomic nuclei, neutrons are unstable, decaying with a half-life of only 12 minutes into protons and electrons. So we would certainly not expect many primary neutrons to survive the long journey to the Moon.

As far as secondary neutrons are concerned, however, 12 minutes is a very long time indeed. Quite enough for practically all of them to be slowed down and captured by successive collisions with lattice nuclei.

But some elements are much more efficient at capturing these slow neutrons than others. Graphite and cadmium, for example, are used to keep the numbers of neutrons in a nuclear reactor down below a critical value. And neutron capture is, in fact, the only means by which really heavy elements (such as uranium) can be created in the first place.

## Neutron reactions in lunar rocks

Two of the most efficient scavengers of these neutrons are gadolinium and samarium, two elements in the rare earth group. And, as the addition of a neutron to an atomic nucleus simply increases the relative abundance of a heavier isotope, what we find is that the ratios of gadolinium-158 to gadolinium-157 and samarium-150 to samarium-149 in lunar rocks may have increased, by as much as 2% where the integrated dose of thermal neutrons was particularly high.

The fluence of secondary neutrons in the lunar regolith, unlike that of cosmic ray protons, should not decrease gradually with depth. Instead, the fastest neutrons should strip more neutrons off target nuclei to produce what is known as a 'neutron cascade', which builds up to a maximum at a depth of 18 inches or so. Confirmation of the neutron cascade in the lunar regolith came from gadolinium and samarium isotope analyses of samples from the Apollo 15 deep drill core. These results also imply that the Apollo 15 regolith (at least where it was sampled by the drill) must have had a very undisturbed irradiation history.

As fast neutrons are being slowed down and captured, other nuclear transmutations must be taking place. Most of the argon-37 being produced in the regolith, for example, is not the result of direct spallation, but rather the product of the reaction between a neutron and an atom of calcium-40, which also results in the ejection from the nucleus of an alpha particle. Such fast neutron reactions are responsible for excesses of other stable rare gas isotopes (such as xenon-131 from barium-130), and for the existence in lunar rocks of the superheavy element neptunium, a transuranic element which was recently discovered in nature for the very first time.

Each neutron reaction requires neutrons in a particlar energy range. So the products from different reactions in the regolith can tell us something about the energy spectrum of lunar neutrons. We would expect fast neutron reactions, for example, to occur closer to the surface than slow neutron capture. But, in order to be able to use these neutron reactions to study regolith dynamics, it is important to investigate the present day neutron density and energy spectrum in the regolith. And it was to this end that an ingenious experiment was devised.

### The lunar neutron probe experiment (LNPE)

The LNPE consisted of a long rod which was inserted into the hole left behind after the extraction of Apollo 17 deep drill core (Figure 2.38). Strategically mounted along this rod were a series of uranium and boron detectors for registering neutrons with different energies. The proportion of uranium-235 induced to fission while inside the regolith, for example, was used as a measure of the total number of neutrons which had passed through the detector in the energy range appropriate for this reaction. So what did the LNPE tell us about lunar neutrons and what have we been able to conclude about the lunar regolith from neutron studies?

Well, for one thing, the measured neutron density profile was similar to that inferred from the Apollo 15 drill core results, and was in excellent agreement with theory (Figure 4.19). But today's neutron flux appears to be somewhat higher than the average over the last 500 million years. Quite why this is so remains to be seen. What is valuable, however, is that we are now in a position to calibrate neutron exposure ages. And, as these ages refer to 'exposure' at depths of 18 inches or so, they should prove to be a useful complement to cosmic ray spallation ages, which refer to exposure at the surface of the regolith. If a rock has spent most of its time buried, for example, it should have a high neutron exposure age but a low proton spallation age. If, on the other hand, it has spent its entire history resting on top of the regolith, neutron capture effects should have been minimal. There is certainly still much work to be done to unravel the evolution of the regolith, and lunar neutrons will surely play a major part here.

This then concludes our brief introduction to cosmic ray effects. But chemical and isotopic anomalies are not the only consequences of cosmic particle interaction. There are also drastic physical effects, in the form of radiation damage. So how much do cosmic ions affect lunar crystal structures and how can we use this damage to investigate lunar history?

## 4.4    Radiation damage

The destructive potential of cosmic rays had been fully appreciated ever since they were first studied directly, using sheets of mica flown into the upper atmosphere with high altitude balloons. It was found that whenever a cosmic ray passed through one of these mica detectors, the lattice atoms were strongly repelled, deforming the crystal structure in the immediate vicinity of the ion path. This deformed region was then attacked by etching reagents more readily than the other parts of the crystal, and a deep track was produced, defining exactly where the cosmic ray had penetrated. These tracks could then be viewed directly

Figure 4.19. *Neutrons in the regolith.* (*a*) The results from the boron detectors mounted on the Lunar Neutron Probe Experiment show that secondary neutrons build up in a cascade, and are most abundant at a depth of 18 inches or so. A similar profile was defined by gadolinium and samarium isotope ratios in the Apollo 15 core, indicating that a very undisturbed regolith had been sampled there. Vertical mixing would certainly have resulted in deviations from this ideal curve. (Courtesy D. S. Woolum, California Institute of Technology, Pasadena.)

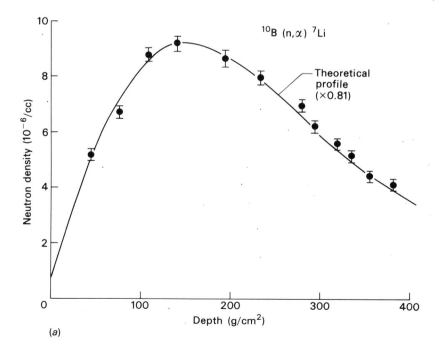

(*a*)

(by either optical or electron microscopy) or a replica could be made, so
that each track is revealed as a plastic spike.

Each dislocation is produced by a single heavy cosmic ray ion (such as
iron), so there is a one-to-one relationship between the number of tracks
in the detector and the number of heavy cosmic rays that have passed
through it. In other words, the longer the detector is exposed to cosmic
rays, the more tracks will be registered.

*Natural cosmic ray detectors*

Any object which has been exposed to interplanetary space will of course
have been subjected to cosmic ray damage. So one serious objection to
extended manned space missions is that astronauts could suffer from
unacceptably high doses of radiation, not just X-rays and gamma-rays,
but cosmic rays as well.

A single cosmic ray can destroy several living cells, and so long
exposure to them could lead to extensive brain damage. Indeed, the
flashing lights reported by several Apollo astronauts were probably just
cosmic rays impinging on their retinae. So any plans for a continuously
occupied space station must include either regular crew turnover or large
amounts of dense shielding. Although more desirable, the latter alternative would be prohibitively expensive, unless, as seriously proposed, the
space station is constructed inside a hollowed out asteroid.

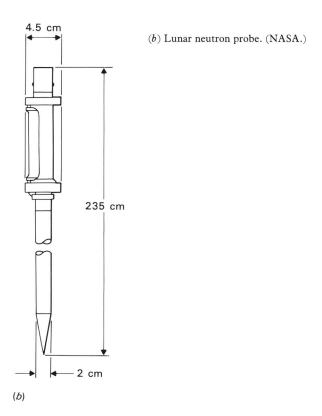

(b) Lunar neutron probe. (NASA.)

(b)

Returning to the subject of cosmic ray tracks, these had already been revealed in extraterrestrial samples (i.e. meteorites) long before Apollo. So we already knew which minerals and etching conditions would be most suitable for track revelation in lunar rocks. The questions to be asked now were how can tracks in lunar rocks be used to learn more about cosmic rays and, conversely, how can these cosmic ray tracks help us to probe lunar history?

The problem with meteorites is that they are not pristine cosmic bodies. For, during their passage through the atmosphere, their outer surfaces were more or less ablated away. And we have no reliable way to establish how far below the original meteorite surface a particular crystal may have been.

Rocks from the Moon, in contrast, do not suffer from this uncertainty. But, unlike meteorites, which received their doses of cosmic rays from all directions as they tumbled about in space, lunar rocks are always shielded on one side by the Moon itself, and may also have had complex exposure histories within the regolith.

### The cosmic ray spectrum

Before we can learn much about the Moon, then, we must discover more about cosmic rays themselves. Just how abundant are the heavy ions necessary to have produced tracks in lunar rocks, and with what energies do these cosmic rays strike the Moon? Well, the best way to measure

Figure 4.20. *Cosmic ray detectors.* This experiment consists of two surfaces (consisting of pieces of mica, aluminium, glass etc.) for registering cosmic ray tracks after etching. (After Price, P. B., Chan, J. H., Hutcheon, I. D., Macdougall, D., Rajan, R. S., Shirk, E. K. & Sullivan, J. D. (1973). *Proceedings of the Fourth Lunar Science Conference*, pp. 2347–61. New York: Pergamon.)

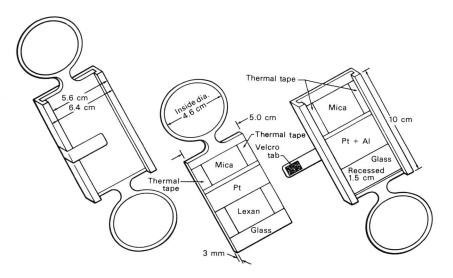

present day cosmic ray activity is to use artificial detectors, just as the Solar Wind Composition Experiment was the best way to investigate the solar wind directly. So a number of special detectors were designed, along the lines of those balloon-borne sheets of mica. These were taken to the Moon by Apollos 16 and 17, and returned to Earth at the end of each mission (Figure 4.20).

These experiments enabled us to measure not only the numbers but also the energies of heavy cosmic rays. For, by stacking the mica detectors on top of one another, only the fastest ones could penetrate through to the deepest levels in the stack. Furthermore, by using plastic and glass detectors (instead of just ones made from feldspar or mica) the ions could be analysed according to their atomic masses. Plastic films, for example, were used to register light slow moving ions, whereas silica glass plates only detected those ions more massive than iron.

The choice of missions on which to fly the two detector packages was indeed fortunate. For there was a small solar flare during Apollo 16 which provided an opportunity to record a really intense burst of solar cosmic rays. The Apollo 17 experiment, on the other hand, yielded a record of normal cosmic ray activity, which turned out to be quieter by a factor of at least 1000.

From these results, then, we can now conclude that most solar cosmic rays must originate in solar flares. For the Apollo 17 'quiet' flux was quite insufficient to account for the density of tracks in a piece of glass (taken from the Surveyor 3 television camera optics) which had been exposed to cosmic ray bombardment on the Moon for $2\frac{1}{2}$ years (Figure 4.21).

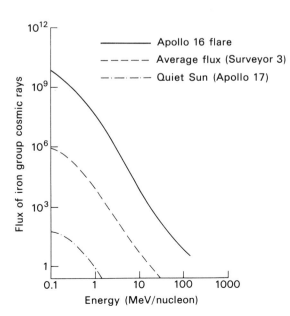

Figure 4.21. *Cosmic ray measurements*. The flux of cosmic rays registered by the Apollo 16 cosmic ray detectors was very much higher than the average flux experienced by Surveyor 3, and higher still than that during Apollo 17. See also Figure 4.15.

*Track profiles*

The Surveyor 3 results represent our most reliable yardstick for converting track densities into exposure ages. For the Surveyor 3 glass clearly sampled a more representative period of cosmic ray activity than either of the two Apollo experiments had done. And if we can only establish how many tracks are recorded in a rock in an average year (or over an average solar cycle) we will be able to calculate how long that rock has spent resting on the lunar surface. So what have we now learnt about the exposure histories of lunar rocks from their densities of cosmic tracks?

Well, first of all it is not just track density that is important here. Because, just as in crater counting, a certain number of tracks per square inch could have been produced either by (*a*) long exposure to infrequent high energy cosmic rays or (*b*) a single short burst of lower energy ones. But high energy cosmic rays are more penetrating than low energy ones. So we would expect to find (and indeed we do find) that there are steep track density gradients close to the surfaces of lunar rocks. Track densities here are invariably very high indeed, frequently more than 100 million tracks per square inch. But as we examine crystals from deeper down, track densities decrease very steeply, perhaps to fewer than one million tracks per square inch at the centre of the sample. And it is these deeper tracks which are produced by the highly penetrating galactic cosmic rays (Figure 4.22).

In order to calculate exposure ages, then, we must ensure that the track density measured and the track production rate which this number is divided by, both relate to cosmic rays in the same energy range. Obviously we must not divide the density of galactic cosmic ray tracks by the production rate that refers to solar flare ions. And the exposure age itself will refer to the penetration depth of the cosmic rays from which it was

Figure 4.22. *Solar flare track gradient.* The steep gradient of solar flare tracks in this etched feldspar crystal in a breccia is apparent in this photomicrograph. The top edge of the crystal must have been unshielded. Width of field = 21 micrometres. (From Macdougall, D., Rajan, R. S., Hutcheon, I. D. & Price, P. B. (1973). *Proceedings of the Fourth Lunar Science Conference*, pp. 2319–36. New York: Pergamon.)

calculated. So galactic cosmic ray exposure ages will always be as high, if not higher, than solar flare exposure ages because of the possibility of shallow burial.

But track density gradients in lunar rocks are found to be shallower than in the Surveyor 3 glass. The absolute track density in the Surveyor 3 glass is, of course, orders of magnitude lower than in any lunar rock. But we would have expected the steepness of the track density gradients to be similar to one another. For we have no reason to believe that Surveyor 3 collected a higher proportion of lower energy cosmic ions than lunar samples did.

It seems that the lunar rock profiles must have been modified by erosion. Because erosion of the surface will have the effect of removing crystals with very high track densities without preventing the interiors of rocks from continuing to accumulate galactic cosmic ray tracks. This would certainly tend to make profiles less steep. It turns out that one thousandth of an inch every 250 000 years would be quite enough to make lunar track density profiles compatible with that in the Surveyor 3 glass. And, interestingly enough, this erosion rate is very similar to that calculated from the depth profiles for spallation radioisotopes in lunar rocks.

But what is responsible for this erosion? Well micrometeorite pitting certainly, but ion sputtering, thermal cycling and physical abrasion must also play their part.

Figure 4.23. *Track exposure age.* This plot shows the track density gradient in a rock chipped from the Apollo 17 Station 6 boulder. The depth dependency is consistent with the boulder having an exposure age of 21 million years, and an erosion rate of 0.05 millimetres per million years. (From Crozaz, G., Drozd, R., Hohenberg, C., Morgan, C., Ralston, C., Walker, R. M. & Yuhus, D. (1974). *Proceedings of the Fifth Lunar Science Conference*, pp. 2475–99. New York: Pergamon.)

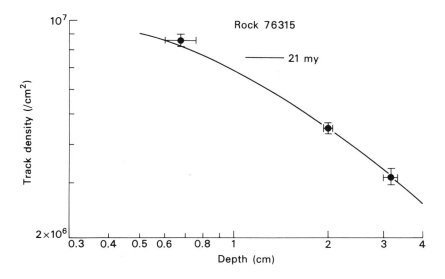

### Track exposure ages

We must clearly refrain from examining crystals close to the surface of a rock when trying to calculate its exposure age. And it is therefore best to concentrate on galactic, rather than solar, cosmic ray tracks. If we do this, track exposure ages usually agree remarkably well with ages based on spallation isotope anomalies. For solar flare protons are just about as penetrating as heavy galactic cosmic rays (Figure 4.23).

Studies of cosmic ray tracks, then, have enabled us to calculate the exposure ages of rocks as well as their surface erosion rates. And they have also helped us to understand regolith evolution. For the tiniest fragments in the soil are subjected to just as much cosmic bombardment as the larger rocks. Cosmic ray track densities in soils are therefore a useful guide to their maturity, the most mature ones having the largest proportion of highly irradiated grains. And there are good correlations

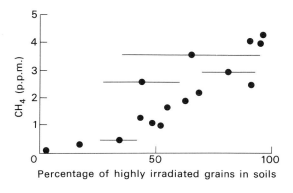

Figure 4.24. *Cosmic rays and solar wind.* Soils with large amounts of solar wind atoms also contain the most tracks. This plot shows a good correlation between solar wind carbon (as methane) and the proportion of very highly irradiated grains. Where a particular soil sample has been examined by more than one investigator the spread in values is shown.

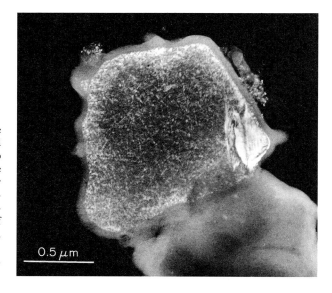

Figure 4.25. *Amorphous coatings.* Under the electron microscope the outermost surfaces of micron-sized dust grains are sometimes found to be so damaged by radiation (the solar wind) as to make them totally amorphous. These coatings facilitate the welding of grains, and contribute to the radiation darkening of soils. (From Dran, J. C., Durrieu, L., Jouret, C. & Maurette, M. (1970). *Earth Planet. Sci. Lett.*, **9**, 391–400.)

with the amounts of solar wind elements and glassy aggregate contents (Figure 4.24). Those deep strata in the Apollo 15 drill core, for example, contain high track densities because they were exposed to cosmic rays for an extended period some 500 million years ago. And the track density profiles in lunar cores, like the rare gas profiles, can best be explained by slow accumulation of debris rather than deep turnover.

The highest track densities of all occur in tiny soil grains rather than in larger rocks. Sometimes, track densities higher than one million million per square inch are revealed here. It is not surprising that minerals cannot always retain their original crystal structure when subjected to this degree of dislocation. Indeed, under the high voltage electron microscope, the outer surfaces of micron-sized grains of dust may be totally amorphous (Figure 4.25). The radiation damage here is so extensive that all long range order in crystals has been erased. And this amorphous coating may play a major role in regolith evolution, radiation darkening the soil and encouraging the tiniest dust grains to be welded back together again.

But what ions are responsible for this amorphous layer, only a few thousand ångströms thick? Well, they must be very low energy ions indeed, so this must be where solar wind ions are initially embedded.

*Exotic cosmic rays*
So far we have assumed that cosmic rays are all ions of the iron group. For less massive ions are incapable of producing dislocation tracks at all, whereas the heavier ones are too rare to make a significant contribution to their total number (Figure 4.26).

But very heavy cosmic rays must exist, so it is just a question of recognising them. And the possibility exists that cosmic rays might include elements not only as heavy as lead and uranium but also superheavy ones not yet known in nature. Theory suggests in fact that there is an 'island of stability' in element building (in the region of atomic number 126) where elements may have half-lives sufficiently long for them to have survived for millions of years. And the conditions in

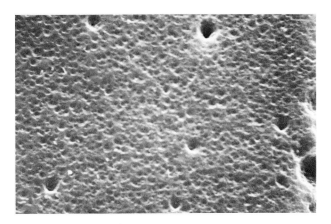

Figure 4.26. *Heavy cosmic rays.* Most of the etch pits on the surface of this glass sphere are due to iron group nuclei. But the larger pits were produced by ions of heavier elements, such as lead. Width of field = 26 micrometres. (From Macdougall, D., Rajan, R. S., Hutcheon, I. D. & Price, P. B. (1973). *Proceedings of the Fourth Lunar Science Conference*, pp. 2319–36. New York: Pergamon.)

exploding stars (from which galactic cosmic rays could have originated) may well be suitable for the synthesis of such extremely heavy elements.

What has been done here has been to measure track lengths under rigidly controlled conditions, because the heavier the ion, the longer the track. And by doing this, tracks due to elements as heavy as lead and even uranium have now been discovered. No one has yet found a track which could conceivably be due to element 126. But the search will, of course, continue.

Ions and electrons, however, may not be the only components of cosmic rays. Experimenters have also been looking for certain other components in cosmic rays, particles which might throw some light on the structure of matter. Quarks and magnetic monopoles are among the theoretically possible particles which have been searched for in lunar rocks, so far without success.

### Spallation recoil

But primary ions are not responsible for all radiation damage in lunar rocks. For, during the spallation process, a target nucleus may recoil a short distance after being struck by a primary cosmic ray proton. And, being electrically charged, it will produce a dislocation which can be made visible as a shallow pit by etching. By counting these shallow

Figure 4.27. *Crater on Apollo 16 boulder.* An Apollo 16 astronaut is here indicating a crater several inches across, with its associated shatter cone, on the Apollo 16 boulder known as House Rock. Note the cuff check list and the sampling bag. (NASA, Apollo.)

spallation tracks we can then calculate a spallation age, an age which should be (and usually is) directly comparable with that based on spallation isotopic anomalies.

This then concludes our brief discussion of radiation damage. But this is not the only damage that is suffered by rocks resting on the lunar surface. For we have heard already how the impacts of micrometeorites must also erode surfaces of lunar rocks. So how can micrometeorite craters be used to learn more about the particulate matter in space, and to probe still further into lunar surface processes?

## 4.5 Cratering by micrometeorites

The distinction between meteorites and micrometeorites is an arbitrary one based on particle size. For the purposes of this discussion, a micrometeorite is defined as a cosmic particle which is small enough to produce a crater on the surface of a rock without shattering it. And in practice this means a crater an inch or so in diameter, and a meteorite size of up to, say, one millimetre (Figures 4.27 and 4.28).

But most lunar microcraters are much smaller than this, and were produced by micron-sized particles. So what did we know about these micrometeorites (also known as cosmic dust) before Apollo, and what have we now learnt about them from examining lunar rocks? Where do they come from and what can they tell us about the exposure histories of lunar rocks?

*Micrometeorite collection*
Well, a number of apparently unrelated studies, including direct collection, have helped us to understand the nature and origin of micrometeorites. They survive entry through our atmosphere remarkably well, and can therefore be collected intact at the earth's surface. As much as 3000

Figure 4.28. *Sub-critical microcrater*. This crater on an Apollo 17 feldspathic melt rock was just insufficient to smash it into pieces. Note the extensive fracturing. The scale is in centimetres. (NASA, JSC–LRL.)

tons of cosmic dust, in fact, may accumulate every year, and so, during its entire history, the surface of the Earth should have been covered to a depth of half an inch or so with micrometeorite debris.

But the Earth, far from being a passive planet like the Moon, is constantly recycling its crust, through erosion, sedimentation and volcanism. As a result, any extraterrestrial material which contaminates its surface rapidly becomes diluted beyond recognition. And the best places to collect micrometeorites on Earth today are therefore wide open spaces far removed from industrial activity, notably the antarctic ice cap and the deep ocean floor, both areas where sedimentation rates are very low indeed. Tiny spheres of metallic nickel/iron have in fact now been discovered in these environments, and their unusual chemistries confirm their extraterrestrial origins. For metallic iron is very rare indeed as a mineral here on Earth.

But if we can only distinguish micrometeorites if they are chemically distinct from terrestrial materials, then it could be that the majority of them might pass unnoticed. A micrometeorite composed of ice, for example, would certainly take some finding in the antarctic. And studies of meteors (or shooting stars) tend to suggest that many micrometeorites may indeed be icy in composition. For, when a micrometeorite enters our atmosphere, it produces a momentarily bright trail across the sky. Studies of these trails suggest that the micrometeorites responsible for them must have remarkably low densities. Only if they have very open structures, in fact, can they possibly be composed of silicate or metal.

This is not really a very surprising conclusion because most cosmic dust is now believed to originate from comets, the nuclei of which are composed primarily of frozen water, methane and ammonia. The reason for this association is that shooting stars generally fall in showers at those times of the year when the Earth happens to be passing through the tail of a dead comet. The November Taurid shower, for example, has been positively associated with the remains of Encke's comet.

To learn much more about cosmic dust, then, we must carry out our collection experiments well above the Earth's atmosphere. And some success has in fact been achieved in this direction, using microparticle collectors on rockets and satellites. But studies of cosmic dust in the solar system have not only entailed experiments performed in space. We can also study the scattering of sunlight by interplanetary dust, which produces the so-called zodiacal light, a dim glow visible in the ecliptic plane just after sunset or just before sunrise. And we conclude from the intensity of this zodiacal light that micrometeorites are moving around the Sun in much the same plane as the Earth.

*Micrometeorite abundances in space*

By the time of Apollo then, we were already quite familiar with micrometeorites, their numbers, masses and velocities. But their collection and detection in space had always been completed within short time periods. In contrast some lunar rocks have been exposed to micrometeorite bombardment for millions of years. So their surfaces can provide us with information about cosmic dust over much longer intervals. Microcraters must also have been produced on the surfaces of meteorites while they were in space, but these were totally destroyed when the meteorites entered our atmosphere. So lunar samples provide us with our first fossil microcraters.

Before turning to lunar rocks, however, let us consider the Apollo spacecraft themselves, for these spent many days exposed to the interplanetary environment. They should not have been struck by any meteorites large enough to have done any damage, but some impacts should have been recorded on suitably smooth surfaces. A meticulous search was therefore carried out on all Apollo command module windows (with the exception of those from Apollo 11 which had to remain intact for the sake of posterity) to see if any microcraters could be found. Also examined carefully were components returned from the Surveyor 3 and Gemini spacecraft (Figure 4.29).

So how many microcraters did we find? Well only fifteen altogether, but this small number is quite consistent with satellite experiments (Figure 4.30) and now enables us to estimate the exposure ages of lunar rocks (from the densities of microcraters which are preserved on their surfaces) with some confidence. The microcrater clock for lunar rocks has therefore now been calibrated, as the density of interplanetary dust today is presumably much the same as it has been for millions of years.

Figure 4.29. *Apollo microcrater.* This photomicrograph shows one of the few microcraters discovered on the command module windows. The small central pit is surrounded by a much larger spall zone, and is about a tenth of a millimetre across. (NASA.)

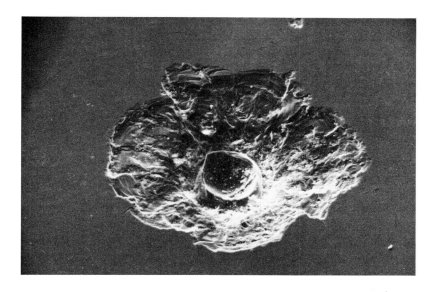

*Lunar microcraters*

So what can lunar microcraters tell us about the exposure histories of lunar rocks and about the nature and origin of cosmic dust? Well the best places to look for them are glassy surfaces, so glass-coated breccias have been extensively studied (Figure 4.31). For it is difficult to recognise the tiniest pits on the more irregular surfaces of crystalline rocks. And some glassy rocks have remarkably high densities of microcraters (Figure 4.32).

When we wish to date an ordinary lunar rock, what we can do is find a glass-lined crater, a millimetre or so in diameter, and count the microcraters there. If we can then find several such glassy craters, we can date the rock rather than just the individual craters. Because we can assume that the craters must have been exposed to the interplanetary environment for about half as long as the rock. In other words, some craters must have formed soon after the rock was exhumed, whereas others must have been created very recently. By doubling up the observed microcrater counts inside glassy craters, then, we can date a rock sample whose surface may not itself register the smallest, and numerically more significant, craters.

Whichever approach we take, we find that, taking erosion processes into account, microcrater ages do indeed agree quite well with those 'sun tan' exposure ages based on solar cosmic ray track densities. The longer a rock has been exposed, the more microcraters are found on its surface at all crater diameters (Figures 4.33 and 4.34).

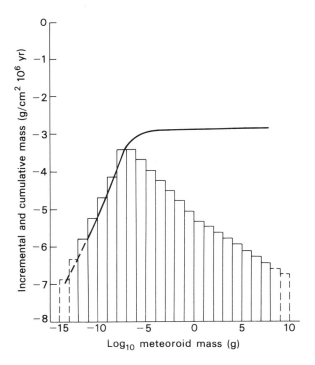

Figure 4.30. *Mass distribution of meteorites.* This plot shows the distribution of meteorites according to mass. From this plot it can be seen that the most abundant particles weigh less than a microgram. For meteorites larger than a gram, the figures are based on seismic data. (From Gault, D. E., Hörz, F., Brownlee, D. E. & Hartung, J. B. (1974). *Proceedings of the Fifth Lunar Science Conference*, pp. 2365–86. New York: Pergamon.)

Figure 4.31. *Glass-coated anorthosite.* This rock, sample 60015, is covered with impact generated glass and is therefore very suitable for dating by microcraters. (NASA, JSC–LRL.)

Figure 4.32. *Highly cratered glassy sample.* This sample of glass has a surface that is covered with clearly visible microcraters. (NASA, JSC–LRL.)

*Micrometeorite masses*

Turning now to investigations of micrometeorites themselves, it is important to be able to relate microcrater diameter to meteorite energy, in order to learn more about the origins of cosmic dust. And this means carrying out laboratory simulations. Accelerators now exist where tiny particles can be made to impact at several miles per second into carefully selected targets. And such experiments have enabled us to relate crater morphology to micrometeorite energy and have also taught us much about the cratering process itself.

But it is one thing to calculate the required energy for a micrometeorite and quite another to calculate its size. For kinetic energy depends on

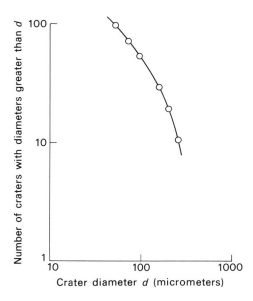

Figure 4.33. *Microcrater distribution.* This plot shows the distribution of larger-sized microcraters on glass coated breccia 60015 (see Figure 4.31). (After Fechtig, H., Hartung, J. B., Nagel, K., Neulkum, G. & Storzer, D. (1974). *Proceedings of the Fifth Lunar Science Conference*, pp. 2463–74. New York: Pergamon.)

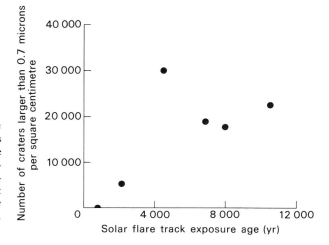

Figure 4.34. *Correlation with track ages.* This plot shows that surfaces with high densities of solar flare tracks also contain the largest numbers of microcraters. (After Hartung, J. B., Hodges, F., Hörz, F. & Storzer, D. (1975). *Proceedings of the Sixth Lunar Science Conference*, pp. 3351–71. New York: Pergamon.)

velocity as well as mass, and the velocities of micrometeorites may vary greatly.

The lowest possible velocity is only one and a half miles per second, or lunar escape velocity. For this is the speed to which a particle being slowly overtaken by the Moon will be accelerated by the Moon's own gravitational field. At the other extreme, however, the solar system is passing at high speed across galactic dust lanes and tiny particles are being blown out of the Solar System by solar radiation pressure at velocities as great as 60 miles per second, almost as fast moving as the solar wind itself.

But most micrometeorites must strike the Moon with velocities between 5 and 25 miles per second. The Apollo 17 Lunar Ejecta and Micrometeorite Experiment (Figure 4.35) was, in fact, designed to establish some firm limits on the numbers of microparticles in a given size range which are striking the lunar surface every year. And by combining these figures with seismic results, we can obtain a reasonable picture of meteorite masses in the Solar System (Figure 4.30). It appears that most extralunar material added to the regolith must be in the form of micrometeorites weighing a microgram or so.

*Origins of cosmic dust*

So where do all these tiny cosmic particles come from? Well, theory tells us that the Solar System should contain very few sub-micron sized micrometeorites. The pressure of solar radiation on these tiniest particles is so great that they should be accelerated out of the Solar System altogether. But, so far, there is no strong evidence to favour this deple-

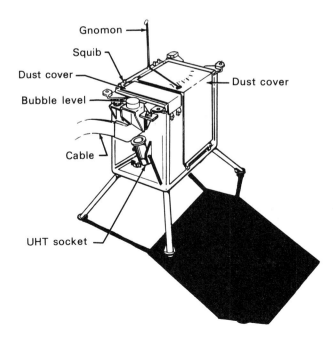

Figure 4.35. *Lunar Ejecta and Meteorites Experiment.* This device was deployed by the crew of Apollo 17 to detect micrometeorites. It was capable not only of counting them but also of measuring their masses and velocities. (NASA.)

tion hypothesis. So there must be some mechanism for the renewal of tiny fragments.

One popular idea is that large micrometeorites gradually spiral in towards the Sun and then break up. This so-called Poynting–Robertson effect should then lead to a higher flux of tiny particles (relative to the larger ones) in the ecliptic plane (i.e. from the Sun) than in the direction of the celestial poles. So a lunar experiment was devised to test this hypothesis.

A suitable site for this experiment was found in the Apollo 17 boulder, one rock from which had two deep cracks in it. The bottom of one crack could only view the ecliptic plane, whereas the bottom of the other was directed well out of the ecliptic.

As expected, larger numbers of microcraters were found on the floor of the first. For studies of zodiacal light had already shown us that most cosmic dust does indeed move in the ecliptic plane. But, strangely enough, we did not find any relative enrichment here in sub-micron sized microcraters (Figure 4.36). In other words, the size distribution of ecliptic micrometeorites is much the same as that of those moving out of the ecliptic. And the size distribution and absolute abundances of micrometeorites does not seem to have changed very much, at least during the last million years or so. So clearly we have some way to go before we fully understand the dynamics of cosmic dust.

But what about the chemical compositions of micrometeorites? Meteor studies may indicate that most cosmic dust is composed of ice, but what do the lunar samples tell us about micrometeorite chemistry?

Well, the density of the micrometeorite responsible for producing a microcrater is reflected in its depth–diameter ratio. For laboratory simulations reveal a strong relationship between this parameter and the

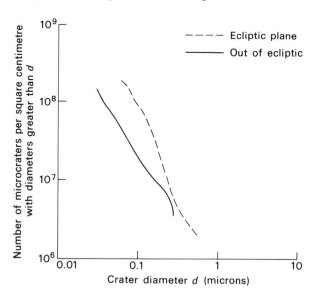

Figure 4.36. *Directions of origin of tiny micrometeorites*. Counts of tiny microcraters in boulder sample 76315 show that the size distribution does not depend on direction, although there are more micrometeorites moving in the ecliptic. The shapes of the two distribution curves are really quite similar with no depletion of the tiniest craters in the ecliptic sample. (After Morrison, D. A. & Zinner, E. (1975). *Proceedings of the Sixth Lunar Science Conference*, pp. 3373–90. New York: Pergamon.)

density of the projectile, for a given impact velocity. And lunar micro-craters have depth–diameter ratios that range from 0.3:1 to 1.2:1, with the most frequent ratio being about 0.6:1. This wide spread is best explained if there are three distinct types of micrometeorites, namely those composed of ice, silicate and metallic iron respectively, with the densest microparticles creating the deepest pits (Figure 4.37).

So far we have seen how microcraters can be used to study the exposure histories of lunar samples and how lunar samples, in turn, may contain valuable clues to the nature and origins of cosmic dust. But micrometeorites do not only produce microcraters and glassy splashes when they strike the Moon. Like the solar wind and cosmic rays, they also leave chemical traces. So let us now consider the chemical remains of meteorites, not only the tiny ones discussed in this chapter, but also the much larger ones responsible for producing lunar craters and basins.

## 4.6  Meteorite remains

Strangely enough, the Earth is more effective than the Moon as a collector of meteoritic debris. This may seem rather surprising in view of the fact that a large proportion of the cosmic debris swept up by the Earth is burnt up in our atmosphere. But, of the meteorites which do survive, all but the very largest have lost all of their cosmic velocity before they hit the ground. For them impact occurs at a few hundred miles per hour, rather than at the several miles per second with which they strike the Moon.

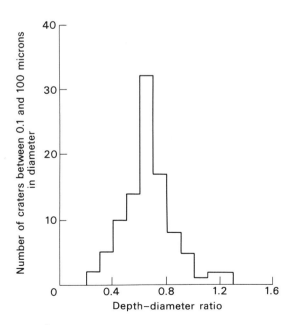

Figure 4.37. *Microcrater depths.* The wide spread in depth – diameter ratios for lunar microcraters must be partly due to erosion effects and differences in impact velocity. But it must also reflect the chemical variability of the projectiles. (After Nagel, K., Neukum, G., Eichhorn, G., Fechtig, H., Müller, O. & Schneider, E. (1975). *Proceedings of the Sixth Lunar Science Conference*, pp. 3417–32. New York: Pergamon.)

*Meteorite remains on Earth*

Only the very largest meteorites, then, produce impact craters on Earth. The famous Barringer Crater, for example, must have been excavated from the Arizona desert by a projectile weighing many thousands of tons. But only tiny fragments of this meteorite (the Canyon Diablo iron) have been recovered, so the main mass must have been totally vaporised on impact. Presumably most of the Canyon Diablo iron is now finely disseminated throughout the ejecta blanket of the Barringer Crater. So finely disseminated, in fact, that it should be clearly recognisable as a chemical fingerprint. In other words, the impact breccias should contain slightly more iron and sulphur, for example, than the surrounding desert soil.

Such chemical anomalies have in fact now been recognised in the impact breccias associated with the still larger Reiss Crater in West Germany. And here, such anomalies provide us with our only chemical clues to the nature of the projectile, because no meteorite fragments whatsoever have been recovered from this crater. This is hardly surprising, however, because the Reiss Crater is several miles across and must have been excavated by a projectile even larger than the Canyon Diablo iron.

We are thus left with the paradoxical situation in which the largest meteorites recovered do not necessarily originate from the largest projectiles. But although small meteorites are regularly picked up from where they first hit the ground, only a few are ever actually seen to fall. Indeed, many must date from prehistoric times, having been recognised by virtue of their unusual appearance. It was the conspicuous iron meteorites, for example, which once provided the basic raw material for the eskimo metallurgical industry. A number of early eskimo tools are found to be composed of a nickel–iron alloy indistinguishable from meteoritic iron.

And the arctic regions are still remarkably good places to collect meteorites today. For, contrary to popular belief, it is not particularly easy to recognise already fallen meteorites, especially the more common stony varieties. To the untrained eye stony meteorites look much the same as any other rock. So they stand out best where other rocks are rare (as in Antarctica) or at least all very similar to one another (as in the Australian desert). And only in such stable environments is the integrity of a meteorite preserved against the ravages of erosion and burial. Meteorite falls are so rare that searching for them is best carried out where they can survive intact for thousands of years. Some meteorites, such as the carbonaceous chondrites, are particularly friable and would certainly not survive long in the bed of a fast flowing river. But there cannot be any place on Earth where black carbonaceous chondrites would show up better than on the pure white snow of the Antarctic.

*Lunar meteorites*

Meteorites certainly can survive on Earth, but active geological processes have meant that the total number recovered so far has been strictly limited. On the Moon however, the situation is reversed. There is no atmosphere there to slow them down, but the absence of tectonic activity on the Moon means that the lunar crust (and particularly its regolith veneer) must have accumulated appreciable quantities of meteoritic debris over the past three or four thousand million years, if not as discrete fragments then at least as chemical traces.

So what evidence do we find for meteorite debris in lunar samples? And what can this evidence tell us about the Moon's early bombardment history? Well, despite the undoubted violence of meteorite impacts on the Moon, we do in fact find some recognisable meteorite fragments in the soil, fragments which are certainly not indigenous to the Moon and which must therefore be extralunar in origin.

Some of these exotic particles can be accurately placed within the established meteorite classification system. One Apollo 15 soil fragment, for example, was positively identified as an enstatite chondrite. For, in addition to orthopyroxene, this particle also contained a magnesium sulphide mineral known as niningerite, a mineral which is never encountered in lunar rocks and is only ever known to occur, in fact, in enstatite chondrites, one of the rarest groups of stone meteorites.

But stone meteorite fragments are not the only recognisable extralunar particles in the regolith. There are also abundant fragments of metallic iron/nickel, not only in the soil but also in highland breccias. Some of this metal must, of course, have a lunar origin. For metallic iron is a minor, yet ubiquitous, component of *mare* basalts. But lunar metal invariably has a high cobalt–nickel ratio (Figure 3.8), whereas most metal particles in the soil are relatively deficient in cobalt, and have cobalt–nickel ratios which are closer to meteoritic values.

Furthermore, we find certain exotic phases in these particles which do not occur in lunar rocks. Minerals such as the phosphide, schreibersite, and the carbide, cohenite, can therefore only be extralunar in origin, and both are indeed commonly found in iron meteorites. Metallic iron is also abundant in other types of meteorite, including chondrites, so metallic iron fragments in the soil provide the most obvious and direct evidence for meteorite debris on the Moon.

*The Moon as a source of meteorites*

Before Apollo it was thought that the Moon might even be the source for at least one type of meteorite. The *maria*, for example, were dark enough for them to have had a carbonaceous chondrite composition and this was closely linked with the idea that the Moon might be a very primitive, undifferentiated cosmic body. Alternatively the lighter coloured highlands could have provided us with the basaltic achondrites, a family of meteorites with undoubted igneous origins. Analyses of lunar rocks, however, soon dispelled these theories and placed the origins of meteorites firmly back in the asteroid belt, or else in the nuclei of comets. Although the basaltic achondrites were superficially similar to certain lunar rocks there were some very fundamental differences between them.

But there was one type of supposedly cosmic object for which a lunar origin once seemed very likely indeed and this was the tektite. Tektites are glassy globules, very rich in silica which, before passing through our atmosphere at high velocity, spent very little time exposed to the interplanetary environment. For they do not contain any cosmic ray tracks or solar wind gases. And so if they are indeed extraterrestrial objects, they cannot have come far and the Moon would seem to be their most likely place of origin. As we have already heard, however, there are no lunar rocks which even approach the silica contents of tektites. So it now looks as though these enigmatic bodies must after all be the products of terrestrial meteorite impacts or explosive terrestrial volcanism.

### Chemical remnants in the regolith

Particulate matter is not of course the only type of meteorite remnant on the Moon. For, just as most of the Canyon Diablo iron must have vaporised on impact, so most meteorites which strike the Moon must end up finely disseminated within the regolith, leaving behind no physically recognisable fragments. So it is to these chemical remains of meteorites that we should now turn. Chemical anomalies in the regolith are what can provide us with a really quantitative measure of the rate of meteorite infall at the lunar surface and the types of meteorite which predominate. So what anomalies might we expect to find and how large is this extralunar component at the various Apollo sites?

Our first attempts to characterise the meteorite component chemically were made with the Apollo 11 soil. And these initial experiments met with remarkable success. For there are a number of elements which are relatively abundant in meteorites, but rare in the Moon. These elemental deficiencies in lunar rocks (volatiles, siderophiles and chalcophiles) have already been mentioned in a different context and are closely related to global processes which occurred early in lunar history.

Figure 4.38. *The micrometeorite component in Apollo 12 soil.* This plot shows the excesses of certain volatile and siderophile elements, normalised to their abundances in carbonaceous chondrite meteorites. The pattern is markedly dissimilar to that which characterises the more fractionated ordinary chondrites, which are relatively depleted in germanium and volatiles. This suggests that the micrometeorite component in lunar soils may have a common origin with carbonaceous chondrites. (After Baedecker, P. A., Schaudy, R., Elzie, J. L., Kimberlin, J. & Wasson, J. T. (1971). *Proceedings of the Second Lunar Science Conference*, pp. 1037–61. Cambridge, Mass.: MIT.)

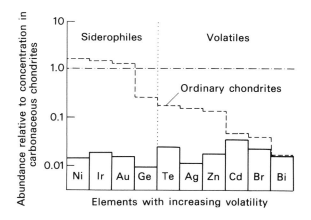

We might expect lunar soils, then, to contain higher concentrations of siderophile and volatile elements than rocks collected nearby. And this is precisely what we found at Tranquillity Base. The amounts of tellurium and bismuth were found to be quite appreciable in the Apollo 11 dust, for example, whereas both elements, being rather volatile, are highly depleted in the local basalts. Similarly, the lunar soil is relatively rich in siderophile elements, such as gold, iridium and germanium.

Rare all these elements may be in absolute terms, but their excess concentrations in the soil enable us to establish what type of meteorite is most abundant in the Solar System today and how abundant the remains of these meteorites are in the lunar regolith.

We can obtain this information quite easily by comparing certain interelement ratios in different meteorite types. If the ratio of excess bismuth to excess gold in the soil is 2:1 for example, then this must be the ratio between these two elements in the parent meteorite type. What we then find is that the best match for the extralunar component in *mare* soils corresponds closely to the most primitive meteorites of all, the carbonaceous chondrites. The more fractionated meteorites (such as the ordinary chondrites) are much more deficient in volatiles relative to siderophiles than is the extralunar component in *mare* soils (Figure 4.38).

Now carbonaceous chondrites are not very common among meteorites found on Earth, largely because they are so friable. Only about twenty, in fact, are represented in the world's meteorite collections. But most meteoritic material is arriving at the Moon in the form of micron-sized dust particles. So there is no reason why this dust should be chemically equivalent to the most abundant group of meteorites, the ordinary chondrites. And it is not really very surprising that micrometeorites should turn out to be tiny carbonaceous chondrites anyway. For this fits in very well in fact with the proposed cometary origin for cosmic dust. We know that comets, like carbonaceous chondrites, did not suffer from the heating to which all other meteorite families were subjected. Comets, then, may well be the source objects for carbonaceous chondrite meteorites as well as micrometeorites.

### The micrometeorite component and soil maturity

If we take the excess concentrations of any tracer element in a soil, and divide this number by the concentration of this element in the carbonaceous chondrites, we end up with the magnitude of the micrometeorite component in the regolith. In the Apollo 11 soil it is as high as 2% but in soils from the other landing sites, which have not been exposed to extralunar influences for so long, it is correspondingly less. The light-coloured KREEP glass returned by Apollo 12, for example, contained very little micrometeorite debris, consistent with its total lack of glassy aggregates and its very low content of solar wind rare gases. At some sites there are, in fact, good correlations between tracer element excesses and soil exposure ages (Figure 4.39).

The micrometeorite component, then, provides us with yet another parameter related to the maturity of the regolith. But, like the others, it must be interpreted with care. For a high meteorite component could well mean a thin regolith, or a poorly mixed one, rather than a regolith which is extremely old. Only by combining all of our exposure history information can we possibly hope to unravel the detailed history of regolith evolution at any particular site.

But as more soils were returned from other sites it became apparent that using a number of volatile and siderophile elements as meteorite tracers was not going to be quite so straightforward. One stratum in the Apollo 12 core tube, for example, was found to contain extraordinarily high concentrations of two volatile elements, cadmium and bismuth, elements which are normally strongly depleted in lunar rocks. And as we have already heard numerous rock and soil samples, including the rusty rocks and the orange and green glasses, have now also been found to contain excesses of these and other volatile metals. Quite where these volatiles originate we cannot be sure, but what is certain is that none can be used any longer as meteoritic tracers. We must therefore concentrate on those elements that are highly deficient in all lunar rocks, namely the siderophiles.

Figure 4.39. *Micrometeorite component as a maturity parameter.* The excess abundances of bismuth and tellurium, both volatile elements, in Apollo 16 soils are correlated strongly with soil exposure age as measured by other methods, e.g. solar flare track-rich grain percentage. The non-zero intercept is indicative of the ancient meteorite component that is present in highland rocks. (From Ganapathy, R., Morgan, J. W., Higuchi, H., Anders, E. & Anderson, A. T. (1974). *Proceedings of the Fifth Lunar Science Conference*, pp. 1659–83. New York: Pergamon.)

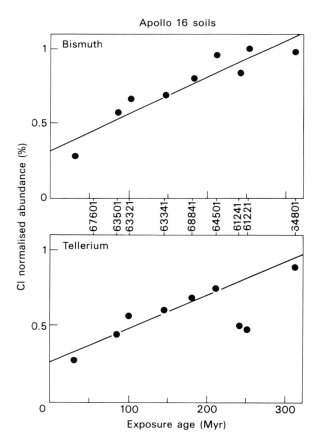

*The ancient meteorite component*

Since Apollo 12 this approach has proved to be the only reliable way to characterise meteorite debris on the Moon. For when we analyse highland breccias we find that they contain an extralunar component that was certainly not chemically indistinguishable from a carbonaceous chondrite. And this is hardly surprising. For highland breccias were formed not by micrometeorites but by huge, basin-forming projectiles, bodies which were more closely related to asteroids than to comets. Even the freshest soils, then, contain this ancient meteorite component (Figure 4.39).

So what have we learnt about the chemistry of these immense projectiles? Were they related to the meteorite families that we recognise today? Or did they constitute a completely separate population of bodies, of which none have survived?

Well to say that we have concentrated on siderophiles to the exclusion of volatiles in our search for the ancient highland meteorite component is not strictly true. Because volatility is a relative term and some siderophile elements are more volatile than others. Antimony and germanium, for example, are not as refractory as gold and iridium.

What we actually find is that the siderophile elements from one highland rock tend to be more or less volatile than those from another. There may, in fact, be several well-defined compositional clusters (corresponding to highland rocks from widely diverse areas) which could be accounted for by a small number of basin-forming projectiles.

How was this dramatic conclusion arrived at? Well the excess siderophile element concentrations are plotted on three element diagrams, one of the most revealing being germanium–gold–iridium. And to ensure that the compositions do not all plot in one corner of the diagram (because of the different absolute abundances of the three elements), the concentrations of all three elements are first normalised to the carbonaceous chondrite composition. In other words, carbonaceous chondrites will plot at the centre of these ternary diagrams.

In contrast, highland breccias are richer in the more refractory siderophiles and tend to cluster into five or six more or less well-defined groups (Figure 4.40). These groups seem to be statistically significant. For we can replace iridium by rhenium and germanium by antimony without destroying them, and some clusters tend to be clearly associated with particular landing sites. But the preferred compositions do not match known meteorite compositions and their significance is by no means unanimously accepted within the lunar science community.

We should remember that bodies the size of asteroids would not necessarily have average compositions matching those of any one meteorite type. Indeed the asteroid-sized bodies must have produced a wide variety of meteorites in the first place, with their cores giving rise to iron meteorites and their crustal material providing the stony ones. So they presumably have intermediate chemical compositions. Perhaps the situation will become clearer when a spacecraft lands on a minor planet.

This concludes the discussion of meteorite remains on the Moon and also rounds off the part of this chapter concerned with the interaction of the regolith with the interplanetary environment. But one aspect of the regolith has not yet been discussed, namely biochemistry. Is there, has there ever been, or might there have been, life on the Moon? This is a question which has always been, and always will be, at the forefront of space exploration.

## 4.7    The search for lunar life

The search for life elsewhere in the Universe will always be one of the major goals of astronomy. For it is difficult to accept that life is the result of a unique chemical accident. And most of us prefer to think that there are intelligent beings somewhere in the Galaxy, perhaps gaining some comfort from the feeling that someone somewhere is looking after our wellbeing.

This is rather strange really because there is no reason to assume that aliens really would show any more respect for our existence than we do for a swarm of locusts. At least locusts and human beings have a common biochemistry. The probability that some form of extraterrestrial life might be based on the same amino and nucleic acids as our own is very remote indeed.

So those distant views of Earth from outer space help us to realise just how precious our planet is, despite its insignificant size compared with the vast reaches of space. It would be more precious still if it happened to be the only oasis of life in the entire Universe.

But however rare Earth-like planets around Sun-like stars may be,

Figure 4.40. *The ancient meteorite components.* The excess siderophiles in lunar highland breccias tend to form a restricted number of compositional clusters, which may indicate that the highland meteorite component is dominated by the chemical remnants of a few large basin-forming projectiles. (From Morgan, J. W., Ganapathy, R., Higuchi, H., Krahenbühl, U. & Anders, E. (1974). *Proceedings of the Fifth Lunar Science Conference*, pp. 1703–36. New York: Pergamon.)

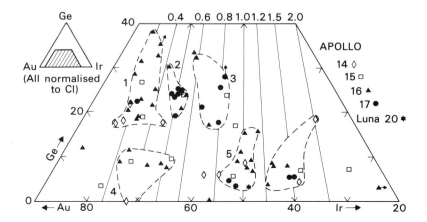

surely there must have been millions of opportunities for life to have arisen elsewhere in the Universe? Some writers, in fact, have taken statistical arguments to extreme lengths to prove that life on Earth is not unique. Until at least one form of extraterrestrial life is found, however, all such arguments are invalid. For no valid statistical prediction can ever be made from a sample of one. And once one lowly extraterrestrial life form is detected the statistical argument is no longer necessary.

It certainly would be strange if life only began on one planet in the entire Universe, but the spontaneous generation of life is so unlikely an event that it seems almost impossible without some form of divine intervention. The probability of life evolving elsewhere in the Universe, then, is the product of a practical certainty and a near impossibility. And the answer to that calculation is anybody's guess.

It should be noted that we are not being narrow minded if we believe that terrestrial life is unique. For, if life only arose on one planet in the entire Universe, then the only beings capable of recognising the existence of that life would have to live on that same planet. This is of course a truism, but it is a consequence that is all too often overlooked.

### The prospects

The discovery of the tiniest microorganism, microfossil or biochemical trace on the Moon would, of course, totally revolutionise our ideas about intelligent life on distant planets. So how could there ever have been life on the Moon and how did we go about searching for it? And what conclusions did we eventually come to, not only about the existence of indigenous lifeforms, but also about the suitability of the lunar environment for colonisation by terrestrial species?

Well there are several points in favour of the possible existence (or former existence) of life on the Moon, some of which are more valid than others.

Firstly, the Moon may once have been part of the Earth and, if it broke away after life had arisen here, then it could have taken some organisms away with it. How they could possibly have survived such a violent event, however, is difficult to imagine, and the fission hypothesis for the Moon's origin is now thought to be improbable anyway.

Secondly, the Moon could have acquired life forms from elsewhere in the Universe. Indeed, one popular theory for the origin of terrestrial life (known as panspermia) is that the primordial germ, instead of evolving spontaneously, arrived here from outer space. If this was indeed the case then the Moon would presumably have accumulated the same life spores. But this panspermia theory just evades the awkward issue of the spontaneous generation of life from inanimate matter. And all it really does is push the most crucial event out to some remote distance and back into the unknowable past.

It is quite possible, however, that some form of life (or protolife) does exist in space. For certain meteorites harbour complex organic molecules (such as amino acids) and the four basic elements of life (carbon, hydrogen, nitrogen and oxygen) are among the most abundant elements in the Universe. So who knows what sorts of organic molecules might exist on interstellar dust grains in the remote depths of space.

Moving further into the realms of fantasy, the Moon could perhaps have been contaminated by an intelligent life form from a distant planet. If alien spacecraft (flying saucers perhaps) once landed on the Moon, then they could conceivably have left some biological traces behind when they left. We are well aware of the crucial importance of sterilising our own space probes in order not to contaminate other planets. But other beings may not have been quite so careful.

### The search for viable organisms

Finally, of course, the physicochemical conditions on the Moon in the distant past could have been eminently suitable for the spontaneous generation of lunar life. And this possibility raised a serious problem. For what if a lunar micro-organism, brought back to Earth with the lunar samples, turned out to be pathogenic. There was a remote, yet finite, chance that Apollo, far from establishing the basis for the exploration of the Universe, might bring about the destruction of the entire human race by infecting it with a deadly virus. No one thought that this was at all likely, of course, but steps had to be taken to ensure that no lunar bug could contaminate the terrestrial biosphere.

And it was this possibility that was behind the elaborate quarantine procedures that were enforced for Apollos 11, 12 and 14. If a lunar micro-organism was discovered and found to be harmful to terrestrial life, then at least it could be contained within the Lunar Receiving Laboratory, where it could presumably be prevented from spreading further. As it turned out, of course, the quarantine procedures proved totally unnecessary. For none of the personnel or test animals contracted any strange diseases, nor did any cultures from the soil reveal any viable micro-organisms, harmful or otherwise.

The materials returned from the Moon were not, however, totally inanimate. Viable streptococci bacteria were discovered on bits from the Surveyor 3 spacecraft returned by Apollo 12, parts of which had been exposed to the harsh lunar environment for 30 gruelling months. It certainly says much for the adaptability of terrestrial life forms when bacteria can survive the vacuum of interplanetary space and the extreme temperatures and high levels of radiation at the surface of the Moon.

The lifting of quarantine restrictions meant a great deal for lunar science. For there are few experiments which can be carried out effectively in a rigidly restricted area. And it also meant that samples returned by Apollos 15, 16 and 17 could be processed much more quickly.

### Fossils

But the search for lunar life did not stop with the lifting of quarantine restrictions. There were a number of other avenues to explore before we could be absolutely sure that no life had ever existed on the Moon. And the first of these involved searching for fossils. Although there are apparently no micro-organisms on the Moon today, such micro-organisms might have existed in the remote past, in which case there could well be microscopic structures in the soil which might be clearly organic in form.

Micropalaeontology is still the only way to investigate the nature of the earliest terrestrial life forms, some of which are known to date back more than 3000 million years. But some structures, including those found in carbonaceous chondrite meteorites, are so simple that they cannot be unambiguously given a biological origin. Indeed, the former existence of micro-organisms in carbonaceous chondrites is a very controversial point.

The lunar situation, however, is more clear cut. No lunar microstructure remotely resembling a bacterium has ever been discovered, so life can never have evolved on the Moon. The physical and chemical conditions must have precluded the spontaneous creation of biologically important organic molecules, or at least the subsequent combination of these molecules into the basic building block of life, the living cell.

## Precursors

But, if life as such never arose on the Moon, how distant was the prospect of life? Biological evolution on Earth must have been preceded by a period of chemical evolution, when hundreds of chemical reactions were taking place in the 'primordial soup', reactions which continued until the necessary combination of atoms came together to form the first self-replicating macromolecule.

In some ways the Moon could be an ideal environment in which to study chemical evolution. For everywhere on Earth, even the Antarctic icecap, has been contaminated to some extent by biological activity. And, as the chemistry of the primordial soup has long since been consumed by the very life which it begat, extraterrestrial organic chemistry could provide our only clues to the origin of life here on Earth. So how far did this chemical evolution progress on the Moon?

Well, both the rarity of carbon and nitrogen and the absence of water on the Moon has prevented even minute quantities of complex organic molecules from forming there. Organic geochemical experiments consuming several grams of lunar soil were quite capable of detecting a billionth of a gram of a compound such as a long chain hydrocarbon. But no large carbonaceous molecules were detected, a result which put the final seal on the search for lunar life, past, present or future.

## Lunar synthesis

Nobody likes performing negative experiments, even where a negative result is just as valuable as a positive one. But there were one or two loose ends to the organic story which took several years to tidy up. We have heard, for example, that solar wind hydrogen converts solar wind carbon into light hydrocarbon gases, such as methane and ethane. And these molecules may react further, whenever the soil is heated, to produce more complex, but biologically insignificant, molecules such as benzene and acetone. More intriguingly perhaps, the lunar soil reacts with water to produce simple amino acids, albeit at vanishingly low levels. So if only the Moon had once been covered by warm seas, the whole lunar life story might perhaps have been very different.

Reports of porphyrin-type molecules in the soil revealed yet another source of organic material on the Moon. But these molecules were not the chemical remnants of chlorophyll or haemoglobin type pigments. Instead, they must have been synthesised from lunar module descent engine exhaust gases. For no such molecules were found in soil samples collected from great distances from the lunar module, or from shielded locations under rocks, or from core tubes.

Whatever we may have thought about the chance of finding evidence for life on the Moon, then, the fact is that there are no traces to be found among any of the returned samples. But what would it take to make the Moon capable of supporting life in the future? Well, the regolith is certainly rich enough in nitrogen and potassium to encourage the rapid growth of terrestrial plants. But we would have to supply the necessary carbon dioxide and water. And this could well prove to be a major undertaking.

But, before we start thinking about establishing a lunar colony, we need to know more about the Moon on a global scale. For how else can we hope to find precious mineral sources or seek out deposits of lunar water.

So what have we managed to learn about the Moon as a whole? What global properties have we measured, what global features have we plotted and how does this data base help us to understand more about the nature, origin and history of the Moon as a whole?

# 5. Global measurements

## 5.1  The shape of the Moon

The global investigation of any planet or satellite must begin with a description of its overall shape. Some bodies, such as Phobos and Diemos (the two tiny moons of Mars) for example, are far from spherical. And one asteroid, in fact, exhibits variations in brightness which can only be accounted for if it is several times as long as it is broad. But these are very small bodies, inside which temperatures are so low that their shapes can only ever be modified by meteorite impact. Stickney, the largest crater on Phobos, for example, is mainly responsible for Phobos's irregular outline, and no amount of structural readjustment is likely to make it disappear. But the larger planets tend to assume more precisely spherical shapes because self-gravitation is more effective on a body which is large, with a hotter, and therefore less rigid, interior.

Even the major planets, however, are not precisely spherical. The very rapid rotation of Jupiter, for example, means that the difference between its equatorial diameter and its polar one is clearly noticeable through a telescope. And the smaller, more solid and less rapidly rotating. Earth is also slightly flattened at the poles.

In addition, the Earth has four subtle bulges, but these were not discovered until our planet was viewed from space. Surveying techniques are simply not accurate enough for such minor deviations from sphericity to be recognised without the aid of satellites.

Finally, of course, planets may have their own topographic irregularities. The surface of the Earth, for example, includes mountains and chasms which respectively reach more than 5 miles above and below mean sea level, a reference horizon that in turn rises and falls in response to tidal forces exerted by the Sun and Moon. So the shape of a planet is modified by external as well as internal forces.

*Early lunar measurements*

To what extent, then, does the shape of the Moon deviate from a perfect sphere? What is its overall shape, how tall are its mountain peaks and how deep are its craters? And what can the Moon's topography tell us about its history and internal constitution?

Well, the science of lunar topography began with the ancient Greeks, who recognised that the Moon must be a sphere rather than just a circular disc. But Galileo was the first to calculate the heights of lunar peaks. He measured the lengths of the shadows which they cast on nearby level terrain, and, although his results may not have been very precise, they did at least demonstrate that the Moon, far from being the perfect orb favoured by the Greeks, was irregular on a vertical scale of several thousand feet (Figure 5.1).

251

But in order to extend Galileo's measurements it was essential to define the lunar equivalent of mean sea level as a reference horizon. Because not all mountain peaks are conveniently situated adjacent to level terrain, nor may all level areas be at the same radial distance from the Moon's centre of figure.

This task is certainly not a simple one, from a vantage point a quarter of a million miles away. And it is not made any easier by the fact that the Moon keeps essentially one face directed towards the Earth. Because, even with the help of geometrical librations, only a small fraction of the lunar surface ever defines the limb, where departures from sphericity can be measured directly (Figure 1.11). During an annular solar eclipse, for example, the deviations of points on the lunar limb from the edge of the slightly larger (and almost exactly circular) solar disc can be measured very accurately. Such eclipses are actually the only times when the entire lunar circumference is visible from Earth. For even at full moon there is a slight phase effect because the Moon must be slightly above or below the ecliptic plane and not all of its Earth-directed hemisphere is illuminated in these circumstances (Figures 1.8 and 1.13).

So what about those parts of the Moon which never appear on the limb? Well, inferior methods have to be used here, which is particularly unfortunate because it is at the centre of the visible disc that any tidal bulge will be pronounced. One such method involves studying the lunar terminator, the imaginary line separating the dark and sunlit hemispheres, the shape of which is related to the Moon's global form (Figure 5.2).

There are problems here, however, not the least of which is that the surface of the moon is everywhere far from smooth. Tall mountain peaks catch the dying rays of the Sun well beyond the true terminator, whereas some craters may be in deep shadow long before sunset. For this reason, terminator measurements have to be carried out photometrically (as opposed to astrometrically), using overexposed photographs. They are therefore of limited accuracy, particularly in the centre of the lunar disc, where the terminator has hardly any curvature.

Here we can again make use of the Moon's geometrical librations and

Figure 5.1. *Shadow height measurement*. The very first estimates of heights of lunar mountains were made by Galileo, based on measurements of the lengths of shadows cast by mountain peaks on nearby level terrain.

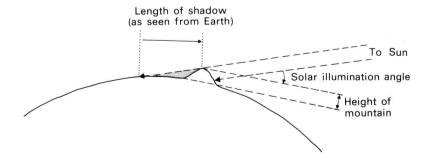

measure the change in angular separation between two lunar features, as the Moon's central meridian swings to either side of the Earth–Moon line. The magnitude of this separation difference depends to some extent on the shape of the lunar globe on which the features are situated (Figure 5.3). Without going into any detail, it can be well appreciated that extremely accurate measurements are required here. In order to obtain the required accuracy, in fact, it is necessary to work within the diffraction limits of the telescope, a questionable practice at the best of times.

So what shape did we think that the Moon had prior to the Apollo programme? Well, the results were conflicting, with some believing in the existence of a small tidal bulge, while others disagreed. The truth is that, within the recognised experimental uncertainties (about two miles) the Moon is essentially a sphere of radius 1080 miles. Any tidal bulge or polar flattening, then, must be within this uncertainty figure. And a much closer survey was therefore necessary before much more could be said about the Moon's global form.

### Topographic mapping

This detailed survey was undertaken by Apollo and is still far from complete. The primary surveying tools were the metric cameras, mounted in the Apollo 15, 16 and 17 Scientific Instrument Module bays (Figure 2.9). Now in order that the metric photography could be used to generate topographic maps of the lunar surface, the precise orientation and altitude of the Apollo spacecraft at the exact moment that each picture was taken had to be ascertained. Because, unlike a ground-based survey using a carefully positioned and levelled theodolite, these lunar ones were carried out from unstable and rapidly moving platforms.

Figure 5.2. *The shape of the lunar terminator.* The exact position of the lunar terminator, as seen from the Earth, depends not only on the solar illumination angle but also on the extent to which the Moon deviates from a perfect sphere. Local topographic irregularities have the effect of diffusing the terminator.

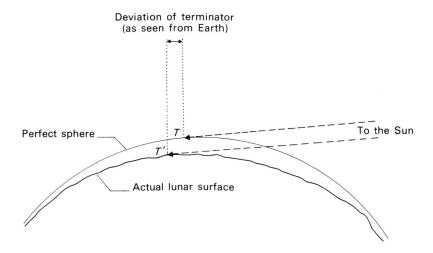

The orientation of the spacecraft was determined by operating a stellar camera in conjunction with the metric one. The calculation of the altitude of the metric camera above the lunar surface was more involved, and was achieved using a laser altimeter. 10 nanosecond laser pulses were sent out every few seconds and reflected back by a 1000 square yard area on the Moon immediately beneath the Service Module. The time taken for the reflected pulse to return, multiplied by the velocity of light, then gives us the distance travelled by the laser pulse, or twice the spacecraft altitude. The accuracy of the measurements, about 30 feet, was solely limited by the accuracy of the time measurements.

Knowing how high the metric camera was above the lunar surface, then, and also how it was orientated with respect to the stars, we could calculate distances and elevations by studying overlapping stereopairs (Figure 5.4).

### The S-band transponder experiment

But this was not all that we learnt from laser altimetry. It was crucial to put lunar topography into a global context by calculating radial distances, from the Moon's surface to its centre of mass.

By accurately tracking the Command Module from Earth this could be achieved, so the laser altimeter was operated over complete orbits, and not just over the sunlit hemisphere in conjunction with the mapping

Figure 5.3. *Stereogrammetry.* The angular separation between two features on the lunar surface when observed under a different libration condition depends not only on the libration angle but also on the heights of the two features, relative to the mean lunar sphere.

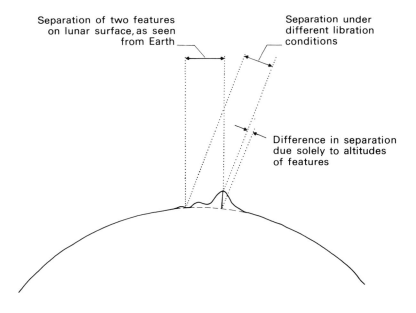

Separation of two features on lunar surface, as seen from Earth

Separation under different libration conditions

Difference in separation due solely to altitudes of features

camera. We were therefore able to obtain topographic profiles for areas not covered photographically.

The radius of the Moon at any particular point cannot be calculated directly from spacecraft altitude because its orbit is not perfectly elliptical, let alone circular. Why? Because irregularities in the Moon's crust may cause the spacecraft to be dragged down slightly towards the lunar surface.

We shall be discussing these mass concentrations, or mascons, later in this chapter. But, as far as lunar topography is concerned, what we needed to know was the extent to which the paths of the command Modules deviated from the elliptical trajectories which they would have followed if the Moon was simply a point mass.

This could be done by making use of the Doppler effect in what was known as the S-band transponder experiment. By transmitting a radio signal to the spacecraft and having it sent straight back, the line-of-sight velocity of the spacecraft at any time could be determined. If it was moving away from the Earth, the frequency of the returned signal would have been less than it would have been had it been stationary. Similarly, if it was moving towards the Earth, there would have been a shift towards higher frequencies. It is just like the change in pitch of an engine whistle when the train passes through a station, except that we are dealing here with electromagnetic, rather than sound, radiation (Figure 5.14).

These frequency shifts have of course, to be corrected for the Earth's rotation, and for the change in the Earth–Moon distance which results from the eccentricity of the Moon's orbit around the Earth. But, after taking these and other factors into account, a precise orbit for the

Figure 5.4. *Profiles based on metric photography.* Metric photography was controlled by stellar photography and laser altimetry. Overlapping pictures were used to generate topographic maps, from which these three profiles of a wrinkle ridge in Mare Fecunditatis (Dorsa Geike) were drawn. The vertical scale has been exaggerated relative to the horizontal by a factor of 16, for the sake of clarity. (After Andre, C. G., Adler, I., Clark, P. E., Weidner, J. R. & Philpotts, J. A. (1976). *Proceedings of the Seventh Lunar Science Conference*, pp. 2649–60. New York: Pergamon.)

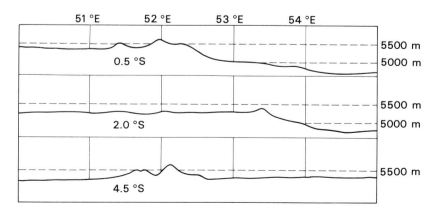

spacecraft can be calculated, with reference to the Moon's centre of mass. The measured altitude of the spacecraft is then subtracted from the calculated distance of the spacecraft from the Moon's centre of mass to yield the Moon's radius immediately beneath the spacecraft. An altimetry profile can then be constructed along the ground track of the orbit. So what do these profiles look like and what can we now say about the Moon's overall shape?

### Altimetry profiles

Well perhaps the most important finding was that, while the Moon's cross section is indeed approximately circular, the centre of the circle (the centre of figure) does not coincide exactly with the Moon's centre of mass (Figure 5.5). This was not totally unexpected, because the impact times of the Ranger probes on the near side had been somewhat later than calculated. It also accounted for the anomalously high Earth–Moon distances which were measured using the Apollo 11, 14 and 15 Lunar Laser Ranging Retroreflectors.

The Moon's centre of mass is, in fact, displaced from its centre of figure by more than a mile in an earthward direction. And a slight flattening (also in an earthward direction), rather than a tidal bulge, means that the moon has not been distorted by tidal forces since it attained its final global form.

But why should there be this centre of mass offset? Well, there are two possible reasons. Either there is an excess of high density material on the near side (in other words a global mascon), or else the far side crust is exceptionally thick. At first sight the former alternative seems more likely, because the near side was clearly flooded with dense *mare* basalts. But these lava flows are far too thin to account for the centre of mass displacement. So the concentration of *maria* on the near side may, in fact, be a consequence (rather than a cause) of the Moon's centre of mass displacement. Because a thin crust would have made it much easier for magma to reach the surface.

Figure 5.5. *Nearside laser altimetry.* The near side of the Moon is characterised topographically by relatively flat *maria*, the surfaces of which are all depressed with respect to the mean lunar sphere. The two profiles shown here were obtained on orbits 15 and 22 on the Apollo 15 mission. (From Kaula, W. M., Schubert, G., Lingenfetter, R. E., Sjogren, W. L. & Wollenhaupt, W. R. (1973). *Proceedings of the Fourth Lunar Science Conference*, pp. 2811–19. New York: Pergamon.)

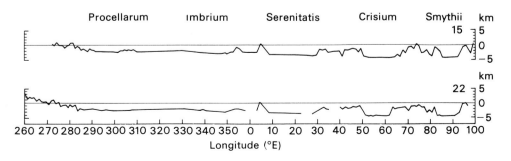

Unfortunately, altimetry data only covered the ground tracks of the last three Apollos. Metric photography allows us to extend this coverage somewhat, but that of the near side is clearly not representative of the hemisphere as a whole, with the Southern Highlands not being over-flown at all by Apollos 15 and 17. It would be highly desirable, then, to mount a laser altimeter on the proposed Lunar Polar Orbiting satellite. We would then be able to calculate the precise direction of the centre of mass displacement and establish whether or not there is any polar flatten-ing. On the basis of the results obtained so far, the centre of mass displacement seems to point 24° or so to the east of the Moon–Earth line rather than directly towards the Earth.

As far as individual features are concerned, the most impressive one was a previously unrecognised 870 mile wide depression (or thassaloid) in the farside highlands, situated on the central meridian and some 30° south of the equator (Figure 5.6). The extra thickness of the farside crust means that this farside basin was never filled with dark *mare* materials. It is therefore more difficult to recognise than smaller basins of comparable age on the near side. Practically all of its ring mountains, in fact, have been totally obliterated by cratering. But its interior was not filled in, so there is still a deep depression, which extends at least 20 000 feet below the surrounding highlands.

Apart from this Big Backside Basin, as it is affectionately called, the most significant discoveries of laser altimetry concern the *maria*, all of which were found to be depressed relative to the mean lunar sphere. This is just what we would expect in view of the centre of mass displacement. But circular *maria* are depressed more than the irregular ones (Figure 5.5).

The *maria* are also extremely smooth, (within a vertical scale of 300 feet) and have characteristically shallow slopes, often as gentle as 1:2000 over distances of many hundreds of miles. These findings confirm that lunar basaltic lavas must have had extremely low viscosities, otherwise gradients as steep as 1:500 would have been inevitable.

Figure 5.6. *Farside laser altimetry.* The far side of the moon is characterised topographically by more rugged terrain and a wide, deep depression. (From Kaula, W. M., Schubert, G., Lingenfetter, R. E., Sjogren, W. L. & Wollenhaupt, W. R. (1973). *Proceedings of the Fourth Lunar Science Conference*, pp. 2811–19. New York: Pergamon.)

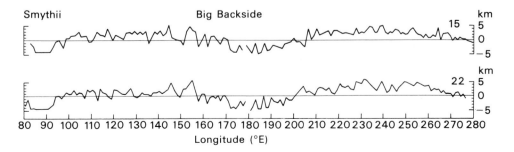

## Radar measurements

The laser altimeter was not, in fact, the only Apollo instrument which was capable of measuring spacecraft altitude. On Apollo 17, the Moon was also probed with radar, although the primary purpose of the Lunar Sounder Experiment was to study the structure of the regolith and upper lunar crust. The radar coverage was of course no more extensive than that of the Apollo 17 laser altimeter, but the radar resolution was much higher, because the Sounder could be operated semicontinuously.

The profiles from both experiments were perfectly consistent, despite the fact that the reflection of radar waves does not all occur from the nadir point, because it depends on the slope of the reflecting surface. The same

Figure 5.7. *Radar altimetry profile*. This lunar profile was obtained using the Apollo 17 Lunar Sounder, an experiment designed primarily to investigate the lunar subsurface. This is essentially the same ground track as that of Apollo 15, except that it is displayed from the north. So Mare Imbrium and Oceanus Procellarum are between 0° and −90° (i.e. 270°E). Exaggeration factor = 391. (From Brown, W. E., Adams, G. F., Eggleton, R. E., Jackson, P., Jordan, R., Kabrick, M., Peebles, W., Phillips, R. J., Porcello, L. J., Schaber, G., Sill, W. R., Thompson, T. W., Ward, S. H. & Zelenka, J. S. (1974). *Proceedings of the Fifth Lunar Science Conference*, pp. 3037–48. New York: Pergamon.)

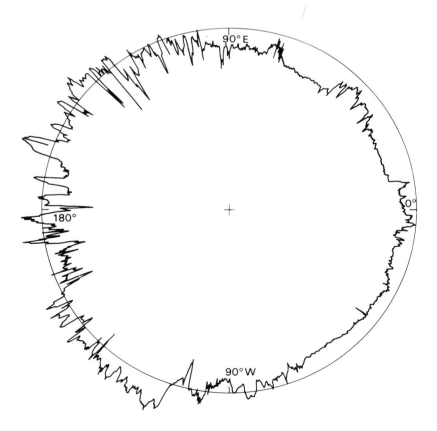

broad scale profile was obtained (showing the centre of mass displacement, the farside basin and the depressed nearside *maria*), and innumerable smaller scale features, beyond the resolution of the laser instrument, were revealed. This consistency does much to inspire confidence in the altimetry results as a whole (Figure 5.7).

Interestingly enough the surfaces of western Oceanus Procellarum, Mare Serenitatis, Mare Crisium and Mare Smythii all define a spherical surface. But the centre of this sphere coincides with neither the centre of mass, nor with the centre of figure. To determine whether or not this surface is shared by other *maria*, such as Humorum and Frigoris, we will have to wait for an altimeter on a Lunar Polar Orbiter satellite. The levels of *mare* surfaces may well tell us something about the Moon at the time of lava eruption.

Our current picture of the Moon, then, is currently based on data from a number of independent sources. Radar ranging, laser altimetry and topographic mapping are complemented (in those areas far removed from the Apollo groundtracks) by limb measurements and long range photography. A smoothed surface through this data allows us to concentrate on the Moon as a whole rather than on individual features.

## 5.2   Gravity, tides and mascons

Having heard that inhomogeneities in the Moon's gravitational field have a measurable effect on the path of a satellite in lunar orbit, what can we infer from the S-Band Transponder Experiment about the distribution of matter within the Moon? Before attempting to answer this question, we must introduce the force of gravity itself, discuss how it controls the dynamics of the Earth–Moon system, and discover why lunar gravity is only one-sixth as strong as that at the surface of the Earth.

Gravity is the weakest of the fundamental forces of nature. The gravitational force between an electron and a proton is, in fact, one thousand million million million million million million times weaker than their electrostatic attraction. So why does gravity seem so strong? Well, it is simply that planets are very massive and the force of gravity depends on mass rather than electric charge. Had the Solar System not been electrically neutral the arrangements of planets and moons would certainly have been very different.

But, like electrostatic attraction, the force of gravity varies with the inverse square of separation distance. In other words double the distance between two bodies and the attractive force between them decreases fourfold. And this gravity force is, of course, what holds the Earth and Moon together. So how does gravitation control the motion of the Moon around the Earth, and how can we calculate the mass of the Moon itself?

*Lunar inequalities*

In the first chapter the Moon's relative orbit was described as being not only elliptical, but also inclined to the ecliptic by between 5° 0′ and 5° 18′ (Figure 1.2). The exact value of this angle depends on whether or not the Sun is in a position to pull the Moon's orbital plane down towards the ecliptic. We also heard that the major axis of the Moon's relative orbit rotates around the Earth every nine years or so (Figure 1.4), and that the lunar nodes rotate in the opposite direction about half as quickly (Figure 1.3). But complex as these consequences of gravitation may be, they alone cannot describe the observed motion of the Moon across the sky to a high degree of accuracy. A number of other factors, largely due to the gravitational pull of the Sun, complicate matters still further.

The most significant of these lunar inequalities concerns the eccentricity (or elongation) of the Moon's orbit and was first recognised by the greatest of ancient Greek astronomers, Hipparchus. Depending on how the major axis of the Moon's ellipse is orientated with respect to the Earth–Sun line, its eccentricity may vary between 0.0432 and 0.0666. In effect the Sun may either be stretching the ellipse out, or else squashing it back closer to a circle (Figure 5.8).

This phenomenon, *evection*, may result in a lunar longitude displacement (an angular displacement from the position that the Moon would occupy if the Sun did not exist) of more than 1° 16′, or greater than twice the apparent diameter of the full moon. It is hardly surprising, then, that evection was noticed in the pre-telescope era, although its origin was of course to remain a complete mystery until the days of Isaac Newton.

A second inequality, *variation*, results in a longitude displacement which, although smaller, may still be nearly 39.5′ arc. Variation arises

Figure 5.8. *Evection*. The eccentricity of the Moon's orbit varies between 0.0432 and 0.0666, depending on the relative direction of the Sun.

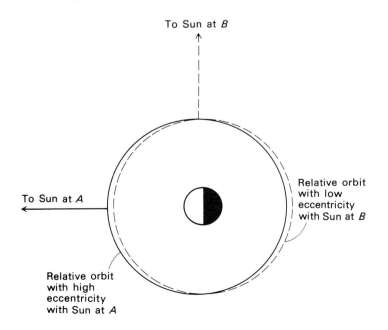

from the fact that the Earth's gravitational pull is enhanced by that of the Sun when the Moon is full, but opposed by it when the Moon is new (Figure 5.9). Variation is clearly also large, compared with the Moon's apparent diameter, and was, in fact, first recognised by a 16th century astronomer, Tycho Brahe.

The third major complication is known as the *parallactic inequality* and arises from the fact that, when the Moon is at first quarter, it is moving away from the Sun and is therefore being retarded by its gravitational pull, whereas at last quarter it is being slightly accelerated (Figure 5.10). As these additional accelerations and decelerations depend on the Moon–Sun distance, the magnitude of this effect can be used to calculate the distance to the Sun.

Finally, the orbit of the Earth–Moon system around the Sun is not a circle but an ellipse, albeit one of small eccentricity. The change in the Sun's distance throughout the year modulates the disturbing effects of its gravitational pull, so that evection, for example, is more significant when the Earth is at perihelion (in other words when it is nearest to the Sun) than it is at aphelion. These modulations are collectively known as the annual inequality, or *annual equation*.

Taking into account all other factors, such as the disturbing effects of other planets, the equations which must be used to describe the Moon's motion to the highest degree of accuracy would now occupy more than 200 pages of text. It is no wonder then that we need high speed electronic computers to track the still more complex orbits of man-made satellites.

But it was as long ago as the 18th century that the major components of lunar motion were first quantified. Only then could the apparent position of the Moon in the sky be used by ocean navigators to establish their

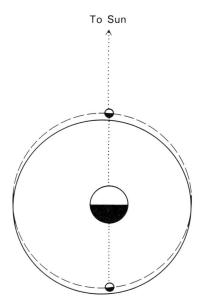

To Sun

Figure 5.9. *Variation*. The Moon's elliptical orbit is further distorted towards the Sun, such that the Moon is subjected to the combined pull of Earth and Sun when it is full.

longitude by lunar parallax. This remarkable feat was accomplished in response to a major financial incentive, but it was around this time that accurate time measurement first became possible, with the invention of the chronometer, rendering the lunar parallax method obsolete. This does not, however, detract from the accomplishments of talented mathematicians of that era, such as Euler and Laplace, who effectively laid the foundations for interplanetary navigation.

### The mass of the Moon

But how can we make use of the effects of gravity to learn more about the Moon as a planet? How can we, for example, calculate the Moon's mass? Well, what we have to do here is study the effect that the Moon has on the motion of a nearby astronomical body. And, before the space program, the only available object was the Earth. For no other planet ever comes near enough for its path to be noticeably deflected by the Moon's gravitational pull. The Moon does not, strictly speaking, revolve around the Earth. Instead, both planets move about their common centre of mass, at distances which are inversely proportional to their respective masses. And it is this centre of mass, rather than the centre of the Earth, which describes an elliptical orbit around the Sun (Figure 5.11). So, by calculating how much the Earth wobbles on either side of this elliptical path every month, we should be able to calculate the ratio of the masses of the two planets.

The most straightforward method used to quantify the Earth's wobble is to make use of the parallax effect. The displacement of a nearby astronomical body whose absolute position is known to a high degree of accuracy depends on the magnitude of the wobble and also on the distance and direction of that body. It is just like the apparent displace-

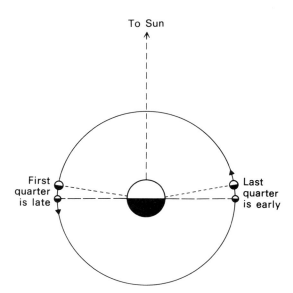

Figure 5.10. *Parallactic inequality*. When the Moon is at first quarter it is being retarded by the Sun's pull, whereas at last quarter it is being accelerated.

ment of an object held at arms length relative to a distant scene, when it is viewed first with one eye and then with the other. This astronomical measurement will clearly be most accurate if the body is very close. And the best possible object for this purpose is an Apollo–Amor asteroid, some of which very occasionally pass within a million miles of Earth.

It turns out that the Earth is some 81.3 times as massive as the Moon, and therefore the centre of mass of the Earth–Moon system lies well inside the Earth. Interestingly enough, the much larger lunar wobble is still so small compared with the Earth–Sun distance that the Moon's actual path in space is always concave towards the Sun. It does not trace out the sort of looping trajectory that one might expect it to follow. No other satellite in the Solar System behaves in this way. And no other Moon, in fact, is so massive compared with its parent planet.

The mass of the Moon has of course been recalculated many times since the dawn of the space age, from the deflecting effects that it has on passing spacecraft. But, had we not already known it to a high degree of accuracy, it is unlikely that spacecraft would ever have reached the Moon in the first place. In absolute terms the Moon's mass turns out to be $7.353 \times 10^{25}$ grams, or more than 70 million million million tons.

Combining this value with the Moon's mean radius we can calculate a mean lunar specific gravity of 3.34, which is significantly lower than that of the earth (5.53) and clearly an important clue to the Moon's internal constitution. If the Moon does have a metallic iron core, for example, it

Figure 5.11. *Determination of lunar mass.* The centre of mass of the Earth–Moon system describes an elliptical orbit around the Sun. The Moon and Earth oscillate on either side of this path, the relative amplitudes of the oscillations being inversely proportional to their respective masses. The size of the Earth's wobble may be calculated from the parallax of a passing asteroid. The mass of the Moon can now be calculated more precisely from its deflection effect on a passing spacecraft. Note that the Moon's path, like the Earth's, is always concave towards the Sun.

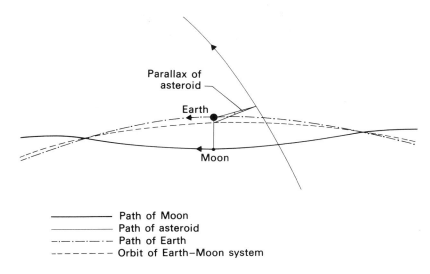

Parallax of
asteroid

Earth

Moon

———————— Path of Moon
———————— Path of asteroid
—·—·—·—·— Path of Earth
– – – – – – – Orbit of Earth–Moon system

cannot be a very large one. And the Moon's mean density is, in fact, already very close to that of the pyroxenite rocks of which the Moon's mantle must be composed. The Moon's density, then, is certainly a firm constraint on all models for the lunar interior.

As the Moon is so much smaller than the Earth, the force of gravity at its surface is by no means 81 times weaker. At 5.31 feet per second, in fact, it is equivalent to about one-sixth of that at the surface of Earth. Avid followers of the moonwalks may recall that an Apollo astronaut performed an experiment to demonstrate the constancy of this acceleration by simultaneously dropping a feather and a geological hammer to the ground. Both were seen to reach the Moon's surface at the same instant, a sight that Galileo would certainly have appreciated.

Figure 5.12. *The raising of tides.* (*a*) The gravitational pull of the Moon is stronger on the oceans on the moonward side of the Earth (*B*) than those on the opposite side (*B*₁). So two tidal bulges are created. The rapid rotation of the Earth is responsible for a lag in time between meridian transit of the Moon and the highest tide. The tidal bulge exerts a force on the Moon which causes it to recede. (*b*) The highest (spring) tides occur when the pull of the Moon is supplemented by that of the Sun, particularly when the Moon is at perigee. As the tibal bulge is essentially in the Moon's orbital plane, rather than in the equatorial plane, one high tide on Earth is invariably higher than the other, a phenomenon known as the diurnal inequality.

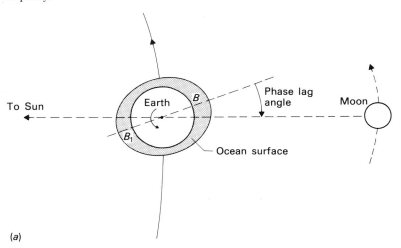

(*a*)

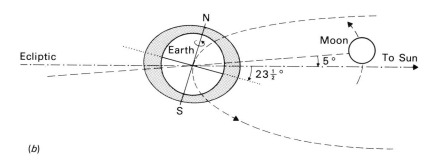

(*b*)

The low gravity on the Moon was a novel experience for astronauts, who had until then been used to the total weightlessness of space. One-sixth *g* did not present any real problems and was in fact found to be quite pleasurable. Not only did it give a sense of up and down, it also provided opportunities for experiments. A well struck golf ball on the Moon would easily disappear over the horizon.

Lunar escape velocity is only one and a half miles per second. So the Moon could be a good place to obtain raw materials to build space stations. A long chain of superconducting magnets could be used like a rifle barrel to blast chunks of iron out into orbit, where they could be collected for further processing. The imagination certainly knows no bounds when it comes to the future exploration of space.

### Secular acceleration and lunar recession

But lunar inequalities are not the only consequences of gravitational forces in the Earth–Moon system. The Moon also acts on the Earth, and its differential attraction produces two tidal bulges. One side of the Earth is closer to the Moon than the other and is therefore more strongly attracted to it. These tides are most obvious in the oceans, but they are also raised in the atmosphere and, to a much lesser extent, in the solid earth.

It is the rotation of the Earth under these tidal bulges which is, of course, responsible for the production of two high tides every day but there is invariably a time lag between the time of strongest pull and the highest tide, for dynamical reasons. In other words, high tide only occurs when the Moon is several degrees past the meridian (Figure 5.12).

Now, if the Earth were a perfect sphere, and if the oceans could flow easily over its surface, the raising of tides on Earth would not affect the dynamics of the Earth–Moon system. But the oceans of the Earth are of variable depth, and are bounded by the highly irregular outlines of the continents. So a considerable amount of energy must be dissipated whenever tides sweep up and down estuaries, or through narrow straits. And this energy must come from the Earth's rotation. In other words, as tidal energy is converted into heat, the rotation of the Earth on its axis must slow down.

Is there any evidence for this slowing down? Well, the number of days in the year back in Devonian times (based on the annual and daily growth ring patterns in certain marine organisms) imply that the year may then have contained as many as 400 days. So the day must have been considerably shorter than it is now. And the rate at which the Earth spins on its axis is currently slowing down by about 16 seconds every million years.

But energy is not the only property that is conserved in a closed system. If the Earth's rotation is slowing down, some of its angular momentum must be transferred to the Moon. In other words, the Earth's tidal bulges should exert a pull on the Moon and cause it to accelerate. But the Moon is not attached to the Earth by a piece of string, so, as the Moon speeds up, it should recede from the Earth, at a velocity of about one and a half inches per year.

Surprisingly enough, this *secular acceleration*, although very small, is easy to recognise, the reason being that total solar eclipse records go back for many hundreds of years. And the timings of these eclipses (and where they were observed) can tell us exactly where the Moon was in its orbit thousands of years ago. To appreciate how sensitive these records are, we need only add that the Moon's secular acceleration has been calculated to be only 10.3″ arc per century.

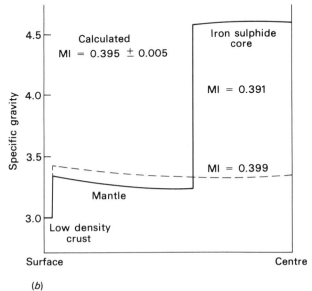

Figure 5.13. *Lunar ranging retroreflector.* (a) This picture shows the Apollo 11 laser reflector. An array of corner cube reflectors returns a laser pulse to Earth and the delay time can be used to measure the Earth–Moon distance. (NASA, Apollo.) (b) In this diagram are drawn two extreme density models for the Moon which both satisfy its moments of inertia. (After Dainty, A. M., Toksöz, M. N., Solomon, S. C., Anderson, K. R. & Goins, N. R. (1974). *Proceedings of the Fifth Lunar Science Conference*, pp. 3091–114. New York: Pergamon.)

The other potentially measurable consequence of tidal friction is, of course, the recession itself. But it is only since the invention of radar and laser ranging that distance measurements have had anything like the degree of accuracy required to confirm the predicted recession rate. We can now obtain distances to the Moon to an accuracy of better than one foot, a truly remarkable achievement. It will be a while, however, before we can isolate the Moon's secular recession from its periodic motions (Figure 5.13a).

It will, of course, be a very long time before total solar eclipses are not just rare, but completely impossible. Eventually, however, the Earth's axial rotation (like that of the Moon) will become synchronised with the Moon's orbital motion. At this point the Moon will recede no further. But by then its distance will have doubled, so it is unlikely that anyone will be around on Earth to witness the event.

But the Moon must presumably have been very much closer to the Earth in the past than it is today. And tidal effects may then have been much more effective at slowing down the Earth's axial rotation. So how close did the Moon get, and might there be a clue here to the Moon's origin?

Well, we cannot simply wind the clock back, because conditions on Earth today are very different from those prevailing in earlier epochs. A long time ago, for example, the Earth had fewer continents, and continental shelves, where most of the Earth's tidal energy is dissipated, were not so extensive. So this accounts for a significant discrepancy. For, based solely on the current recession rate, the Moon should have been in close proximity to the Earth only 1800 million years ago. There is, however, no evidence to support such a close encounter during the last 4000 million years, in the form of widespread melting of the lunar surface or distortion of crater shapes. The chemistries of the *mare* basalts preclude them from having such a secondary origin. So, if the Moon ever was subjected to extreme tidal forces, this can only have happened during the very first 700 million years of its history.

### Physical librations

Most of our lunar ranging data comes from the Apollo 11, 14 and 15 Lunar Laser Ranging Retroreflector Experiments, operated in conjunction with Earth-based telescopes. But laser reflectors were also deployed on the Moon by the Russian Lunokhods, as part of a joint Franco–Soviet ranging program. Not only do these experiments enable us to increase the accuracy and precision of lunar distance measurements, they also allow us to study the phenomenon of gravitation itself. For the anticipated recession rate of the Moon, based on Einstein's general relativity theory, is slightly different from that predicted by classical Newtonian gravitation. So the recession rate of the Moon may one day provide evidence to support Einstein's theory of gravitation.

But lunar ranging is not only concerned with the Moon's rate of recession. We also want to study irregularities in the motion of the Moon about its centre of mass, known as its physical librations. These forced oscillations of the Moon allow us to determine an extremely important lunar property, its moment of inertia, our only real clue to density variations within the Moon.

So just how big are these physical librations? Well, they are very minor indeed compared with the geometrical librations discussed in the first chapter at the very most amounting to barely 2' of lunar longitude. Seen from the Earth the entire lunar disc has a diameter of only half a degree, so angular displacements arising from the Moon's physical librations amount to 0.5" arc at the very most. That such incredibly small angular distances have been detected at all says much for our current understanding of lunar motion, for the dedication of lunar astronomers and for the precision of their instruments. Recently acquired laser data should improve the results still further.

But what causes physical librations and what can they tell us about the Moon? Well, their very existence indicates that the distribution of matter within the Moon is not spherically symmetrical, and we now know that this asymmetry must arise from the greater thickness of the farside crust. In effect, the Earthward displacement of the Moon's centre of mass provides a handle on which the Earth's gravity may pull and make the Moon oscillate.

Figure 5.14. *Apollo S-band transponder experiment and mascons.* Changes in line-of-sight spacecraft velocity indicate the presence of superisostatic masses associated with the circular *maria*, mascons. Although the dense *maria* are depressed relative to the mean lunar sphere above which the highlands are elevated, isostasy cannot prevail. One theory is that *mare* lavas rose until the molten lava column balanced the equivalent column under the highlands $(P = P')$ and stood at level $L$. But, as this lava solidified, it sank to level $L'$, allowing a superisostatic layer $(L - L')$ to be extruded over it. The strength of the crust then prevented this extra load from sinking to isostasy. A mantle plug could also contribute to mascons.

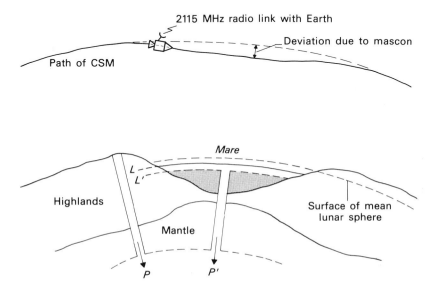

A much more important piece of information, however, can be extracted from the frequency of the Moon's physical librations and this is its moment of inertia. For, when combined with its total mass, the moment of inertia of the Moon (namely $0.395 \pm 0.005 \, M_m R_m^2$, where $M_m$ and $R_m$ represent the mass and radius of the Moon respectively) tells us that it is effectively a sphere of uniform density. The small uncertainty in this figure cannot accommodate a thick low density crust, or a dense iron core of any appreciable size. So once again we have a firm constraint on the structure of the lunar interior (Figure 5.13*b*).

### The discovery of mascons

But how does the distribution of mass within the Moon affect its topographic outline? And what has now been learnt about this distribution through satellite tracking? Well, the S-band transponder experiments (Figure 5.14) flown with the Lunar Orbiter satellites, enabled us to quantify changes in line-of-sight velocity as small as 1 millimetre per second. The Lunar Orbiters were initially inserted into high altitude parking orbits in order to investigate the Moon's gravity field in preparation for low altitude photographic reconaissance. By taking this precaution, all survived to complete their appointed tasks, and were only crashed on to the lunar surface after their film packs had been exhausted.

After removing the low order terms in the Moon's gravitational field (those due to its irregular shape and centre of mass displacement) a number of circular positive gravity anomalies, or mascons, can be clearly distinguished (Figure 5.15). So what do these excess masses consist of, where did they come from and what can their existence tell us about the Moon's internal structure?

Well, first of all, mascons are invariably associated with circular *maria*, such as Mare Imbrium and Mare Serenitatis. Smaller ones are embedded in Mare Crisium, Mare Smythii, Mare Nectaris and Mare Humorum. As spacecraft cannot be tracked from Earth beyond the lunar limbs, we cannot yet say whether there are any on the far side. The Apollo 16 subsatellite crashed on the far side, as a direct result of our lack of detailed knowledge of the farside gravity field. So an S-band transponder experiment would clearly be desirable on a Lunar Polar Orbiter, with a second satellite in equatorial orbit acting as a relay link when the Polar Orbiter is out of direct radio contact with Earth.

### Origins of mascons

At one time it was naïvely believed that mascons are remnants of the dense projectiles responsible for the excavation of basins. But it was soon realised that they would have been totally vaporised. Furthermore, the largest mascons only occur in those basins which were filled with *mare* basalt.

But if mascons are simply layers of *mare* basalt lying on top of low density crust, how was this excess mass brought to the surface in the first place?

Well, if a planet is in isostatic equilibrium, the overburden pressure will be constant at any given distance from the centre of the planet (Figure 5.14). Terrestrial continents are much thicker than the oceanic crust simply because granitic-type rocks have lower densities than oceanic basalts. Glancing at the lunar altimetry profiles, it might seem that isostasy prevails on the Moon as well (Figure 5.5), because the nearside *maria* are all significantly lower than the surrounding highlands. But the proximity of dense *mare* basalt to an orbiting spacecraft is not in itself sufficient to produce much of a mascon. And the large mascons of Imbrium and Serenitatis would certainly not have survived on a Moon in isostatic equilibrium. They would have quickly sunk without trace.

So what exactly produces mascons and why have they survived for so long? Well, the most compelling theory is based on the fact that lava, unlike water, will shrink when it solidifies. Imagine a vertical crack in an

Figure 5.15. *Lunar gravity map.* The lines on this map of the Moon are gravity contour lines, showing the strength of lunar gravity at the lunar surface (in units of 100 milligals) over and above the mean value, as compiled from radiotracking data. The mascon associated with Mare Orientale is not shown, but is a more complex feature. (NASA.)

1 Mare Imbrium
2 Mare Serenitatis
3 Mare Crisium
4 Mare Tranquillitatis
5 Mare Nectaris
6 Mare Humorum

otherwise rigid lunar crust, extending all the way down to a *mare* basalt source region more than 100 miles inside the Moon. Basaltic lava rises up this crack and floods the basin floor until the column of fluid just balances the pressure appropriate to the depth of the source region, at which stage isostatic equilibrium is attained. This is directly comparable to mercury rising up a barometer tube until its height corresponds to atmospheric pressure.

But the fluid lava will then freeze and shrink, causing the *mare* surface to subside. And this subsidence could then allow a little more lava to rise up the crack and spread out over the original surface, because the equilibrium fluid level will now be above the original solid surface. This extra layer of *mare* basalt, the amount required to compensate for the subsidence on solidification, is then responsible for the mascon. And the deeper the basin, the larger we would expect the mascon to be (Figure 5.14).

How does this theory stand up to scrutiny, and what other possibilities exist? Well, there is certainly some evidence for *mare* subsidence. For sinuous rilles are the ruptured remains of lava drainage channels (Figure 1.35), and drainage implies a decrease in lava volume. It is also possible to distinguish high lava marks around the borders of some *maria* (Figure 1.33).

But there are other possible solutions to the mascon problem. Dense metallic iron may have sunk to the base of the lava lake, for example, or pressures may have been so great at depth that the dense rock type known as eclogite was formed.

A more popular alternative is that some mascons may consist of dense mantle materials which rose up during basin excavation (Figure 5.14). This last hypothesis is favoured for the small mascon in Orientale, a basin which was not extensively filled with *mare* lavas.

But the Orientale mascon is rather complex, with the outer parts of the basin exhibiting negative, rather than positive, gravity anomalies. This suggests that significant amounts of material were removed and deposited elsewhere, without later crustal readjustment inside the basin itself. We find smaller negative gravity anomalies in young unfilled craters, such as Copernicus, and we might expect to find a very large one in Big Backside Basin. The crater Grimaldi, on the other hand, is filled with *mare* lavas and is therefore the site of a small positive anomaly.

*Survival of mascons*

Perhaps the most remarkable aspect of mascons is their very survival. The fact that they still exist today implies that the Moon's crust must have been remarkably rigid for a very long time. But it is interesting to note that Mare Smythii and Mare Crisium are significantly depressed relative to the somewhat younger *maria*, Imbrium and Serenitatis (Figure 5.5). This could mean that they have sunk further towards isostatic equilibrium, but the masconless Mare Tranquillitatis is possibly the oldest *maria* on the Moon, and yet it is elevated with respect to Mare Smythii.

It is possible that we are dealing with underlying *mare* basalts of very different ages and chemical compositions, a factor that is not normally stressed in mascon discussions. In the author's Gargantuan Basin hypothesis for explaining the distribution of natural radioactivity in Oceanus Procellarum, it is proposed that the entire *mare* was flooded by KREEP basalts, which were later covered by a thin veneer of iron-rich lavas. Now the significant point to note here is that KREEP basalts, being aluminous and iron-poor, are much less dense than ordinary *mare* basalts. So even if the lunar crust was strong enough to support mascons at the time of their eruption, there may not have been much of a mascon to support. And Oceanus Procellarum is, in fact, elevated relative to Mare Imbrium. A similar argument can be used to explain the lack of mascons in the irregular eastern *maria*, because mascons only form when there is a significant density contrast between lava and crust.

But mascons and topography are not the only lunar properties that can be mapped using satellites. The Moon can also be characterised by means of electromagnetic radiation. And such studies can reveal much about the chemical and thermal properties of the Moon, not just at a handful of landing sites, but over the entire globe.

## 5.3   Remote sensing

There is more to remote sensing than simply recording and interpreting images on photographic film, important as this task may be. Electromagnetic radiation extends over a vast range of wavelengths, and each band in the spectrum yields its own particular clues about the physical and chemical state of the radiation source. We will begin by considering infrared and radio frequencies, and then move across the optical window to ultraviolet and the even shorter wavelength X-rays and gamma-rays.

### Thermal radiation

Although the full moon seems very bright, the surface of the Moon is really a very poor reflector of light. It absorbs more than 90 % of incident sunlight, and so it is in fact as black as charcoal. But, because of this absorption, the regolith heats up and re-emits at longer wavelengths, mainly in the infrared region at around 10 microns.

An infrared image of the Moon, then, reveals thermal (rather than topological) contrasts on the lunar surface, because we are, in effect, measuring thermal radiation. But such an image cannot be captured on photographic film, because photographic emulsions are insensitive to all but the shortest wavelength infrared rays. So they must instead be built up using special instruments such as the infrared scanning radiometer flown in the Apollo 17 Scientific Instrument Module. Less sophisticated Earth-based devices, with lower resolving power, had already proved the usefulness of this technique.

So what do these infrared pictures tell us? Well, the images which we obtain of the full moon are certainly very different from those taken in visible light. For one thing, the brightest region is the centre of the lunar disc, where the Sun is directly overhead and where surface temperatures may easily exceed 100 °C. The limbs are cooler and therefore darker. And, during the long lunar night, the regolith may be as cold as liquid air. So not much infrared radiation is emitted from the dark face of the Moon.

But temperature changes on the Moon can never be instantaneous, whether at sunrise or sunset, or when the Moon is eclipsed by the Earth. Infrared brightness is therefore not solely a function of Sun angle. Large boulders, having been heated to high temperatures during the lunar day, for example, take an appreciable time to cool when suddenly plunged into darkness. So infrared pictures of the Moon taken during the night or a lunar eclipse reveal those regions which can retain heat most efficiently, such as the young rayed craters Copernicus and Aristarchus (Figure 5.16), which presumably harbour high concentrations of heat retaining boulders. These same boulders also take a long time to heat up after sunrise, so young craters show up as dark spots when they are sunlit but still close to the sunrise terminator.

But thermal radiation from the Moon is not confined to the infrared. It also reaches the Earth's surface through the radio window, at wavelengths of a few millimetres. And, unlike the infrared rays, this longer wavelength radiation yields information about the Moon's subsurface, where temperatures are not influenced by the wildly fluctuating diurnal heat wave.

At depths of only a foot or so, temperatures do not vary much from one month to the next, because the lunar soil is such a bad thermal conductor that surface heat cannot penetrate down to such depths. Microwave radiation might therefore enable us to detect the Moon's internally generated heat. So a microwave scanner may one day be flown on the proposed Lunar Polar Orbiter.

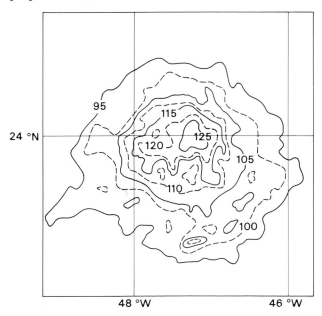

Figure 5.16. *Thermal map of Aristarchus*. Temperatures inside the young crater Aristarchus were mapped using the Apollo Infrared Scanning Radiometer. Temperatures (in degrees absolute) are as much as 30 degrees higher during the night than in the surrounding terrain. (After Mendell, W. W. (1976). *Proceedings of the Seventh Lunar Science Conference*, pp. 2705–16. New York: Pergamon.)

### Spectral reflectance

Investigations of the Moon at long wavelengths have not been restricted to thermal measurements. Certain elements (such as iron and titanium) absorb strongly in the near infrared (Figure 5.17). Indeed, spectral reflectance at such wavelengths is now a powerful tool for remote chemical analysis. And not just of the moon. Asteroids and some satellites of outer planets can also be studied, because they too lack an atmosphere.

The iron absorption band, for example, will occur at a wavelength which characterises the chemical state of iron at that particular point on the Moon's surface. So, by locating the peaks in the absorption spectrum we have been able to map out the areal distributions of known rock types, such as titanium-rich basalt. We have even identified some rock types which are not represented among the returned samples. So there may still be some important petrological discoveries to be made on the Moon by future prospectors.

The fact that some rocks are coloured means that the absorption of solar radiation must occur in the visible part of the spectrum as well as the infrared. And this absorption does in fact also extend into the ultraviolet. So this is why the *maria* all appear darker than the highlands. They are simply that much richer in iron and titanium. But there are even colour differences between *mare* rocks. And these contrasts may be suffficiently large for them to be recognised at a distance, particularly if colour difference photography is employed. By combining photographs taken through infrared and ultraviolet filters, for example, individual basalt flows can be readily identified. And the heavily cratered shores of Mare Serenitatis are found to be distinctly bluer than its younger interior (Figure 5.18).

### Lunar albedo and polarisation

Moving to shorter wavelengths, one Apollo 17 orbital spectrometer mapped parts of the Moon in the far ultraviolet, and at these wavelengths the *maria* were found to be more, rather than less, reflective than the highlands.

But the Moon's albedo is determined not only by the structure, but also by the chemistry, of the lunar surface. The Moon is, in fact, eleven

Figure 5.17. *Spectral reflectance curves.* The reflectivity of bright craters in the *mare* (open circles) is very similar to that of crushed *mare* basalt (solid curve). All data is normalised to a standard (Apollo 12 soil), to highlight subtle differences. (After Adams, J. B. & McCord, T. B. (1971). *Proceedings of the Second Lunar Science Conference*, pp. 2183–95. New York: Pergamon).

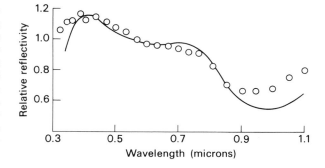

times brighter when it is full than when it is at first or last quarter, rather than just twice as bright. And the lunar limbs are just as reflective as the centre of the lunar disc. In other words, those parts of the Moon which are illuminated by a low Sun appear equally as bright as those which are experiencing the full force of the Sun overhead (Figure 5.19).

This phase effect is in marked contrast to the Moon's reflectance at radio wavelengths. Of the total energy in a reflected radar pulse, half comes from an area amounting to barely 1 % of the nearside hemisphere. The absence of such limb darkening at optical wavelengths can be attributed to the existence of tiny cracks and crevices in the regolith, crevices which will always appear partly shadowed unless illuminated from directly behind the observer. So the lunar surface effectively consists of innumerable corner cube reflectors, each of which reflects much of the incident light in the same direction from whence it came.

Figure 5.18. *Colour difference map.* By combining photographs taken through infrared and ultraviolet filters, colour differences on the Moon may be enhanced visually. Note in particular the young blue lavas in the west. (Courtesy E. Whitaker, University of Arizona.)

These crevices, however, are too small to be very effective at radio wavelengths. Radar waves will therefore be reflected elsewhere, except from those areas of the Moon which are essentially perpendicular to the Earth–Moon line on a scale comparable to or larger than the radar wavelength.

Radar, then, as well as enabling us to determine distances, can also provide us with clues about the structure of the regolith. It was radar experiments, in fact, which first indicated that the uppermost layer of the Moon's crust is not only fine grained and porous, but also is several metres thick. And these studies were extended by the Apollo 14, 15 and 16 Bistatic Radar Experiments.

The highly sophisticated Lunar Sounder Experiment was capable of extending this analysis of the Moon down to depths of hundreds of feet, because radar reflectance also depends on the electrical properties of the target. It was possible on Apollo 17 to detect a distinct boundary between the thick regolith and the underlying *mare* basalt in the Taurus–Littrow valley.

Another effect which can provide valuable information about the nature of a planet's surface is the extent to which the light reflected by it becomes polarised. What we find is that the brightest regions of the Moon tend to polarise sunlight the most, but the degree of this polarisation is also dependant on the phase of the Moon (Figure 5.20). The observed polarisation curve for moonlight was, in fact, what indicated that the surface of the Moon is not only very fine grained but also extensively radiation damaged. Fine-grained amorphous silicates were what could best mimic what was observed. This technique has now found application elsewhere in the Solar System, on moons from which samples are not likely to be returned to Earth for many years.

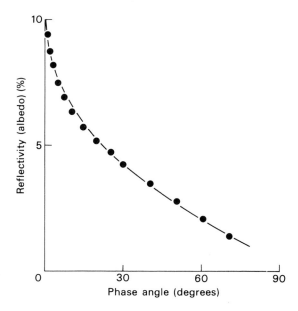

Figure 5.19. *The phase effect.* The albedo of the Moon at optical wavelengths increases very sharply towards full moon, because of tiny crevices in the regolith surface.

## Luminescence and fluorescence

We have already noted that much of the radiation absorbed by the Moon is re-emitted in the infrared and microwave region of the spectrum. But this is not the whole story. For some of the radiation incident upon the Moon is in the form of powerful X-rays. And the energy absorbed at these short wavelengths may be re-emitted promptly and in several discrete stages, a process known as luminescence. Such luminescence, particularly in the red part of the visible spectrum, may well account for certain variations in lunar brightness which have been observed to correlate with solar activity. Luminescence could also explain certain types of Transient Lunar Phenomena, namely those reddish glows occasionally associated with lunar craters, such as Aristarchus, Grimaldi, Alphonsus and Gassendi.

But luminescence is not the only consequence of solar X-rays interacting with the lunar surface. Atoms in the outermost few microns of the regolith may be excited, and emit secondary X-rays with wavelengths which are characteristic of those atoms. This process, known as X-ray fluorescence, allows us to measure the surface concentrations of aluminium, magnesium and silicon along those parts of the groundtracks that were sunlit. This is very useful, because the aluminium and magnesium contents of lunar rocks vary widely, with high aluminium contents being characteristic of highland rocks, whereas high magnesium contents are typical of *mare* basalts (Figure 5.21).

The X-ray fluorescence experiments performed chemical analyses in

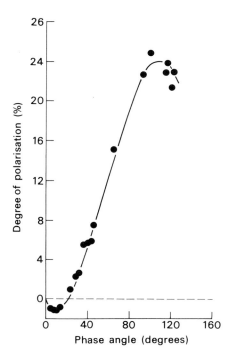

Figure 5.20. *Lunar polarisation curves.* The manner in which light from the Moon is polarised, as a function of phase angle, was used to characterise the lunar surface before the Apollo program. The points are telescopic measurements, the solid line Apollo 11 fines at the same wavelength. (After Gold, T., O'Leary, B. T. & Campbell, M. (1971). *Proceedings of the Second Lunar Science Conference*, pp. 2173–81. New York: Pergamon.)

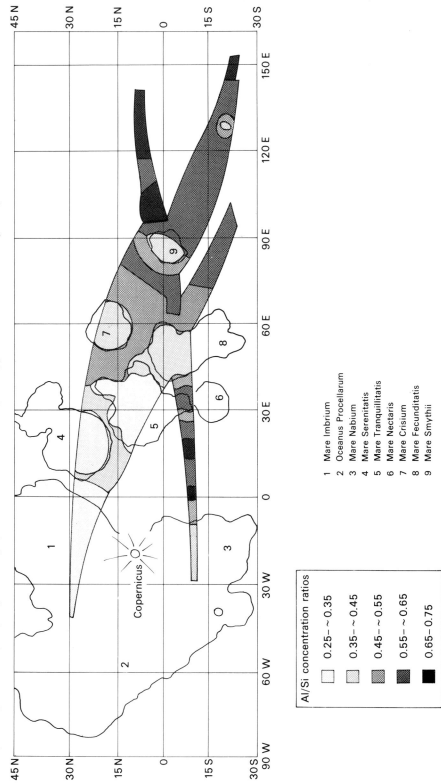

Figure 5.21. *X-ray fluorescence map*. The spectrum of fluorescent X-rays from the lunar surface can be used to measure aluminium–silicon ratios in the regolith. In this map the *maria* (including the floor of the far side crater Tsiolkovsky) are much less rich in aluminium than the highlands. (After Adler, I., Trombka, J. I., Schmachebeck, R., Lowman, P., Blodget, H., Yin, L., Eller, E., Podwysocki, M., Weidner, J. R., Bickel, A. L., Lunn, R. K. L., Gerard, J., Gorenstein, P., Bjorkholm, P. & Harris, B. (1973). *Proceedings of the Fourth Lunar Science Conference*, pp. 2783–91. New York: Pergamon.)

1 Mare Imbrium
2 Oceanus Procellarum
3 Mare Nabium
4 Mare Serenitatis
5 Mare Tranquillitatis
6 Mare Nectaris
7 Mare Crisium
8 Mare Fecunditatis
9 Mare Smythii

Al/Si concentration ratios

0.25 – ~ 0.35
0.35 – ~ 0.45
0.45 – ~ 0.55
0.55 – ~ 0.65
0.65 – 0.75

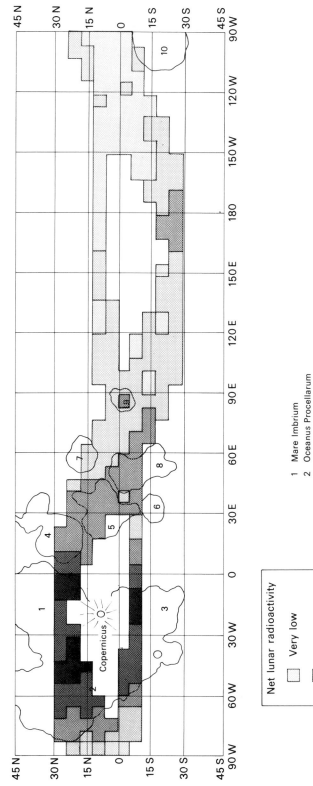

Figure 5.22. *Map of lunar radioactivity.* This map of gamma ray intensity reveals those areas which contain the most thorium, uranium and potassium. Note in particular the high levels in the Imbrium/Procellarum area, especially associated with the Fra Mauro formation, and the craters Aristarchus and Archimedes. To a large extent this is a map of KREEP distribution. Note also the anomaly associated with Big Backside Basin. (After Trombka, J. I., Arnold, J. R., Reedy, R. C., Peterson, L. E. & Metzger, A. E. (1973). *Proceedings of the Fourth Lunar Science Conference*, pp. 2847–53. New York: Pergamon.)

Net lunar radioactivity

- Very low
- Low
- Medium
- High
- Very high

1  Mare Imbrium
2  Oceanus Procellarum
3  Mare Nubium
4  Mare Serenitatis
5  Mare Tranquillitatis
6  Mare Nectaris
7  Mare Crisium
8  Mare Fecunditatis
9  Mare Smythii
10 Mare Orientale

areas that were far removed from the landing sites, confirming, for example, that the dark floor of the farside crater Tsiolkovsky must indeed consist of *mare* lava. The interior of this crater has distinctly more magnesium and less aluminium than the surrounding highlands. But an area to the east of the nearside crater, Ptolemaeus, was less easy to characterise, being rich in both magnesium and aluminium. This is odd because the concentrations of these two elements in lunar rocks are usually inversely correlated with one another.

### Gamma ray spectrometry

The other important remote chemical analysis experiment flown by Apollo was a spectrometer to monitor the emission of the even shorter wavelength gamma rays. Most lunar gamma rays result from the decay of long-lived radioisotopes, particularly thorium and uranium. Gamma ray maps of the Moon, then, are basically maps of lunar natural radioactivity.

Now we have already heard that uranium and thorium, like zirconium, yttrium and the rare earths, must have been strongly concentrated in the last liquids to crystallise during the Moon's chemical differentiation. This is why we find concentrated deposits of these elements on the Moon, in rare granitic fragments and in the much more widespread rock type called KREEP. So gamma ray maps of the Moon reveal those areas where KREEP is now most abundant at its surface.

It turns out that Oceanus Procellarum and Mare Imbrium are the areas which are richest in KREEP, with a few specific features, such as Fra Mauro, Archimedes and Aristarchus being distinct hotspots (Figure 5.22). Nowhere else on the Moon are higher levels of natural radioactivity to be found, at least among those areas overflown by Apollos 15 and 16. Other hotspots will doubtless be revealed if we can fly a gamma ray spectrometer on a Lunar Polar Orbiter satellite.

That such sharp chemical discontinuities do exist is itself a useful indicator of lunar activity, or rather the lack of it. Because, if there were extensive horizontal movements of regolith materials on the Moon, such discontinuities would have been washed out long ago. So lateral transport, although indicated unequivocally by the presence of exotic anorthositic and KREEP particles in *mare* soils, cannot be very significant. And this argues against the hypothesis that highland light plains consist of Orientale ejecta. Material excavated from this basin must certainly be widely disseminated over the lunar surface, but it cannot exist today as a coherent rock unit.

But why is the western near side of the Moon so abnormally rich in KREEP basalt? One theory postulates that KREEP was excavated by the Imbrium Basin and not erupted as lava at all. But why, then, is the distribution of radioactivity not symmetrical? Some highland areas close to Mare Imbrium, such as the Haemus Mountains, are not abnormally radioactive at all.

In another hypothesis KREEP lavas were erupted long before the excavation of the Imbrium Basin, so the present day surface distribution of KREEP may be a consequence of a single global convection cell in the primordial magma ocean. Not only could this explain how radioactive elements were concentrated on the nearside hemisphere, we can also then interpret the thicker farside crust as being an incipient lunar continent.

The author has proposed a third alternative, namely that a truly vast crater was excavated, defined to the south and west by the borders of Oceanus Procellarum and to the north by the shores of Mare Frigoris (Figure 5.23). This huge depression, known as the Gargantuan Basin, was then filled with KREEP basalts (just as other basins were later flooded by *mare* basalts) to give rise to the observed broad distribution of natural radioactivity. The radioactive hotspots are then simply those regions where KREEP basalt was exhumed from beneath a thin veneer of the less radioactive *mare* basalt which later covered the KREEP lavas to produce Oceanus Procellarum and Mare Imbrium. The Jura Mountains to the northwest of Mare Imbrium should therefore also be strongly radioactive, but this prediction cannot of course be tested until a gamma ray spectrometer is flown on the LPO. This Gargantuan Basin theory is certainly supported by remote chemical analyses of the Apennine Mountains. Because the conventional view is that these hills are typical highland crust. But if the Imbrium Basin was excavated from inside the Gargantuan Basin, its third ring should be composed of KREEP rather than anorthosite. And this is exactly what we find. Aluminium – silicon ratios in the Apennines are characteristic of non-*mare* basalt, and very few anorthosites were present among the Apollo 15 samples. Of the Apollo 15 rocks with highland affinities, KREEPy breccias predominated.

Gamma ray spectrometry from lunar orbit also enabled us to map a

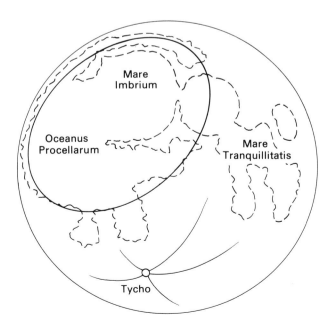

Figure 5.23. *The Gargantuan Basin*. This sketch map shows the proposed outline for the Gargantuan Basin, inside which KREEP lavas may have been erupted, later covered by *mare* basalt and later still excavated to result in the observed radioactive hotspots.

number of non-radioactive elements on the Moon. Because some atoms in the regolith are converted into gamma ray emitting short-lived radioisotopes when they interact with high energy cosmic rays. As gamma ray energy depends on the particular nuclear transmutation, measurements of gamma ray intensity at different energies can then provide us with elemental concentrations.

Iron, magnesium, titanium and two more abundant, but less variable, elements, oxygen and silicon, have now been measured, although the areal resolution for all these elements is not as sharp as that for thorium. But some intriguing features have been revealed, and the magnesium results allow us to make cross correlations with the X-ray fluorescence results.

Perhaps the most significant finding was that the western half of the Moon's far side is distinctly poorer in iron than the eastern half. This is a curious anomaly, and one which still defies an adequate explanation. Perhaps it is linked in some mysterious way to the Moon's near-side–far-side asymmetry.

### Alpha particle spectrometry

Electromagnetic radiation is not however the only means by which we can explore the Moon from a distance. Because, just as we can investigate the Sun by collecting solar wind atoms, so we can also search for alpha particles (or helium nuclei) originating from the Moon.

Like gamma rays, alpha particles are produced by natural radio-activity, but in this case the important parents are not isotopes of uranium and thorium, but much shorter-lived intermediates. One of the elements created during the decay of both uranium and thorium is, in fact, the rare gas radon. And it is the very fact that radon is a gas which makes alpha particle spectrometry from orbit so worth while.

Now it so happens that one isotope of radon is rather longer-lived than the other. Radon-222, the uranium decay product, has a half-life of 3.8 days, whereas radon-220 from thorium has a half-life of only 55 minutes. And, whenever an atom of radon decays, it emits an alpha particle of characteristic energy.

Unless the parent radon happens to be very close to the lunar surface,

Figure 5.24. *Radon emission from Aristarchus*. The Apollo 15 alpha particle spectrometer showed a pronounced peak in radon-222 activity above the crater Aristarchus, where transient lunar phenomena are sometimes observed. The lack of abnormal polonium-210 activity implies episodic gas release. (After Gorenstein, P., Golub, L. & Bjorkholm, P. J. (1973). *Proceedings of the Fourth Lunar Science Conference*, pp. 2803–9. New York: Pergamon.)

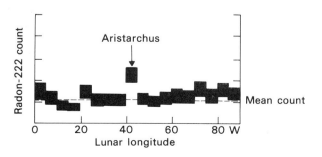

however, the alpha particles produced will be unable to escape from the regolith. This is because, unlike gamma rays, they have very little penetrative capability. So only if alpha particles are emitted in the right direction and from the very surface of the Moon can they possibly be expected to reach orbital altitudes.

But radon is a gas. And gases can diffuse out of solid grains and escape through pores in the regolith to form a tenuous lunar atmosphere. So an alpha particle spectrometer in lunar orbit could detect alpha particles from any atmospheric (or surface-adsorbed) radon-222. The only real question was whether or not the half-life of radon-222 was sufficient for its diffusion out of the regolith to be effective. Over most regions on the Moon the numbers of alpha particles from radon-222 (and its two very short-lived daughter isotopes) were found to be consistent with the uranium concentrations measured by the gamma ray spectrometers. The natural radioactivity of the lunar surface was simply being mapped using a different method.

But usually high concentrations of radon-222 were found by Apollo 15 in the immediate vicinity of Aristarchus Crater, one of the few spots on the Moon where Transient Lunar Phenomena are frequently observed. So could it be that TLPs are gaseous discharges? It certainly looks that way, although we cannot yet say whether this gas comes from the regolith or from much deeper inside the Moon (Figure 5.24).

Radon can only be a transient component of the lunar atmosphere, of course, because of its relatively short half-life. But it must take days, (rather than hours) to reach the lunar surface, otherwise we would have been able to detect the even shorter-lived radon-220 as well. For a radon excess to have been observed at Aristarchus, then, there must have been tectonic activity there shortly before the Apollo 15 mission. Otherwise the alpha particle activity would have dropped back to the mean global level. Unfortunately, the Apollo 16 ground track was too close to the equator for the alpha particle activity above Aristarchus to be re-measured.

*The fate of lunar radon*

Radon is much too heavy to escape from the Moon completely, so it must bounce across the lunar surface until it either decays or finds somewhere cool enough to be adsorbed. We would therefore expect a migration of radon towards the dark side. And, sure enough, a slight enhancement in radon radioactivity was detected in the vicinity of the sunrise terminator, where adsorbed radon is presumably boiling off (Figure 5.25).

But after radon decays we are no longer dealing with a gas. And the next long-lived alpha-emitter is, in fact, polonium-210. Although not volatile, the polonium derived from atmospheric radon should also reside on the regolith surface, so it should be possible to detect its decay there as well. Fortunately for us alpha particles from polonium have slightly higher energies than those from radon and its prompt daughters. So it is possible to count polonium alpha particles separately.

Interestingly enough, no Aristarchus polonium anomaly was found.

How can this best be explained? Well, the production of polonium-210 from radon-222 is delayed by a long-lived beta-emitting intermediate, lead-210. So insufficient time must have elapsed since the beginning of gas emissions at Aristarchus for polonium levels there to have built up above the average polonium background. The absence of a polonium excess at Aristarchus, then, can only mean that the release of radon from the Moon is episodic.

Like the occurrence of moonquakes and Transient Lunar Phenomena, it could in fact be triggered by tidal stresses. The Moon is certainly most active seismically when it is close to perigee, and it is then that most transient phenomena are observed.

The alpha particle spectrometer results are certainly consistent with this view, but more data is urgently needed. It is interesting to note, however, that polonium levels are enhanced around the perimeters of circular *maria*, and in the dark-floored crater Grimaldi. And these are also sites where Transient Lunar Phenomena are sometimes observed (Figure 5.26).

As there are no radon anomalies in these areas, however, we must conclude that radon emissions here must have occurred several months, if not years, before the flight of Apollo 15.

But how can gases be vented from the Moon, and to what extent do they remain as a lunar atmosphere? When we have discussed what other gases exist above the lunar surface, we will be in a much better position to tackle the subject of lunar seismology, and the structure of the Moon's interior.

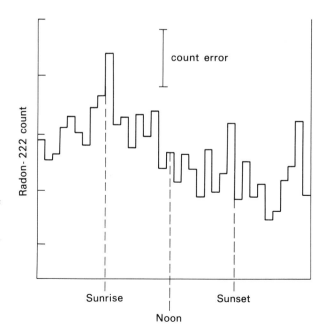

Figure 5.25. *Radon condensation.* Radon-222 activity was found to be enhanced in the vicinity of the sunrise terminator, indicating that previously absorbed radon must be boiling off under the heat of the sun. (After Gorenstein, P., Golub, L. & Bjorkholm, P. J. (1973). *Proceedings of the Fourth Lunar Science Conference*, pp. 2803–9. New York: Pergamon.)

## 5.4   The lunar atmosphere

Although radon may be a very rare and short-lived element on the Moon, the great mass of the radon atom ensures that it is retained most effectively, and the lunar atmosphere is such a diffuse gas that radon could even be a significant component.

But exactly how thin is the lunar atmosphere today? What does it consist of chemically, what can it tell us about the Moon as a planet? And could the Moon, like the Earth, once have been surrounded by a thick reducing atmosphere that was subsequently stripped away by solar action?

Well, it had been known for a long time that the total pressure of gas at the surface of the Moon must be incredibly low. No-one had ever observed clouds or aurorae above the limb, for example, and, during a lunar occultation, the extinction of the light from a star, unless it happens to be a close binary, is always instantaneous. Such occultations are, in fact, excellent opportunities for mapping astronomical sources which are too small to be resolved by a telescope in the normal way. The Crab Nebula, for example, was first mapped as an X-ray source during a lunar occultation.

There is also never any refraction of sunlight around the limb during a total eclipse of the Sun, nor is the lunar terminator anything less than crystal sharp. So we would never experience twilight on the Moon if we were to stay there after sunset. Long before Apollo, then, it had been established that the lunar atmosphere must exert a pressure of less than one-thousand-millionth of the Earth's at sea level, by all accounts a very thin gas indeed.

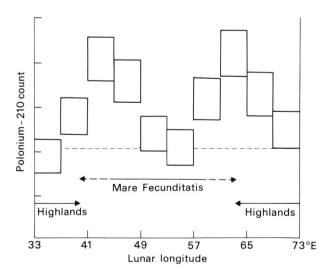

Figure 5.26. *Polonium profile*. There is a tendency for polonium activity, like TLPs, to be enhanced around the perimeters of *maria*, again indicative of gas venting. (After Gorenstein, P., Golub, L. & Bjorkholm, P. J. (1973). *Proceedings of the Fourth Lunar Science Conference*, pp. 2803–9. New York: Pergamon.)

### *The prospects for a lunar atmosphere*

But nature abhors a complete vacuum, and even interplanetary space consists of gas at a pressure of $10^{-19}$ Earth atmospheres. Obviously a lunar atmosphere cannot possibly be more diffuse than interplanetary space, where most of the gas consists of solar wind atoms originating from the Sun. These are able to reach the surface of the Moon largely unchecked.

If a lunar atmosphere of any sort exists, then, it must be collisionless gas where many familiar concepts become meaningless. Gas temperatures, for example, refer to a characteristic distribution of particle velocities in a collection of atoms or molecules, a distribution that can only be maintained if the particles can exchange kinetic energy with one another. As this situation will not occur in a collisionless gas, it is more appropriate to refer to particle densities and velocities rather than to overall gas pressures and temperatures.

But what sorts of atoms and molecules might we expect to find in the lunar atmosphere, and what instruments were deployed by Apollo to investigate them directly? Well, first of all we should find two rare gases other than radon. And also as a result of natural radioactivity. For helium is a byproduct of uranium and thorium decay, whereas the disintegration of potassium-40 gives rise to an isotope of argon, argon-40.

These gaseous decay products provide a means for dating lunar rocks because their radioactive clocks will be reset by melting. In other words the amounts of argon and helium that accumulate inside a rock are related to the time which has elapsed since these gases were last removed from the rock. And the gas that is lost must eventually reach the lunar atmosphere.

The existence of helium in the lunar atmosphere has already been demonstrated. For those alpha particles from radon and polonium are nothing more than energetic helium ions. They are very transient components of the atmosphere, however, having more than enough energy to escape immediately from the Moon's gravitational pull. But those alpha particles emitted within the regolith are invariably slowed down and

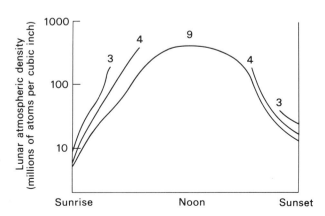

Figure 5.27. *The cold cathode gauge.* This plot shows the density of the lunar atmosphere during the third, fourth and ninth lunar days following the Apollo 14 landing. (After Johnson, F. S., Carrol, J. M. & Evans, D. E. (1972). *Proceedings of the Third Lunar Science Conference*, pp. 2231–42. Cambridge, Mass.: MIT.)

neutralised to become helium atoms, which may then slowly diffuse out of the regolith, entering the atmosphere at only thermal velocities. The anticipated amounts of such radiogenic helium in the regolith are of course quite negligible compared to the amounts of solar wind hydrogen and helium. But the same cannot be said of argon-40, which is not present in the solar wind at all. Any argon-40 in the atmosphere must be of radiogenic origin.

Now surface temperatures during the lunar day are so high that atmospheric hydrogen and helium will rapidly escape from the Moon altogether. The Moon's gravitational field is simply not strong enough to retain atoms as light as these, even though they must be vented into the lunar atmosphere in appreciable quantities.

If the Moon does have an atmosphere, then, it must consist of atoms such as argon, krypton and xenon, which are so massive that, even at lunar surface temperatures, only a small fraction will have speeds that exceed lunar escape velocity. Other possible atmospheric constituents, also ultimately derived from the solar wind, are reactive species, such as carbon oxides, methane and molecular nitrogen. So which of these gases really do exist and what is their ultimate fate?

*Total gas pressure and 'orphan' argon*

The first direct measurements of lunar atmospheric pressure were made by the Apollo 12 cold cathode gauge. And these measurements indicated gas leakages during, and immediately after, the mission itself, not only from astronauts' spacesuits, but also from the experimental hardware left behind at the site. During the first lunar night, however, particle densities fell by three orders of magnitude, to only 3 million atoms per cubic inch. And even during the next day they did not rise much above a few hundred million particles per cubic inch. This means, in fact, that the entire lunar atmosphere cannot weigh more than a few tons (Figure 5.27).

Why is it so thin? Thermal escape should only be really effective for hydrogen and helium, so where are those heavy rare gases and carbon molecules which are constantly escaping from the regolith? The degassing of radiogenic argon alone should be capable of producing an atmosphere with the measured density in the space of a single year. So there clearly must be some other mechanism by which argon and other heavy gases are lost.

Our first real clue here came from the rare gas analyses of lunar samples. For not only is argon-40 present in lunar soils in quantities which greatly exceed those expected from the radioactive decay of indigenous potassium, this argon is also surface correlated. In other words, like those solar wind isotopes mentioned earlier, it is located very close to the surfaces of soil grains, and is therefore most abundant in fine particles with high surface areas per unit mass (Figure 5.28). And the most likely explanation for this surface concentration is that atmospheric argon atoms are being reimplanted into soil grains.

What we think happens is that some of the argon atoms vented into the lunar atmosphere become ionised by solar ultraviolet photons and are then accelerated away by interplanetary electric fields. But permanent escape from the Moon depends not only on the velocity but also on the direction of motion of the ion. And theoretical calculations suggest, in fact, that as much as 8 % of the argon in the lunar atmosphere at any one

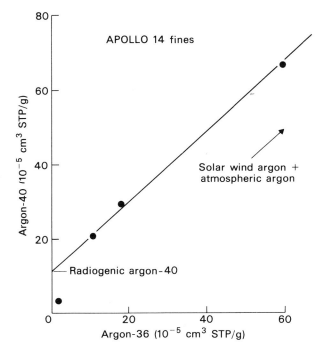

Figure 5.28. *Orphan argon*. Some of the argon-40 in lunar soils is the result of *in situ* decay of potassium-40. But the remainder correlates well with argon-36 from the solar wind, and this component must have been reimplanted from the lunar atmosphere. (After Pepin, P. O., Bradley, J. G., Dragon, J. C. & Evans, D. E. (1972). *Proceedings of the Third Lunar Science Conference*, pp. 1569–88. Cambridge, Mass.: MIT.)

Figure 5.29. *Reimplantation from the lunar atmosphere*. If atoms in the lunar atmosphere become ionised, they will be accelerated along curved trajectories by the electric field associated with the solar wind, the lighter the ion the more curved the trajectory. But the heavier atoms tend to be confined closer to the Moon's surface at a given temperature. About 8 % of the argon in the lunar atmosphere is accelerated back into the regolith from whence it came. (After Manka, R. H. & Michel, F. C. (1973). *Proceedings of the Fourth Lunar Science Conference*, pp. 2897–908. New York: Pergamon.)

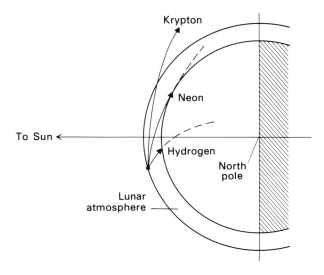

time may be accelerated towards the Moon, where it becomes firmly embedded into the regolith (Figure 5.29). Not only can this mechanism explain the measured excesses of argon-40 in the soil, it can also account for the low density of the lunar atmosphere as a whole. Because it must also operate on heavy gases, such as krypton and xenon, which would otherwise accumulate to appreciable levels.

*Rare gases in the lunar atmosphere*
So what steps were taken to find out how much argon exists above the Moon's surface? And what other neutral and ionic species are present as well? Theory is fine, but it really is no substitute for direct measurement.

The atoms and ions of the atmosphere had to be collected separately, so some sort of mass spectrometer had to be employed, the first being the Suprathermal Ion Detectors flown by Apollos 12, 14 and 15. Only on Apollo 17 was a proper mass spectrometer set up on the lunar surface, although orbital analyses had been performed during the previous two missions.

The Suprathermal Ion Detector Experiments were not strictly speaking atmospheric analysers, because they could not register neutral atoms, only those that were already ionised. A mass spectrometer ionises neutral atoms so nothing is then left unmeasured. One important early result,

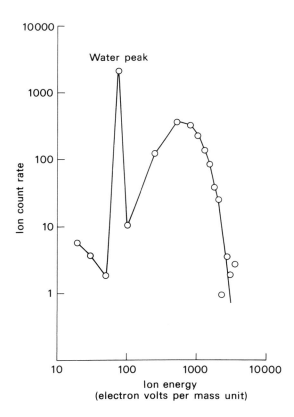

Figure 5.30. *Lunar water?* This spectrum from the Apollo 14 suprathermal ion detector experiment exhibits a low energy peak, also detected by the Apollo 12 SIDE, that seemed to correspond to water vapour. The event occurred on 7 March 1971. (After Freeman, J. W., Hills, H. K. & Vondrak, R. R. (1972). *Proceedings of the Third Lunar Science Conference*, pp. 2217–30. Cambridge, Mass.: MIT.)

however, was that gases are occasionally emitted from the Moon, and that these emissions seem to coincide with natural seismic events, at times of maximum tidal stress. A cloud of gas (possibly water) was detected by both Apollo 12 and 14 instruments. So here again there was some compelling evidence for Transient Lunar Phenomena (Figure 5.30).

The Apollo 15 and 16 mass spectrometers revealed, for the first time, the unequivocal presence of the neon in the lunar atmosphere. Even at a height of some 60 miles there were some 130 000 atoms of neon-20 (the major isotope) per cubic inch. These neon atoms must originally have been solar wind ions, which were captured by (and subsequently released from) the lunar soil. So once again the Sun is a major contributor to the lunar atmosphere.

But it was the Apollo 17 instrument which taught us most about the lunar atmosphere. Close to the lunar surface, night-time concentrations of neon are some ten times higher than at orbital altitudes. But it was not the only rare gas detected. Also measured were two other solar wind isotopes, helium-4 and argon-36, and the other expected radiogenic species, argon-40.

Of particular interest is how these three elements vary in abundance throughout the lunar day. Helium is completely incondensible so its lifetime in the atmosphere during the day depends solely on surface temperature. In other words the highest helium concentrations occur during the intensely cold lunar night (Figure 5.31). Neon exhibits similar behaviour. But neon is found to be more abundant than helium because it cannot escape from the Moon's gravitational pull so easily, on account of its higher mass. The behaviour of argon-40, on the other hand, is rather unexpected, because it seems to condense out during the night. Unlike helium and neon levels, which increase sharply after sunset, then, the

Figure 5.31. *Helium behaviour*. The concentrations of helium in the lunar atmosphere are highest during the lunar night, because of thermal escape during the day. (After Hoffman, J. H., Hodges, R. R., Johnson, F. S. & Evans, D. E. (1973). *Proceedings of the Fourth Lunar Science Conference*, pp. 2865–75. New York: Pergamon.)

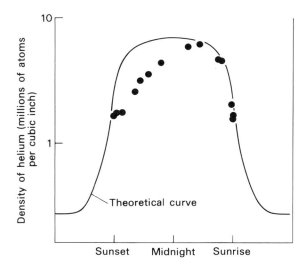

concentrations of argon drop steadily, through three orders of magnitude, reaching barely 1500 atoms per cubic inch before sunrise. This low level shows just how sensitive this mass spectrometer was (Figure 5.32).

But just before dawn there is a significant enhancement in argon-40. In other words, there must be an influx at that time of argon atoms which are boiling off those nearby regions (such as the tops of the surrounding massifs) which are just being warmed up for the first time in two weeks. This is certainly the behaviour that we would expect for a condensible gas, but not what we would have anticipated for argon.

But what was even more intriguing about the behaviour of argon-40 was that minimum concentrations were not the same from month to month. During its nine months of successful operation, in fact, the Apollo 17 mass spectrometer showed that the concentrations of atmospheric argon-40 at lunar midnight may vary by a factor of two. And the only reasonable explanation for this variation is that the venting of argon into the lunar atmosphere must be episodic, up to 3 tons (plus 15 tons of helium) must be released every year. This argon must originate from deep inside the Moon (rather than just from the regolith) because the measured ratio of argon-40 to argon-36 in the atmosphere (10:1) is significantly higher than this ratio in lunar soils (typically 1:1). So radiogenic gases may be seeping up from the Moon's partially molten deep interior, their release possibly being triggered by seismic disturbances.

### Other atmospheric components

Unfortunately, the outgassing of Apollo hardware (particularly the mass spectrometer itself) precluded the unambiguous detection in the lunar

Figure 5.32. *Argon behaviour*. Argon concentrations decrease steadily during the long lunar night, because of condensation. Minimum levels vary from one month to another because of episodic release of radiogenic gases from the lunar interior. (From Hodges, R. R. & Hoffman, J. H. (1975). *Proceedings of the Sixth Lunar Science Conference*, pp. 3039–47. New York: Pergamon.)

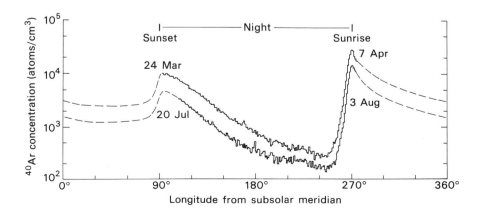

atmosphere of such gases as carbon dioxide, nitrogen and methane, and it also made the measurement of all rare gases heavier than helium impossible during the lunar day. So if we ever want to measure day-time gas concentrations we will have to find some other way of doing it.

But a mass spectrometer was not the only instrument flown on Apollo 17 to analyse the Moon's atmosphere. Up on the Service Module was a far-ultraviolet spectrometer, designed to detect the absorption of reflected ultraviolet rays by such species as atomic hydrogen. The atmosphere is so diffuse, however, that all that this experiment could do was establish upper limits on the concentrations of certain components. For atomic hydrogen this upper limit was a mere 150 atoms per cubic inch.

Of the 3 million or so atoms per cubic inch which are present in the lunar atmosphere during the lunar night, then, about 40 % must be neon, 20 % helium and 5 % argon, the remainder presumably being largely molecular hydrogen. But it is carbon-containing and nitrogen-containing molecules which could well be the dominant constituents during the day, because these species freeze out completely when the Sun sets.

The Moon's atmosphere may, however, contain dust as well as gas. This may seem surprising, but dust storms of global extent occur on

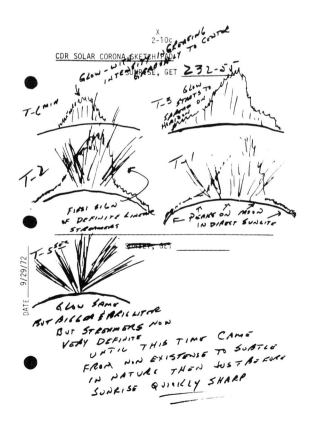

Figure 5.33. *Pre-dawn streamers.* The Apollo 17 crew noticed and sketched solar streamers immediately before sunrise. They grew in size so rapidly that they had to have a local origin, possibly the scattering of sunlight by fine dust in the lunar atmosphere. (NASA.)

Mars, where surface pressures are very much lower than on Earth. The first clue here came from a Surveyor 1 picture, which showed a narrow band of light above the horizon just after sunset. It looked as though the dying rays of the Sun were being scattered by levitated dust particles.

And, during the Apollo 17 mission, two astronauts sketched what seemed to be solar streamers, a few seconds before sunrise (Figure 5.33). They grew in size very rapidly and so, unlike the solar corona or zodiacal light, they could not have been at astronomical distances. The only reasonable explanation is that lunar particles are somehow being suspended in the lunar atmosphere. The most likely levitation process is an electrostatic one, because strong electric fields are known to be generated close to the lunar terminator. Alternatively, tiny fragments of soil may be ejected very rapidly during micrometeorite impacts and the Moon may have a permanent cloud of such secondary particles in orbit around it.

So what is there left to know about the lunar atmosphere? Well, by studying fossil soils from deep cores, we might one day be able to discover what the Moon's atmosphere was like in the remote past. So far there is no evidence in favour of the ancient Moon ever having had an appreciable atmosphere, with the total absence of water and lack of other volatiles in all lunar rocks and soils implying that an atmosphere has always been essentially non-existent.

But we also want to know more about how lunar seismic activity drives or at least triggers the venting of gases into the atmosphere. So what do lunar seismograms tell us about the Moon's internal activity and structure?

## 5.5 Seismology and lunar structure

The study of rocks at the lunar surface can only provide us with indirect clues about the inside of the Moon. Only by studying the passage of shock waves through the Moon can its interior be investigated directly. Compositional discontinuities can then be characterised, because different rock types transmit seismic waves at different velocities, and these waves are reflected and refracted, at, for example, the crust–mantle boundary, just as light is bent when it passes through a pane of glass. And such seismic studies can not only reveal the structure of the lunar crust. Distant seismic sources can also provide information about the Moon's deep interior.

### Lunar seismometry

But what are seismic waves, and how do they travel through rock? Well, there are two basic types, compressional (P) waves and transverse (S) waves. P waves are always faster moving than S waves, so they are invariably detected first. But, not only are S waves slower than P waves, they are also more likely to be strongly attenuated. For they are soon dissipated in rocks which are partially molten or rich in volatiles. If we cannot detect any S waves, then, it is likely that a partially molten zone exists between seismic source and seismometer.

But seismometry is not just a question of timing first arrivals. For there are many possible paths that seismic waves can take from source to seismometer, and the first S waves may well arrive long before the more devious of P waves. So the vibrations in a seismogram which are due to the first S waves could well be superimposed on later P wave arrivals. But these more devious P waves can also provide us with valuable clues to the structure of the Moon, if only they can be interpreted correctly. In other words, the more seismic data that can be collected from the Moon, the more tightly can models for the lunar interior be constrained.

So how did we set about detecting seismic waves on the Moon? How were lunar seismometers designed, and where were they deployed? Well, the very first attempts at lunar seismometry were, in fact, made before Apollo. Because seismometers were optimistically included on the earliest Ranger probes, none of which succeeded in reaching the Moon, let alone in softlanding.

So it was on Apollo that the first seismic data were returned and the Apollo instruments had to be handled very gently. This was because the smallest seismic vibrations on the Moon have amplitudes of less than a millionth of a millimetre.

So how do lunar seismometers work? Well, their essential component is a weight suspended inside a light framework (Figure 5.34). When the ground vibrates, the framework vibrates with it, whereas the weight suspended inside this framework tends to remain where it is, its high inertia preventing it from responding quickly to external forces. When it eventually does move, its motion is artificially damped. And it is the motion of the framework (with respect to the weight) that is amplified to produce the seismogram.

Each lunar instrument did in fact consist of four seismometers, three for registering seismic vibrations in each of three mutually perpendicular directions (to provide more information about the direction of origin of the seismic signal) and a short period seismometer, for registering higher frequency vibrations in a vertical direction. This was because certain seismic sources on the Moon may be characterised by higher frequency vibrations, as rapid as one per second.

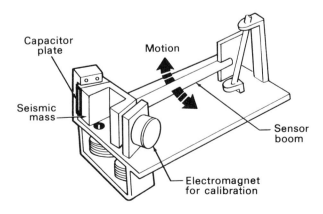

Figure 5.34. *Apollo seismometer.* The vibration of the supporting framework with respect to the seismic mass is amplified to produce the seismogram. (NASA.)

*The Apollo seismic network*

The first lunar seismometer was set up on the Moon by Apollo 11, but, driven by solar cells, its operational lifetime was rather limited. In contrast (Figure 5.35), the Apollo 12, 14, 15 and 16 instruments were powered by Radioactive Thermoelectric Generators, and all four are in fact still in an operational condition today. No such passive seismometer was flown on Apollo 17 because on this occasion the upper crust was probed using small controlled explosions and arrays of geophones, the so-called Lunar Seismic Profiling Experiment. Active seismometry was also performed on Apollos 14 and 16, when the seismic sources consisted of a thumper device and explosive mortars, and the resulting vibrations could be recorded by the passive seismometer as well as by geophones (Figure 5.36).

So deep profiling of the Moon has been achieved using a triangular network approximately 700 miles along each side, with the Apollo 12 and 14 seismometers only 110 miles apart at one apex of this triangle (Figure 5.37). In order to locate a seismic source precisely such a network is

Figure 5.35. *The Apollo 16 passive seismic experiment.* Powered by the RTG, this instrument was the most sensitive of all lunar seismometers. (NASA, Apollo.)

Figure 5.36. *Explosive grenades at Cayley–Descartes*. This mortar was used to launch explosive grenades to provide local seismic sources for the Apollo 16 geophone array after the astronauts had left and on remote control from Earth. (NASA, Apollo.)

Figure 5.37. *Apollo seismic network*. The four long-lived passive seismometers defined the apexes of an approximately equilateral triangle, enabling the precise location of seismic sources. (NASA.)

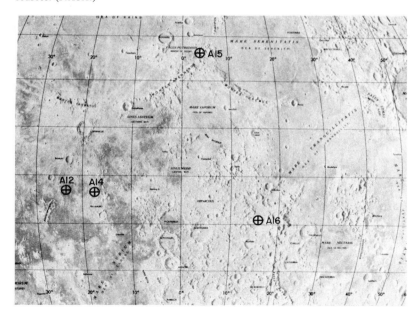

essential because we must be able to record its seismic waves from at least three widely separated locations. It is just like triangulating with a theodolite, except that seismic waves (unlike light waves) do not travel in straight lines, nor at constant velocity (Figure 5.38). And moonquakes may occur deep beneath the crust, adding a third dimension to the problem of location. The existing network was certainly of sufficient areal extent to locate all major seismic sources to within a few miles, but only those sources that were strong enough to be detected by the weakest station could be fixed precisely. For some strange reason, the Apollo 16 seismic station turned out to be much more sensitive than the other three.

So what do we now know about the structure of the Moon from lunar seismograms? Well, we certainly could not take any chances when it came to providing the necessary sources for the lunar seismic network. If the Moon were totally inactive seismically, we would have to generate some large artificial seismic sources ourselves and rely on these to probe the Moon.

Fortunately, it did not take much energy to adjust the trajectories of the discarded Saturn 5 third stage boosters (SIVB) in order to guide them towards preselected impact points on the lunar surface. The first such usable artificial impact was, interestingly enough, that of the Apollo 13 third stage, so this otherwise ill-fated mission was not completely wasted after all, although there was at that time only one seismometer active on the Moon, that set up by Apollo 12. The impact of the Apollo 17 booster, in contrast, was recorded by all four seismic stations (Figure 5.39).

But these were not the only artificial projectiles to be used as seismic

Figure 5.38. *Propagation of seismic waves.* Seismic waves may follow many different paths before reaching the seismometer. As seismic velocity tends to increase with depth on the Moon, the first P wave arrival may be the one which penetrated deepest. Scattered surface waves may take much longer. Waves from deep moonquakes pass at a steeper angle through the lunar crust, but S waves are strongly attenuated in the partially molten lower mantle. Several seismometers are required in order to locate a seismic source in space and time.

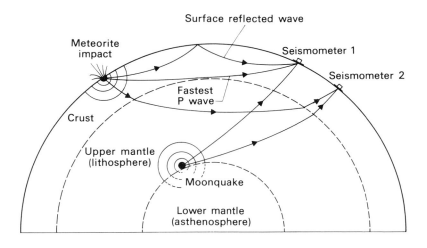

sources. After the landing phase, each lunar module ascent stage was jettisoned and crashed onto the lunar surface as soon as its two astronauts had safely transferred back to the orbiting command module. The Apollo 13 LM was an exception here, of course, because it was used as a lifeboat to ferry the shipwrecked astronauts safely back to Earth. It says much for the flexibility of the whole Apollo concept that the explosion in the Apollo 13 service module (which brought about the premature end of this mission as far as science was concerned) did not have more disastrous human consequences.

The Lunar Module impacts were certainly not as violent as those of the more massive SIVBs. But they could at least be recorded by the seismometer which had just been activated. By crashing the Apollo 17 LM on to the South Massif, for example, there was a relatively distant, yet strong, source for the Lunar Seismic Profiling Experiment. Unfortunately, this particular impact was just too far from the primary network to be of much use in probing the Moon's interior (Figure 5.39).

The great advantage of employing artificial impacts in this way is that we knew so much about them from radio tracking data, their times of impact to the nearest millisecond and their impact points to within a mile. The more information that has to be extracted from the seismograms in order to locate and time the seismic event, the less tightly will models for the Moon's internal structure be constrained.

Figure 5.39. *Artificial impacts.* Artificial seismic sources for the Apollo seismic network were produced by crashing Saturn IVB boosters and spent Lunar Modules (S and L respectively) on to the lunar surface at selected locations. Most impacts were in the Mare Cognitum area. Not all combinations of impact and seismometer were fruitful, with the Apollo 17 LM impact being too far from the Apollo 12 and 14 stations to be registered. The usable seismograms correspond to the paths shown by dotted lines. (After Toksöz, M. N., Dainty, A. M., Solamon, S. C. & Anderson, K. R. (1973). *Proceedings of the Fourth Lunar Science Conference*, pp. 2529–47. New York: Pergamon.)

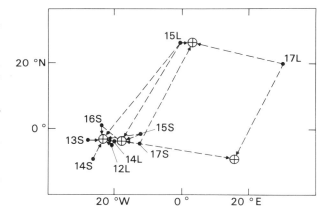

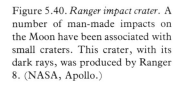

Figure 5.40. *Ranger impact crater.* A number of man-made impacts on the Moon have been associated with small craters. This crater, with its dark rays, was produced by Ranger 8. (NASA, Apollo.)

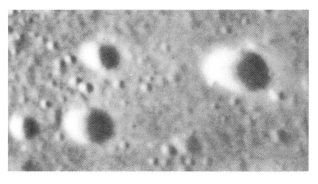

Some of the craters produced by these impacts have even been identified in Apollo photographs, as have those of Ranger before them (Figure 5.40). We also know how much kinetic energy was dissipated during each of these impacts, which in turn allows us to calculate the energy released during natural impact events.

But the use of artificial impacts was not wholly sufficient, because they all had to be reasonably close to the network for the first P wave arrivals to be detectable. Apart from the Apollo 15 LM crash, in fact, all usable artificial impacts occurred within 225 miles of the Apollo 12 and 14 stations. And even the longest usable distance between source and seismometer was only 706 miles.

With such relatively short distances, few seismic waves can penetrate into the Moon more than about 60 miles, so artificial impacts have really only allowed us to probe beneath eastern Oceanus Procellarum to about this depth. The other disadvantage is that the total number of such artificial impacts was restricted. Rockets could not be sent to the Moon simply to provide extra sources for the seismic network.

But boosters and spent lunar modules have not been the only projectiles to hit the Moon in recent years. There have also been several large meteorite impacts, several of which occurred at considerable distances from the seismic network, some even on the far side. These provided opportunities to investigate the Moon's core, because seismic waves from farside impacts must necessarily travel through the centre of the Moon in order to reach the seismometers.

But there have been other natural seismic sources in the Moon in addition to meteorite impacts. Indeed, some such moonquakes are much stronger than impact events, and are even better as deep lunar probes. For, despite the fact that still more seismic data have to be wasted in order to calculate the depth of the moonquake focus, the seismic waves from them will only spend a short time passing through the crust. Those from distant impacts, on the other hand, always have to traverse the crust twice, and usually do so at more acute angles (Figure 5.38). Mantle seismic velocities calculated from moonquake data will therefore be less dependent on the depth of, and seismic velocity assumed for, the lunar crust.

*The upper lunar crust*

So how thick is the lunar crust and of what does it consist? Well, perhaps the most intriguing aspect of all lunar seismograms is that they last for so long. In some cases, seismic waves were still being recorded a full hour after the first P wave arrival (Figure 5.41). In other words, the Moon rings like a bell when it is struck, and this can best be explained if its crust is highly fractured down to a depth of several miles.

The necessity for the existence of such a 'megaregolith' is quite simple. If the crust were solid throughout, seismic waves would travel through it very rapidly, and along well-defined paths. In a fragmented crust, on the other hand, they are slowed down and may be reflected off internal surfaces and spread out in all directions, just like a beam of light passing through a pane of frosted glass.

The result of such a diffusion process is that some seismic waves travel much farther than others. So the scattered waves are greatly delayed compared to the more direct arrivals (Figure 5.41). But the most surprising fact is that the scattered waves arrive at all, rather than being totally absorbed by the megaregolith. And this can only mean that the pores in the crust are completely dry. The signals would have been strongly attenuated over even short distances if any water had been present. So the very length of the seismic wave train yields a valuable piece of information about the chemistry as well as the structure of the lunar crust.

But just how thick is this scattering layer, and how can its existence be accounted for? Well, it must be at least 1000 feet thick because the Apollo 15 LM impact occurred on the far side of Hadley Rille and, despite the fact that Hadley Rille is 1000 feet deep, the normal scattering behaviour was still observed in the Apollo 15 seismogram. If the scattering layer had been of comparable depth to the rille some abnormal effects should have been observed here.

Not until the source is more than 90 miles from the seismometer, in fact, do we observe any significant change in lunar seismograms. And theory tells us that lunar shock waves arriving from this distance should be penetrating as much as 12 miles into the lunar crust. Interestingly enough, this is the depth at which the pressure exerted by the overlying rocks should squeeze out any pore spaces that might exist. In other words, the megaregolith is about as thick as possible on a planet where the surface gravity is only one-sixth that on Earth.

Figure 5.41. *Lunar seismograms*. This picture shows seismograms from the three main classes of natural seismic event. Note the extreme lengths of the wave trains, the long rise times and the S wave first arrival in the short-period seismogram from the High Frequency event. (From Nakamura, Y., Dorman, J., Duennebier, F., Ewing, M., Lammlein & Latham, G. (1974). *Proceedings of the Fifth Lunar Science Conference*, pp. 2883–90. New York: Pergamon.)

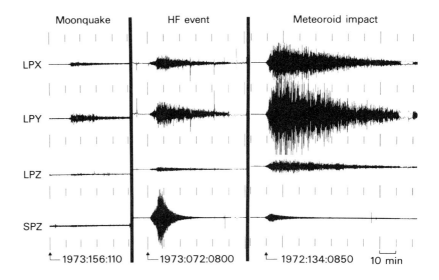

Intense fracturing of the upper lunar crust during major impact events and the highly vesicular nature of certain types of highland breccia and *mare* basalt must help to keep seismic velocities down in this megaregolith. But where does the regolith stop and the megaregolith begin, or do these two zones grade into one another?

Well, most of our knowledge about the vertical structure of the regolith has been obtained through active seismometry, particularly by means of the Apollo 17 Lunar Seismic Profiling Experiment. These studies have enabled seismic velocity, and thus the degree of consolidation, to be determined down to depths of a few miles. The more compact the crust, the more rapidly will seismic waves travel through it.

In the fine surface regolith, P waves travel at little more than 300 feet per second, two or three times slower than in the zone immediately below. So the Apollo 16 Active Seismic Experiment was able to confirm the conclusion, based on crater morphology, that the highland regolith is about 30 feet deep.

On this occasion it was not possible to penetrate much deeper than 100 feet, but during the final mission, seismic signals from the LM impact on the South Massif were of sufficient intensity to be recorded by the extensive array of surface geophones after having been reflected off more deep-seated layers. Strangely enough the Taurus–Littrow regolith was found to be thicker than that at the Cayley–Descartes site (as much as 100 feet in places), and this may be due to the fact that the valley floor has tended to accumulate debris from the surrounding massifs.

Below the regolith, P wave velocities are initially about 900 feet per second, but they increase sharply to some 3400 feet per second at a depth of about 700 feet (Figure 5.42(*a*)). These two layers must consist of fractured and coherent *mare* basalt respectively, and they, in turn, must overlie the highland crust (in which P wave velocities are as high as 13 000 feet per second) at a depth of some 3500 feet. Interestingly enough, this last discontinuity was also revealed by the Apollo 17 Lunar Sounder Experiment. So, all in all, the results indicate that seismic velocities in the upper lunar crust must increase in a stepwise fashion, rather than uniformly.

*The lunar crust and other seismic phases*

But what happens as we penetrate farther into the crust? Well, seismic velocities continue to rise steadily until a depth of about 17 miles, where P wave velocities reach more than 4 miles per second and remain constant for the next 20 miles or so. As this velocity matches that for anorthositic gabbro, the Moon must indeed have a feldspathic crust, and one that is about 37 miles thick (Figure 5.43).

Strictly speaking, this value is only appropriate to eastern Oceanus Procellarum. The farside crust must be as thick as 90 miles, in order to account for the marked displacement of the Moon's centre of mass towards the Earth. And, in the centre of the nearside circular *maria*, the *mare* basalt fill may be several miles in thickness, rather than just a few thousand feet as in Oceanus Procellarum.

Figure 5.42. *The upper crust.* (*a*) The LM impact on the South Massif produced seismic waves of sufficient intensity to enable the Apollo 17 Lunar Seismic Profiling Experiment to investigate the structure of the lunar crust down to a depth of several thousand feet. An even slower velocity zone, the fine-grained surface regolith, is too shallow to show up on this scale. The evidence suggests that seismic velocities in the lunar crust increase in a stepwise fashion. (Courtesy R. L. Kovach.) (*b*) Travel times increase with distance, but the rate of increase becomes less as seismic waves penetrate further into the lunar interior, where seismic velocities are higher. (After Dainty, A. M., Goins, N. R. & Toksöz, M. N. (1975). *Proceedings of the Sixth Lunar Science Conference*, pp. 2887–97. New York: Pergamon.)

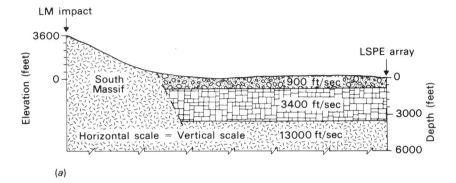

(*a*)

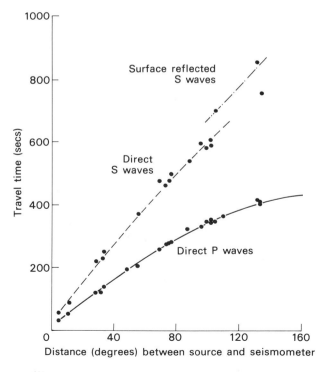

(*b*)

But what about the slower S waves? So far we have only considered the very first P wave arrivals. Well, S wave detection is severely compromised by scattering, because multiply reflected P waves are still arriving with the first S waves. But it so happens that S waves have shorter rise times than P waves and are more significant at high frequencies. So, in short period seismograms, the first S wave arrivals can show up clearly on top of the P wave envelope (Figure 5.41). It has therefore been possible to calculate that S wave velocities in the lower lunar crust are about 2.3 miles per second.

Another method that has been used to disentangle lunar seismograms is known as polarisation filtering, a complicated process designed to sort out the directions of the various vibrations which arrive at the seismometer. It is valuable to find out which vibrations are going up and down, and which are going from side to side, for example, because pure S waves only vibrate in a transverse horizontal direction with respect to the source. Scattered P waves, on the other hand, do not vibrate in any preferred direction and can therefore be filtered out mathematically.

The technique of polarisation filtering has enabled us to pick out later P and S wave arrivals. And this is important because only a small fraction of the seismic energy from a surface impact takes the fastest route to the seismometer. Instead, the strongest waves may be those reflected off, say, the crust–mantle boundary, possibly changing from a P wave to an S

Figure 5.43. *The crust and below.* S and P waves increase sharply down to a depth of nearly 40 miles, which is taken to be the thickness of the Moon's feldspathic crust. P waves continue to increase and remain at about 5 miles per second to the Moon's centre. S waves, in contrast, move more slowly with depth (indicating chemical zoning and the steady rise in temperature) and are strongly attenuated in the partially molten lower mantle. Seismic velocities in the mantle are consistent with a pyroxenitic composition.

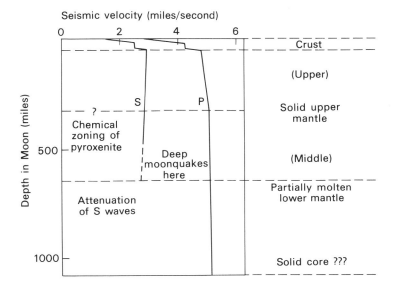

wave in the process. And this reflected phase may arrive later than the first P wave because it may have spent more of its journey in the low velocity crust (Figure 5.38). If these modified phases can be recognised for what they are it may be possible to construct models for the Moon's internal structure which are not solely dependent on timings of the weak first P wave arrivals, which in some cases may be undetectable. And even when the first P wave arrival is clearly recognisable the modified phases are still valuable (because they provide extra constraints on all models of interior structure) and a number of reflected and inverted phases have now been identified unequivocally.

### The Moon's deep interior

To what extent has lunar seismology allowed us to penetrate deeper into the Moon? Well, first of all, immediately below the crust there is a marked increase in P wave velocity to about 5 miles per second (Figure 5.43). The slower S waves also accelerate here, to about 3 miles per second. And the rock type responsible for this deeper zone? Well, these seismic wave velocities certainly match those for pyroxenite, which just happens to be the rock type from which *mare* basalts can be derived most readily by partial melting. So, as far as the crust and mantle are concerned, seismology is certainly consistent with petrology.

But seismic signals from deep moonquakes and farside impacts have

Figure 5.44. *Distribution of moonquakes.* This projection shows the epicentres for locatable deep moonquakes and shallow HFT events. The relative proximity of moonquakes to the seismic network reflects the fact that HFT events are generally stronger than moonquakes. (Courtesy Y. Nakamura.)

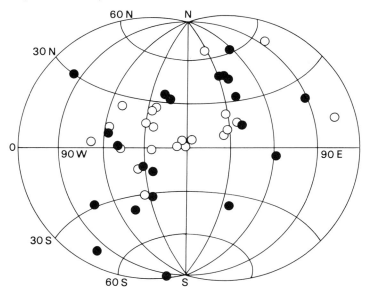

● High Frequency Teleseism (HFT)

○ Deep moonquake

revealed other, still deeper, seismic discontinuities within the Moon. There is some disagreement about the precise depth of the first one (which may simply mean that it is not a very distinct boundary), but below a depth of 200–300 miles, S waves begin to move more slowly, whereas P wave velocities are unaffected. This effect must in part reflect the gradual increase in temperature with depth. But temperature alone cannot be responsible, because otherwise the mantle would melt. The only reasonable explanation is that, as predicted by lunar petrology, the mantle must be zoned chemically, becoming increasingly rich in magnesium with depth.

The other major discontinuity occurs at a depth of 560–620 miles, below which S waves are strongly attenuated. We can reasonably conclude from this that the Moon has a partially molten asthenosphere, or lower mantle, with a radius of 500 miles. Only such a partially molten zone could possibly account for the complete loss of S waves from farside impacts. There is also now some tentative evidence in favour of a small (solid?) core in which P waves travel quite slowly, but more farside impacts are necessary before the existence of this core can be proved beyond doubt.

Our current picture of the Moon, then, consists of a thin feldspathic crust, a thick and chemically zoned pyroxenitic mantle, a partially molten lower mantle, and possibly a small (metallic iron?) core. This model may still be crude, but it will no doubt be refined when more seismic data become available, particularly from farside sources.

### Moonquakes

So much, then, for the structure of the lunar interior, but what about the moonquakes themselves? How strong are they, whereabouts do they occur and what causes them, on a planet which is essentially dead?

Well, there are three types of moonquake, the first of which is really just a soilquake, a tiny movement of the regolith within a few hundred yards of one of the seismic stations. These events are most intense at around noon but most numerous at sunset. So thermal expansion of the regolith is presumably causing small scale soil slumping, on the sides of hills and on the walls of craters.

The weaker variety of true moonquakes invariably occur deep down inside the Moon, at depths between 400 and 600 miles, or close to the base of the solid mantle. And what is particularly intriguing about these deep moonquakes is that many hundreds of the seismograms obtained can be accounted for by a mere 22 foci. So, for some strange reason, stresses must be localised at a small number of weak points inside the Moon. Some seismograms are so similar, in fact, that they can be combined to improve the ratio of signal to noise, and ensure that the vital first P wave arrivals are not missed. This would certainly not be possible if foci had moved more than a few miles (Figure 5.44).

But why should stresses build up inside the Moon at all, and why should they be released at such great depths? Well, like TLPs and radon gas emissions, tidal regularities are also apparent in the seismic records.

And stresses presumably build up where they do because of the great difference in rigidity between solid upper mantle and the partially molten lower mantle. It is perhaps significant that the epicentres of these moon-quakes occur along two narrow belts (one running east–west, the other north–south), so they might be the outward expression of deep mantle convection.

Examples of the third type of moonquake all occur within 200 miles of the lunar surface, and may even just be confined to the crust. They are more intense than the deep ones (up to strength 4 on the Richter scale) and because they are characterised by more rapid vibrations they are collectively referred to as High Frequency Teleseisms (Figure 5.41). Only about 20 HFTs have so far been recorded and, unlike the deep moonquakes, their occurrence follows no obvious pattern in time or space. So they must be caused by the release of crustal stress, possibly indicating mascon readjustment.

### Natural impacts

Moonquakes may well be numerous as seismic sources on the Moon, but there are also abundant natural impact events. So at what rate is the Moon being bombarded by large meteorites, and how does today's flux compare with that in the remote past, as inferred from the observed crater densities?

Well, as many as 150 meteorite impacts have been recorded in a single year, with the projectiles responsible ranging in size from a few ounces to as much as a ton. The largest impacts must have produced craters as much as a hundred yards in diameter. But Earth-based estimates of meteorite abundances in space imply impact rates on the Moon that are many times higher than this. And the measured impact rate is certainly insufficient to explain the observed crater densities on the *maria*.

So it could well be that the impact rate is variable over a timescale of a few years and that the average rate has decreased significantly since the period immediately following the eruption of the youngest *mare* basalts. But the rate at which the Moon is struck by the largest meteorites seems to vary even within a single year, with most falling between April and July. So these may be fragments from a shattered asteroid, or long-dead comet, the orbit of which the Earth–Moon system may be crossing at that time of year.

Although seismology is invaluable for counting meteorite impacts, however, it is not our only tool for probing the deep interior of the Moon. We can also investigate the Moon's magnetic field, not only today but also in the remote past. So let us now turn to the subject of lunar magnetism.

## 5.6 Magnetic studies

The magnetic records preserved in lunar rocks must reflect conditions prevailing during the earliest epochs of lunar history. And global magnetic studies should tell us much about the Moon's present day constitution. But, before we can begin to investigate lunar magnetism, we must appreciate what causes magnetism and discuss how it can best be studied, in space as well as in the laboratory.

The Earth's natural magnetism first became apparent when freely suspended pieces of lodestone (a mineral now referred to as magnetite) were always found to point in the same geographical direction. This discovery led, of course, to the invention of the magnetic compass, without which many early explorers of uncharted territories would doubtless have perished.

But magnetism is also intimately associated with electricity, because, whenever electricity flows, a magnetic field is generated. In an electromagnet, for example, an electric current passes along a wire which has been wound many times around a piece of iron. And the magnetic field generated by the flow of electrons down this wire is enhanced by the magnetism induced in the iron. If this iron is then allowed to cool through the Curie temperature (about 770 °C) while magnetised in this way, its magnetisation will become permanent. In effect, the field produced by the electromagnet becomes frozen into it.

The magnetic field of the Earth must also be due to electricity, but in this case immense electric currents inside the Earth must be responsible, presumably driven by convective motion deep within the Earth's core. And the reversals in magnetic polarity which occur on Earth with some regularity may well be due to instabilities in this convection.

So, when molten rocks cool at the Earth's surface, their remanent magnetisation will reflect the direction and intensity of this field. On Earth we can use the magnetisation directions in rocks to tell us something about sea floor spreading, and the rearrangements of the continents relative to the magnetic poles. In other words we can use magnetised rocks as fossil compasses. So how useful has palaeomagnetism been in unravelling the Moon's history? How did we set about measuring the Moon's magnetism, and what have magnetic studies now told us about our sister planet?

*Magnetic field measurements*

The easiest way to measure the intensity of a magnetic field is simply to move a conducting wire through it and then measure the electric current that flows down the wire. The size of this current will reflect the strength of the field. The first lunar probes to do just this were the crash lander, Lunik 2, and the first lunar satellite, Luna 10. More refined analyses were later performed by a 230 pound American satellite called Explorer 35.

Explorer 35 was launched in July 1967 with the sole purpose of measuring particles and fields several thousand miles above the lunar surface, and it continues its lonely vigil to this day. But, despite its added sophistication, Explorer 35 told exactly the same story as its Russian predecessors. In other words, the global magnetic field of the Moon was found to be effectively zero, and all that could be detected was the weak interplanetary magnetic field associated with the solar wind. So the Moon's global field must be more than a million times weaker than that of the Earth. The future for lunar magnetic studies certainly looked unpromising, but this did not last for long. There were a number of surprises still in store, but men were needed on the Moon first.

Magnetometers were among the most numerous Apollo instruments, each capable of measuring magnetic field strength in three mutually perpendicular directions. They were set up on the lunar surface by Apollos 12, 15 and 16 (Figure 5.45) and also flown in orbit, on the Apollo 15 and 16 subsatellites (Figure 5.46). In addition, local intensity variations at the lunar surface were monitored by portable magnetometers, and by instruments on board the Lunokhod roving vehicles (Figure 2.48). So with this impressive investment in instrumentation, it is perhaps just as well that the story of lunar magnetism was not a simple one after all. But why did we need so many magnetometers, and what did they tell us about the Moon?

Figure 5.45. *Apollo 16 magnetometer*. Three-axis magnetometers, such as this one at Cayley–Descartes, were set up on the lunar surface to monitor surface magnetic fields. The cable connecting the instrument to the ALSEP central station and the RTG can be seen clearly. (NASA, Apollo.)

## The Moon's induced field

Well, first of all a great deal can be learnt about the chemical composition and internal temperature of the Moon even though it has no magnetic field of its own. Because iron-bearing minerals behave characteristically when subjected to a magnetic field. Ferromagnetic minerals such as metallic iron, and to a smaller extent paramagnetic ones (such as olivine and pyroxene), will be magnetised in the same direction as the interplanetary field. So they should add to it, just as an electromagnet is increased in strength by the presence of an iron core.

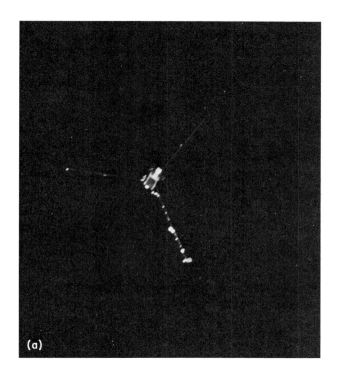

(a)

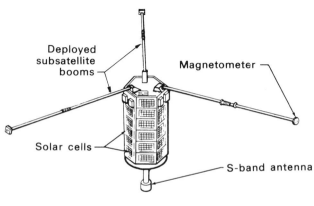

(b)

Figure 5.46. *Apollo 15 subsatellite.* (*a*) This photograph shows the subsatellite immediately after its launch from the Scientific Instrument Module. (NASA, Apollo.) (*b*) Subsatellite magnetometers were mounted at the ends of three booms. Electrical power was provided by solar cells. (NASA.)

But not all substances behave in this way, others may be magnetised in the opposite direction, so as to decrease the induced field, and these materials are referred to as diamagnetic. The strength of the induced field divided by that of the applied field is known as the magnetic permeability of the material so, if we can establish whether the Moon as a whole is diamagnetic or paramagnetic, it should be possible to establish some firm constraints on its chemical composition.

Explorer 35 was in such a high orbit that at all times it effectively monitored the interplanetary magnetic field, unmodified by the presence of the Moon (Figure 5.47). So it was the surface magnetometers, and those on the low flying subsatellites, which were used to measure the induced field. By comparing these surface (or near-surface) fields with those measured by Explorer 35, then, the magnetic and electrical properties of the lunar interior could be investigated.

Before discussing the results of these experiments, however, something should be said about the nature of the interplanetary magnetic field, which is far from constant in the vicinity of the Earth. The solar wind fluctuates wildly in intensity and direction, so it is by no means an ideal environment in which to investigate the Moon's magnetic permeability. Fortunately there is a much better place to carry out these measurements.

Now, as far as electromagnetic radiation is concerned, the Moon is only shielded from the Sun during a lunar eclipse. But, when it comes to being shielded from the solar wind, the Earth's shadow is very much larger, and the Moon passes through this shadow once every month. Solar wind ions interact strongly with the Earth's magnetic field, stretching it out in the antisolar direction like a comet's tail, and compressing it on the sunward side to only a few Earth radii (Figure 5.48). So, for an appreciable fraction of every month, the Moon is bathed in this so-called geomagnetic tail, where the magnetic field is relatively constant (about

Figure 5.47. *The induced field.* The magnetic properties of lunar materials determine the extent to which the interplanetary field (as measured by Explorer 35) is modified by the presence of the Moon. So the Moon's interior can be probed by surface magnetometers and those on the Apollo 15 and 16 subsatellites.

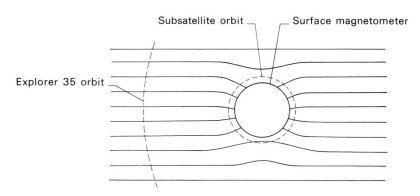

$10^{-4}$ oersted) and is directed towards the Earth above the ecliptic, but away from it to the south.

So what is the magnetic permeability of the Moon as a whole? Well, magnetic fields at the lunar surface imply that, after discounting the diamagnetic effects of the lunar ionosphere, the Moon has a magnetic permeability of $1.012 \pm 0.010$, and is therefore paramagnetic or weakly ferromagnetic in nature.

So what constraints does this result place on the Moon's iron content? Well, despite the relatively large uncertainty, some reasonable limits can be established. But the validity of these limits is complicated by the fact that metallic iron is ferromagnetic only below the Curie point. Above this temperature it is paramagnetic, like the ferromagnesian silicates. And the problem here is that we have no way of knowing where within the Moon any metallic iron might be concentrated. If it were all in the crust then it would certainly be ferromagnetic, but below a depth of 150 miles it would just as certainly be paramagnetic.

But the Moon must contain a minimum of 0.8 % of ferromagnetic iron, and it could have a total iron content as high as 13.5 % if the remainder were all in the form of orthopyroxene. The maximum possible metallic iron content, on the other hand, is 4.8 % (assuming that it is all above the Curie temperature and therefore totally paramagnetic) and the Moon's total iron content could be as low as 5 %. So once again here are some important constraints for any lunar models to accommodate.

Figure 5.48. *The geomagnetic tail.* The solar wind interacts with the Earth's magnetic field to produce what is known as the geomagnetic tail, through which the Moon passes once every month. Magnetic fields are more constant here than in the free streaming solar wind. (After Dyal, P., Parkin, C. W. & Dailly, W. D. (1974). *Proceedings of the Fifth Lunar Science Conference*, pp. 3049–62. New York: Pergamon.)

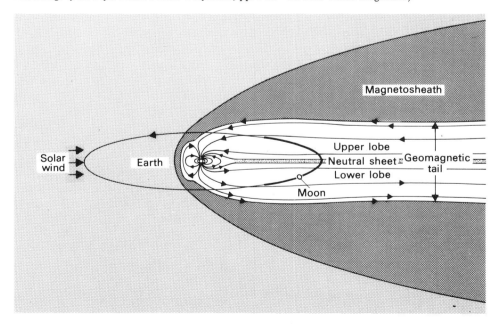

### The ancient field

The Earth's magnetic tail is also a good place to detect any permanent magnetic fields that may exist in the Moon. Because the 5° inclination of its orbital plane to the ecliptic means that the Moon sometimes passes through the northern lobe of the tail and sometimes through the southern lobe. By combining lunar surface measurements made in each lobe for equal periods of time, the induced field can then be made to decrease to zero, leaving just the permanent lunar field. When this is done the Moon's global dipole field is found to be even weaker than we had at first thought, at least 30 million times weaker than the Earth's. So there is certainly no convective dynamo inside the Moon today.

But what of the past? Did the Moon ever have its own magnetic field? Well, as soon as lunar rocks were studied, some were found to have been so strongly magnetised that they must have cooled in a magnetic field that was several orders of magnitude stronger than that which exists in the lunar environment today.

Detailed laboratory studies of highland rocks (involving their demagnetisation by subjecting them to high temperatures and strong alternating magnetic fields, and their subsequent remagnetisation under controlled conditions) have revealed that the magnetic carrier (metallic iron and not magnetite in lunar rocks) must have acquired its natural remanent magnetisation in magnetic fields that were at least comparable to that of the Earth today. By the time that the youngest *mare* lavas were extruded, however, the strength of the Moon's magnetic field must have decreased by a factor of ten or so, although it would have been still quite appreciable. Some results indicate a global field as high as 1.2 oersted some 3900 million years ago, which decreased over the next 700 million years to about 0.05 oersted before disappearing altogether (Figure 5.49).

So what could have produced this ancient magnetic field, and why does it no longer exist? Well, a number of intriguing theories have been proposed, but there is still no universally accepted one. Perhaps the most attractive, if only for its elegant simplicity, is that the Moon's core was once magnetised. For this to have happened, the Moon must have accumulated metallic iron grains below the Curie temperature during the earliest phase of its formation. In the presence of a strong and relatively constant solar magnetic field, these iron grains could then have become aligned as in a giant bar magnet.

Later partial melting of the outer layers of this magnetic Moon would have resulted in magnetised rocks at its surface. But as the deep interior started to heat up, through the decay of natural radioisotopes, its metallic iron could no longer have remained below the Curie temperature, and would therefore have become demagnetised. So the magnetised rocks at the lunar surface would now provide the only clues to its former existence.

But attractive as this theory may seem, there is simply not enough metallic iron inside the Moon to carry the required magnetisation. Even with 4.8%, a magnetic field as high as 75 oersted would have been necessary, and it is difficult to see how this could have been generated in the early Solar System. For any reasonable solar field, the amounts of metallic iron necessary would be incompatible with the Moon's measured density, moments of inertia and magnetic permeability.

But at least the permanent magnet theory can account for longevity of the ancient magnetic field, which is more than can be said for some of the alternatives. One of these attributes the magnetisation of lunar rocks directly to an intense solar magnetic field, because it is known that young stars may eject vast quantities of electrically charged particles into space, a solar hurricane rather than a solar wind. As the Sun presumably also went through this so called T Tauri phase, the interplanetary magnetic field could well have been orders of magnitude stronger than it is today. It is difficult to see, though, how this intense field could have been maintained for very long, because the T Tauri stage is but a brief interlude in a star's evolution, before it joins the main sequence.

A similar argument can be made to exclude the Earth's magnetic field, because even if the Moon was once very close to the Earth it would not have remained there for very long because of tidal friction.

### The magnetised crust and its implications

So perhaps the Moon really did have its own self-sustaining dynamo, when much of its interior was molten and convection was still possible on a large scale. The strength of this field would have decreased steadily to zero as the Moon cooled inwards. So the lack of a present day global field

Figure 5.49. *The decay of the ancient field*. Studies of magnetised lunar rocks of known age indicate that the ancient lunar field responsible for their magnetisation may have decayed from more than 1 oersted, 4000 million years ago, to less than 0.05 oersted, 3200 million years ago. (After Stephenson, A., Runcorn, S. K. & Collinson, D. W. (1975). *Proceedings of the Sixth Lunar Science Conference*, pp. 3049–62. New York: Pergamon.)

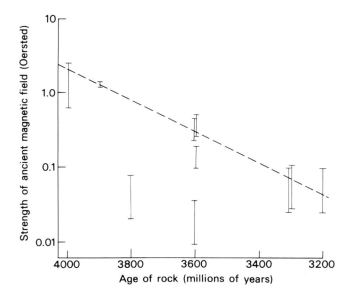

can be accounted for. But the most compelling evidence in favour of a dynamo is negative. Because if the rocks of the lunar crust had been magnetised externally, rather than internally, then there should be a global field of measurable intensity today, due solely to this magnetised crust. Only if the crust was magnetised by an internal field, which has since disappeared, would we expect the magnetic lines of force to be confined within the Moon in this way (Figure 5.50).

But to say that the Moon has no global magnetic dipole field does not mean that no magnetic forces at all can be detected from lunar orbit. Indeed, the strongly magnetised rocks of the highland crust can produce quite appreciable magnetic anomalies, in some places as strong as 0.003 oersted. But such local fields may not extend very far, as demonstrated by the Apollo 14 portable magnetometer, which recorded a complete reversal of field direction over a distance of less than a mile.

So highland cratering must certainly have done much to destroy and randomise the orientations of local fields. But this randomisation process has certainly not been complete, because the Apollo 15 and 16 subsatellite magnetometers (Figure 5.46) revealed an impressive magnetic anomalies (or magcon) in the farside craters Van de Graaff (Figure 5.51) and Aitken. Smaller magcons occur elsewhere and more may be revealed if our magnetic coverage of the Moon can be increased further by means of a Polar Orbiter satellite.

The very existence of these magcons means that large magnetising fields must have existed in the past which could not, for example, have

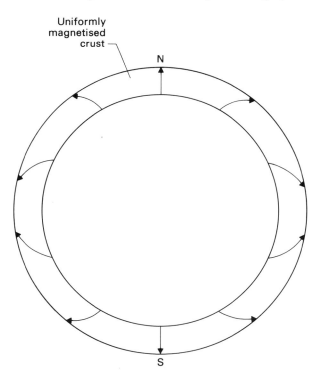

Figure 5.50. *Consequence of a dead dynamo.* If the Moon's crust was magnetised uniformly by an internal source (such as a dynamo) which later disappeared, the Moon would have no net magnetic dipole today. (After Stephenson, A., Runcorn, S. K. & Collinson, D. W. (1975). *Proceedings of the Sixth Lunar Science Conference,* pp. 3049–62. New York: Pergamon.)

Figure 5.51. *The Van de Graaff magcon.* (*a*) The most impressive magnetic anomaly detected from lunar orbit was associated with the farside crater Van de Graaff. (After Russel, C. T., Coleman, P. J., Lichtenstein, B. R., Schubert, G. & Sharp, L. R. (1973). *Proceedings of the Fourth Lunar Science Conference*, pp. 2833–45. New York: Pergamon.)

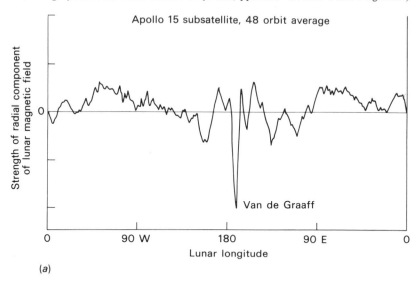

(*a*)

Figure 5.51. *The Van de Graaff Magcon.* (*b*) Van de Graaff is unusual in that it is a double crater. (NASA, Apollo.)

just been frozen into the crust during meteorite impacts. Magnetic features as small as a mile across would simply not be detectable from orbit if they were randomly distributed. And magnetic measurements at different altitudes imply that crustal magnetisation may in places be coherent down to depths of at least several thousand feet. Only a global magnetic field, long since disappeared, could possibly account for this coherence. The magcons may simply represent disturbed areas in an otherwise uniformly magnetised crust.

*Electrical conductivity*

Magnetic studies in the geomagnetic tail have certainly provided us with useful information about the Moon's magnetic fields and the nature of the lunar interior. But, for most of its orbit, the Moon is not bathed in such a uniform field. So most surface magnetometer data reflect conditions either in the solar wind itself, or totally shielded from it during the lunar night.

These environments have been described above as being less than ideal for magnetic studies. But as far as the Moon's response to change is concerned, the interplanetary environment is more suitable. Such changes in magnetic field intensity can also occur within the geomagnetic tail (when the magnetosphere is distorted by solar storms) and our most valuable data, in fact, comes from the solar disturbance of 20 April 1970. So how then does the Moon respond to a changing magnetic field and what can be learnt from this interaction?

Well, the Moon is a barrier to solar particles, so there is a shadow behind the Moon from which the solar wind is excluded. But this cavity is not a perfect shadow, because magnetic fields in the vicinity of the terminator tend to disturb the path of the solar wind and the magnetic field associated with it. And these so-called limb compressions can be used to investigate the magnetisation of the lunar crust.

But the manner in which the Moon responds to change (in other words the way in which the induced field varies with changes in the interplanetary field) is much more important. Because this response cannot be instantaneous, large electric currents must flow around inside the Moon to oppose the driving field. And these so-called eddy currents should be able to tell us something about the electrical conductivity, and hence the physical and chemical state, of the lunar interior.

These, like the permeability measurements, are based on comparisons between the driving field (as measured by Explorer 35) and the induced field (as measured by surface magnetometers). In this case, however, it is not the strength of the induced fields that is important, but rather the time taken for it to respond, of the order of several minutes (Figure 5.52).

What has to be done in practice is to invent a model to describe how the Moon's electrical conductivity might vary with depth and then see how the theoretical response time for this model compares with the measured response times. The model is adjusted accordingly until a good fit is achieved. The model parameters would then approximate to reality.

The calculated electrical conductivity profile for the Moon has much in common with the seismic one, not surprisingly perhaps, because both are intimately related to chemical composition and temperature. But the electrical conductivity changes with depth are more extreme, covering many orders of magnitude. The fragmental upper crust, as well as having a low seismic velocity, is a very good electrical insulator. But the electrical conductivity of basalt rock is reached at a depth of a few thousand feet and it then continues to rise steeply down at a depth of 125–200 miles, and, after that, more steadily to a depth of 450 miles. Finally, there could even be a conducting core. Although such a core is not actually required to explain the results, its maximum possible radius is some 340 miles.

These transition depths should be compared with the seismic discontinuities, the postulated transition between the upper and middle mantle (Figure 5.43), and somewhat deeper down the upper boundary of the asthenosphere, or partially molten lower mantle. So what do these conductivity measurements mean in terms of temperature and chemical composition?

Well, we are now becoming involved in the question of the lunar heat budget, and there are some aspects of this subject to be discussed before a clear picture of the Moon's thermal history can be obtained. So let us now consider the flow of heat through the lunar crust and postulate on likely temperature profiles within the Moon, not only today but also in the remote past.

## 5.7 The Moon's heat

The measurement of lunar temperature has already been discussed in the context of remote sensing, when we heard how the infrared and microwave radiation emitted by the Moon can be used most effectively to map

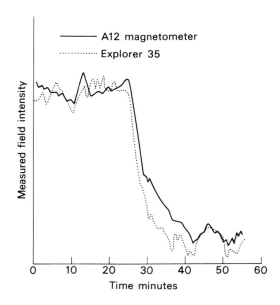

Figure 5.52. *Magnetic response times.* The induced field does not respond immediately to changes in the external field. Studies of response times enable us to investigate the electrical conductivity, and hence temperature, of the lunar interior. (After Dyal, P., Parkin, C. W. & Dailly, W. D. (1975). *Proceedings of the Sixth Lunar Science Conference*, pp. 2909–26. New York: Pergamon.)

regolith temperatures (Figure 5.16). At the very surface these may vary over short distances by as much as 35 °C, because some areas on the Moon (particularly those heavily strewn with boulders) are more efficient than others at retaining the Sun's heat.

But, so far, the emphasis has been on the re-irradiation of solar energy. There has not yet been any discussion of the dissipation of the heat which is generated within the Moon itself. And this is a subject of the utmost importance. For it is the Moon's internally generated heat which is behind such phenomena as magnetism and moonquakes, and was directly responsible for the production of *mare* basalts. For lunar volcanism can best be accounted for in terms of partial melting of the Moon, at depths of more than a hundred miles.

So what are the questions to be asked about the Moon's heat and thermal history? And how adequately have these questions now been answered? Well the most important measurable quantity is of course the amount of heat which is currently flowing out through the lunar crust. And, since the successful emplacement of the Apollo 17 heat flow probes, this crucial measurement has now been performed at two widely spaced locations on the Moon.

Then there are the constraints on the nature of the Moon's deep interior, imposed by the seismic and electrical conductivity measurements just described. These studies add that crucial third dimension to the heat problem. And thirdly, there is the question of the heat sources

Figure 5.53. *Heat flow probes.* Heat flow probes were inserted into the regolith at the Apollo 15 and 17 sites, enabling measurements of heat flow out through the crust to be determined. Difficulties in drilling meant that the Apollo 15 measurements were not as deep as those from Apollo 17. (After Langseth, M. G., Keihm, S. J. & Peters, K. (1976). *Proceedings of the Seventh Lunar Science Conference*, pp. 3143–71. New York: Pergamon.)

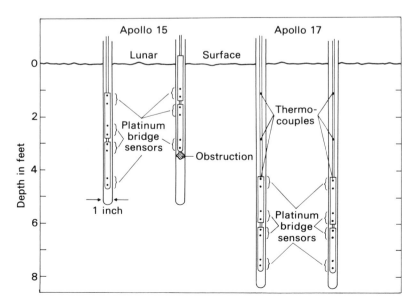

themselves. Where did the Moon's interior heat come from and where are these heat sources today, if indeed they still exist? Finally, all the results, constraints and conjectures must be incorporated into a coherent model for lunar thermal evolution. In other words, what does the Moon's interior temperature profile look like, not only today but also in the remote past.

*Heat flow probes*

The first direct measurements of lunar subsurface temperature were undertaken on the Apollo 15 mission, when two heat flow probes were inserted into the 1 inch diameter, 5 feet deep, holes which were left behind after the extraction of the deep drill strings. The Apollo 17 drill cores were even deeper still, so it was possible to measure temperatures in the Taurus–Littrow regolith down to depths of more than 8 feet (Figure 5.53).

Unfortunately, the Apollo 16 heat flow experiment was rendered inoperative when an astronaut accidentally tripped over one of the cables. This was particularly unfortunate because a heat flow measurement in purely highland terrain would have been an extremely valuable constraint.

All four heat flow probes contained two distinct types of temperature sensor. Mounted along the lower 4 feet of each probe were 8 platinum bridge sensors, capable of measuring absolute temperatures to an accuracy of better than 0.015 °C (Figure 5.53). Temperature measurements in the uppermost four feet of the regolith are less critical because temperatures here are liable to be strongly influenced by the diurnal heat wave and are consequently of limited use for heat flow studies. The upper sections of the probes were therefore just equipped with unsophisticated thermocouples.

*Regolith temperatures*

As far as absolute subsurface temperatures are concerned, these were found to vary from −24 °C, at a depth of 18 inches, up to −16 °C at the base of the Apollo 17 probe. The anticipated monthly variation (the diurnal heatwave) does not, in fact, extend much below a depth of two feet. And even here its amplitude amounts to less than one degree (Figures 5.54 and 5.55).

The effects of solar heating do, however, penetrate much deeper than this. Temperature variations having a yearly, rather than a monthly, period are readily detectable well below three feet. This longer period modulation is due to the change in the effectiveness of solar heating which arises from the eccentricity of the orbit of the Earth–Moon system around the Sun. The temperature on the Moon's surface at noon, in fact, is some 3 °C higher at perihelion than it is at aphelion. This temperature variation is, of course, much smaller than the diurnal one, but there is much more time available for it to propagate downwards. Consequently, it is detectable at considerably greater depths (Figure 5.55).

### Thermal conductivity of the regolith

The rapid attenuation of the diurnal heatwave with depth implies that the lunar regolith must be an extremely efficient thermal insulator. And this is perhaps not too surprising. For lunar soil is made up of highly irregular rock particles and its *in situ* bulk density is such that as much as half of the regolith must consist of empty space. Heat flow through the regolith therefore occurs by a combination of conduction through point contacts and radiation across pore spaces, neither of which are very efficient heat transfer processes.

But just how good an insulator is the lunar regolith? In order to determine the Moon's global heat flow it is necessary to know very accurately the thermal conductivity of the regolith in the immediate vicinity of each heat flow probe. For, the lower this conductivity is, the steeper will be the temperature gradient up the probe.

The experimental approach was simply to provide an artificial heat source at the base of each probe. Measuring the temperature gradient up the probe, with and without this head source switched on, is quite sufficient to enable the thermal conductivity in the vicinity of each probe to be determined. Some of this heat, of course, is transmitted up the heat flow probe itself, but the conductivity of the probe is well known so this effect can be readily corrected for. And when this is done the regolith thermal conductivity works out to be about 250 millionths of a watt per square inch, for a temperature gradient of 1 °C per inch. This value is 100 times less than that for solid lunar rock.

The regolith is clearly not an efficient thermal conductor. But this is not sufficient to explain the longest term temperature variation. For it appears that, from a thermal point of view, the soil in the vicinity of each probe is still recovering from its initial disturbance during probe emplacement. Only now, several years after the return of the last Apollo, is the thermal conductivity of the regolith finally approaching equilibrium. So only now are mean temperatures (that is those corrected for diurnal

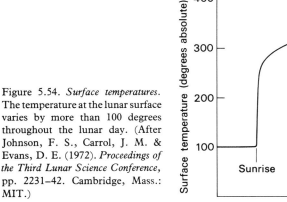
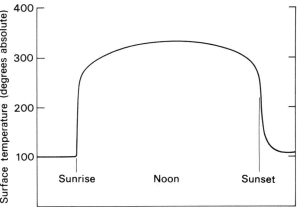

Figure 5.54. *Surface temperatures.* The temperature at the lunar surface varies by more than 100 degrees throughout the lunar day. (After Johnson, F. S., Carrol, J. M. & Evans, D. E. (1972). *Proceedings of the Third Lunar Science Conference,* pp. 2231–42. Cambridge, Mass.: MIT.)

and annual variation) settling down (or rather rising up) to constant values (Figure 5.55). And it was this long term change in thermal conductivity, in fact, which led to a reevaluation of the experimental results, and thus led to a reduction in what were embarrassingly high heat flow values.

*Lunar heat flow measurement and natural radioactivity*
What then is the rate at which heat is flowing out through the lunar crust at each landing site and what might the global value be? Well the measured heat flow values were, in fact, in reasonable agreement with one another. The experimentally determined temperature gradients (of between 0.2 °C and 0.8 °C per foot, Figure 5.56) correspond to heat flows of 13.7 and 9.1 millionths of a watt per square inch.

The problem of extrapolating to the entire Moon from just two isolated measurements is clearly fraught with difficulties. The main reason being our inevitable uncertainty about the nature and location of the necessary heat sources. It is significant, however, that the Apollo 15 site, where the higher heat flow value was obtained, is appreciably richer in natural radioactivity than the Taurus–Littrow area. The gamma ray spectrometer results indicated, in fact, that the entire Oceanus Procellarum – Mare Imbrium area is mantled by the rock type known as KREEP (Figure 5.22). But just how representative is this high surface concentration of radionuclides? Do the gamma rays originate from a thin veneer of KREEP dust, or does the high thorium concentration extend deep down into the crust?

The relatively higher Apollo 15 heat flow certainly does suggest the

Figure 5.55. *Subsurface temperatures.* The amplitude of the diurnal heat wave is little more than one degree at a depth of only 18 inches. If this diurnal component is removed mathematically, the annual variation is clearly apparent, despite the fact that its amplitude at the surface is much smaller than the diurnal one. It can be seen that temperatures have risen steadily at this depth since probe emplacement. (After Langseth, M. G., Keihm, S. J. & Peters, K. (1976). *Proceedings of the Seventh Lunar Science Conference*, pp. 3143–71. New York: Pergamon.)

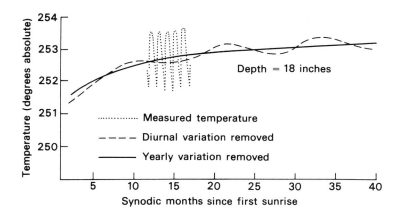

existence in Mare Imbrium of more than just a superficial deposit of KREEP. But, equally well, the measured concentrations (4 parts per million) cannot be typical of the entire crust. For, if they were, the Moon would be molten at a depth of only 40 miles and this is certainly not the case. Similarly, if the Moon's uranium, thorium and potassium was evenly distributed throughout the entire Moon, it would be partially molten below a depth of 150 miles or thereabouts. So there seems to be no way to escape the conclusion that all three elements (plus many more) must have been segregated into the crust early in lunar history.

### Heat sources and thermal modelling

The assumption that radioactive elements are the prime sources of heat within the Moon today is certainly a valid one. But it is not just their global distribution which determines the natural flow of heat at the Moon's surface, important as this distribution may be. For there may be heat within the Moon which is left over from its original formation. This possible source of heat must be seriously considered in models for the Moon's thermal evolution in the remote past.

But, staying with the present, the other important unknowns here are the effects, if any, of convection and the thermal conductivity of the Moon's outer layers. The upper regolith may be a poor conductor of heat,

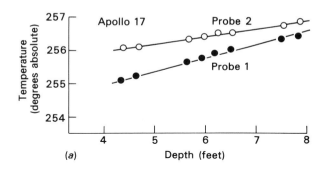

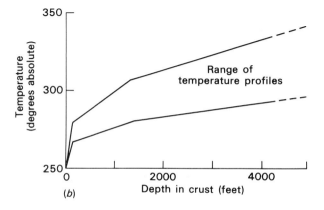

Figure 5.56. *Temperature profiles.* (*a*) Temperature gradients up the Apollo 17 probes differ because of differences in regolith thermal conductivity. (*b*) The steep temperature profile that characterises the regolith must level off, otherwise melting temperatures would be reached at very shallow depths. This graph is an estimate of the true temperature profile, based on seismic and heat flow studies. (After Langseth, M. G., Keihm, S. J. & Peters, K. (1976). *Proceedings of the Seventh Lunar Science Conference*, pp. 3143–71. New York: Pergamon.)

but what about the *mare* lavas and the anorthositic crust? The seismic results imply a rapid decrease in porosity (and hence a steady increase in conductivity) with depth in the upper crust. And this is to be expected. For, if the temperature gradients measured in the regolith remained constant with depth, melting temperatures would be reached at a depth of less than a mile! It seems likely, in fact, that temperature gradients immediately below the regolith are more of the order of 1 °C per 100 feet (Figure 5.56).

Another factor which can affect the heat flow at a particular site on the Moon is local topography. Hills and rilles may have the effect of enhancing the heat flow in one area at the expense of another, particularly if one area is covered by a thinner regolith blanket. The Apollo 17 heat flow measurement, in fact, was adjusted downwards to allow for local topographic effects.

Larger scale differences (farside highlands compared with nearside circular *maria* for example) are more difficult to evaluate because of the lack of adequate seismic and heat flow data. It is unlikely, however, that the Moon has any of the sort of hotspots that, on Earth, are associated with tectonic plate margins. Variations in lunar heat flow must be largely just a function of crustal thickness and chemistry.

Considering all of these factors together, it is now possible to estimate a value for the mean global heat flow of about 12 microwatts per square inch, of which 2.5 microwatts originate from the mantle and core, while 1.3 microwatts come from the KREEP layer. The remainder must be generated within a 40 mile thick crust containing about 1 part per million of thorium. These values reflect the known enrichment of radionuclides in the crust, as well as numerous geochemical studies of lunar rock types. As far as the overall uranium content of the Moon is concerned, this works out to be about 50 parts per billion which, if slightly more than Earth's, is at least consistent with the Moon's slight enrichment in refractory elements.

### The deep temperature profile

So far, only the uppermost few miles of the Moon's crust have had to be considered. For when it comes to the Moon's deep interior it is necessary to turn to seismic and electrical conductivity studies in order to progress much further. Because surface heat flow measurements cannot really provide useful information about the location of deep heat sources, without making assumptions about the thermal conductivity of the mantle. It is rather like attempting to study the ocean floor by simply examining waves on the surface.

The electrical conductivity measurements have been interpreted as indicating a steep rise in temperature down to a depth of 125 miles or so, by which depth temperatures must at least exceed the Curie temperature for metallic iron (Figure 5.57). Thereafter the temperature rise is thought to be less extreme.

The conversion of electrical conductivity to temperature depends to some extent on exactly what chemical composition (e.g. olivine or orthopyroxene) is chosen for the lunar mantle. But, whatever chemistry is chosen, the rise in temperature with depth in the lower mantle is certainly not so steep as it is in the crust. And this conclusion is also consistent with the negative shear wave velocity gradient discussed earlier. The decrease in seismic S wave velocity in the upper mantle can best be explained in terms of a steady temperature rise with depth. The seismic data also suggest that the Moon's interior is only molten below a depth of 600 miles or so, which again implies a shallow temperature gradient.

### Other heat sources

But what of the remote past? Today's lunar temperature profile must clearly have changed quite drastically since the time of the Moon's formation. For one thing the Moon used to be much richer in radio-nuclides, particularly uranium-235, the less long-lived of the two naturally occurring uranium isotopes. And, in the very early days, there must also have been an appreciable quantity of the now extinct radioisotope plutonium-244, the chemical and physical remnants of which are clearly detectable in lunar rocks.

There may even have been some of the very much shorter-lived isotope aluminium-26, which has a half-life of only 700 000 years, a brief period indeed on the cosmic time scale. The decay products of this last isotope have in fact now been found in the oldest meteorites. And this suggests that it may have played a very major part in the thermal evolution of the early Solar System. It is certainly the best way to account for the early

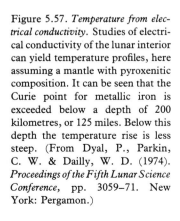

Figure 5.57. *Temperature from electrical conductivity.* Studies of electrical conductivity of the lunar interior can yield temperature profiles, here assuming a mantle with pyroxenitic composition. It can be seen that the Curie point for metallic iron is exceeded below a depth of 200 kilometres, or 125 miles. Below this depth the temperature rise is less steep. (From Dyal, P., Parkin, C. W. & Dailly, W. D. (1974). *Proceedings of the Fifth Lunar Science Conference*, pp. 3059–71. New York: Pergamon.)

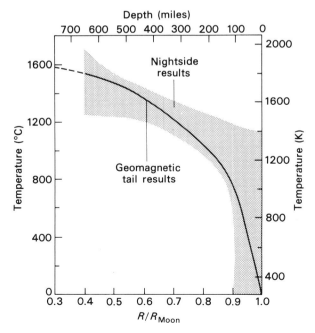

melting of meteorites and, by inference, the asteroids and perhaps the larger terrestrial planets as well.

But radioactivity may not have been the only major mechanism for heat production. There are a number of other possible sources as well. And the most important of these, at least as far as the Moon is concerned, was the gravitational energy of planetary accretion. If the Moon was formed by the gradual accumulation of solid bodies, and this conclusion now seems to be inescapable, then during the final stages of its formation the protomoon would effectively have experienced an extremely intense period of bombardment. And the shorter this accretion period was, the hotter the outer layers of the Moon would have become, as the kinetic energy of the incoming bodies was all converted into heat. The inevitable conclusion is that the outer part of the Moon, possibly down to a depth of 200 miles or so, was at one stage almost completely molten, a conclusion which has, of course, already been described in the context of the origins of lunar rocks. This 'magma ocean' idea has certainly caught on in the lunar scientific community and will no doubt remain viable for some time to come.

Kinetic energy conversion is, however, not the only alternative to natural radioactivity as a possible early lunar heat source. There are other more or less exotic mechanisms, including tidal disruption (particularly important if the Moon ever came really close to the Earth), magnetic induction heating, and gravitational core separation. But the widespread belief is that the Moon's early heat budget was indeed controlled by its own accretion, possibly supplemented by the decay of short-lived radionuclides. Only later did long-lived radioactive isotopes begin to be effective.

*Lunar thermal history*

The starting point for all models of lunar thermal history, then, would seem to be an undifferentiated protomoon with a hot, probably molten, outer layer. It has been suggested alternatively, however, that the Moon accreted cold. This could conceivably have come about if it had a thick gaseous atmosphere, like that of Jupiter perhaps, which could readily absorb the kinetic energy of the incoming projectiles. But there is certainly no evidence today for the former existence of such an atmosphere. And the very early formation of the lunar crust, as discussed in Chapter 3, certainly suggests a deep primordial magma ocean. But just how deep was it and how did it evolve with time?

Well, first of all, it must have been deep enough to result in a feldspathic crust some 40 miles thick when it finally cooled. On the other hand, there must have been sufficient natural radioactivity left behind in the interior (that is not segregated upwards in the form of KREEP) to generate the *mare* basalts and to remain hot to this day. The green glass, the source region for which was barely affected by the primordial chemical differentiation, appears to have originated from a depth of about 200 miles, so this would seem to be an upper limit on the depth of the magma ocean.

But perhaps the strongest constraint comes not from petrology but from photogeology. For the absence of evidence for major lateral crustal movement implies that the Moon cannot have expanded or contracted appreciably throughout its entire history, or at least during that period which is reflected in its exterior features.

If the Moon's radius had increased by, say, 5 miles over the past 4000 million years then we would expect great rift valleys to have ruptured its crust, valleys comparable in magnitude if not in detail to the giant Coprates canyon on the surface of Mars. Contraction features should no doubt be just as dramatic. So it is possible to conclude from the absence of such features that the magma ocean must have been less than 190 miles deep, otherwise it would have contracted too much as it solidified. An ocean only 70 miles deep, on the other hand, would not only have failed to produce the necessary 40 mile thick feldspathic crust, it would have left far too much radioactivity deep inside the Moon. And these radioactive isotopes would have heated the interior up so much that the Moon as a whole would have expanded appreciably. In other words, in order to account for the Moon's essentially constant radius throughout its history, the contraction of the magma ocean must have been more or less compensated for by the expansion of the interior.

The overall picture of the Moon's thermal evolution then consists of a hot exterior which cooled down and a cool interior which heated up (Figure 5.58). As time passed, the partially molten zone gradually migrated inwards, generating the various *mare* basalts on the way. It must by now have reached a depth of some 600 miles, for it is only here

Figure 5.58. *Thermal model for the Moon.* In this model for the Moon's thermal history, the uppermost 200 miles were melted by the energy of accretion, and subsequent heating was due to 30 parts per billion of uranium and corresponding concentrations of thorium and potassium. As time passed, the zone of melting gradually moved inwards, producing basaltic magmas on the way. (After Hubbard, N. J. & Minear, J. W. (1976). *Proceedings of the Seventh Lunar Science Conference,* pp. 3421–35. New York: Pergamon.)

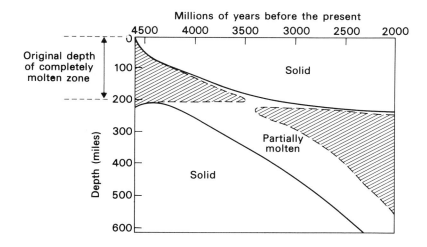

that seismic S waves start to be strongly attenuated. As this central core is all that remains of the original molten zone, it is small wonder that the Moon is outwardly such a dead planet.

This discussion of lunar thermal history has yet again implied knowledge of the absolute ages of lunar events. The chemical differentiation of the crust, for example, has been placed more than 4000 million years before the present, while it has been stated that *mare* volcanism continued for at least a further 1000 million years. But, so far, these ages have had to be taken on trust. Methods for establishing relative ages, such as crater counting and morphology, are just not capable of providing absolute ages. And it is therefore appropriate that, before going on to discuss the most important question of all, namely the origin of the Moon itself, the various methods of dating lunar rocks and the results that these methods have yielded should be discussed in detail. Just how old are the various lunar rock types? And how can we account for their ages in a global context? This is an important subject with which to begin the final chapter.

# 6. The Moon's history and future

## 6.1 The age of the Moon

Any discussion of lunar chronology must begin with the most fundamental question of all. Just how old is the Moon? Is it as ancient as the other terrestrial planets or was it formed much later on, perhaps by the fission of a primitive Earth? The determination of the Moon's age should certainly establish a major constraint on its possible origin. But what exactly is meant by the age of the Moon? It is important to be able to differentiate between the various ages which have been calculated. For some dates refer to the Moon as a whole, while others only relate to individual rocks or surface features.

*Formation ages*
First of all, it is the age of the planet itself that is of prime importance here, not the ages of its constituent rocks or atoms. For the atoms themselves date back to long before the Sun was born, having been laboriously built up from hydrogen and helium nuclei in stars which have long since exploded as supernovae and been disseminated as gas and dust throughout the Galaxy. So by dating lunar atoms we would be dating stars rather than the Moon itself.

But it so happens that there were element-building stellar explosions just before Solar System formation, at least on a cosmic time scale. Indeed, some astronomers believe that the shock wave generated by the expanding gas from one of these so-called supernovae may in fact have been directly responsible for the birth of the Sun and its family of planets. For young stars are commonly observed wherever such shock waves are seen to be passing through gas clouds elsewhere in the Galaxy.

The reason why it is believed that there was at least one supernova just before the birth of our Solar System is that the oldest known samples of Solar System material, namely the chondrite meteorites, contain chemical and physical traces of long-extinct short-lived isotopes. Two of these radionuclides, aluminium-26 and plutonium-244, have already been mentioned in the context of possible early lunar heat sources. But there is another one, namely iodine-129, which was also still abundant when the chondrites formed some 4600 million years ago. And the relative abundances of its gaseous daughter isotope, xenon-129, are such that all of these meteorites must have formed within a few million years of each other at the very most.

But the amounts of iodine-129 and plutoniun-244 that must have been present by that time were so low that their parent supernova must have occurred some 100 million years earlier, in other words about 4700 million years ago. So why were the chondritic meteorites formed within such a short period? And how could there possibly have been any aluminium-26 still around after all that time? Well it now looks as if there was a second supernova 100 million years after the first. And although this second stellar explosion failed to yield (or at least contribute to the Solar System) any more iodine-129 or plutonium-244, it did synthesise some other novel isotopes, including an excess of oxygen-16 and that all-important short-lived radionuclide, aluminium-26. The expanding shock wave from this second supernova must then have brought about the rapid collapse of a nearby gas cloud to form our Solar System.

So much, then, for the Solar System as a whole, but where does the Moon fit into this picture? Well it now seems very likely that the Moon did indeed form at the same time as the other planets, about 4600 million years ago, and that it too must have accumulated its fair share of aluminium-26. But how is this conclusion arrived at and how strong is it? There is certainly no firm evidence as yet for the former existence of aluminium-26 in lunar rocks. And this is really the crux of the problem. How can we possibly date the formation of a planet when there may be no rocks at all which still survive from that earliest period in its history?

*Model ages*
The answer to this problem is that strictly speaking we cannot. The best we can possibly do is attempt to date such physico-chemical effects as the early differentiation of the Moon. And such dates, because they depend to some extent on physical processes which even today are by no means well understood, are usually known as model ages. In other words they relate to a particular physical model for the Moon's formation and early evolution. So what are the various model ages for the Moon as a whole, what assumptions are they based upon and how do they compare with one another?

Well first of all the only clocks that can possibly be used here are based on natural radioactivity. For, unlike all other physical and chemical processes, radioactive decay proceeds at a rate which is to all intents and purposes unaffected by external conditions. And the half-life of a radioactive isotope, the time taken for any amount of that isotope to be reduced by half, is not only constant but can be readily measured in the laboratory. Non-radioactive absolute dating methods, such as those based on crater statistics, rely much too heavily on uncertain physical processes (e.g. past meteorite fluxes, crater erosion rates) to be of any real value here, and are certainly not applicable to the very earliest period of lunar history.

Fortunately there are a number of long-lived radionuclides which can be used to date the Moon (Table 6.1). These isotopes have sufficiently long half-lives (several hundred, or thousand, million years) for appreciable quantities of them to have survived to the present day. For these dating methods (unlike studies based on extinct radionuclides) require the measurement not only of the products of radioactive decay (or fission), but also of the parent isotope. Effectively it is the ratio of daughter isotope to parent which is meaningful here. The longer the

Table 6.1. *List of naturally occurring, long-lived primordial lunar radionuclides used for chronological purposes. The last three isotopes are now extinct, but evidence for the former existence of plutonium-244 and iodine-129 can be found in lunar rocks. Aluminium-26 may have been a significant source of heat in the early Moon*

| Parent isotope | Daughter isotope(s) | Half-life (million years) |
|---|---|---|
| Uranium-238 | Lead-206 | 4500 |
| Uranium-235 | Lead-207 | 713 |
| Thorium-232 | Lead-208 | 13900 |
| Potassium-40 | Argon-40 (and calcium-40) | 1300 |
| Rubidium-87 | Strontium-87 | 52000 |
| Samarium-147 | Neodymium-143 | 108000 |
| Plutonium-244 | Various fission products | 82 |
| Iodine-129 | Xenon-129 | 16 |
| Aluminium-26? | Magnesium-26 | 0.74 |

elapse time since the event being dated, the more daughter isotope there will be, relative to its parent.

One of these long-lived radioisotopes, potassium-40, decays to an isotope of an inert gas, argon-40. This makes potassium-40 very valuable for dating individual rocks, but totally useless for studying the Moon as a whole. Because some, if not all, of the argon formed during the very first epoch of lunar history may well have escaped from the Moon completely as the rocks in the Moon's crust were successively remelted.

*Initial strontium*

Another long-lived natural radioisotope, however, is rubidium-87, which decays, even more slowly, to an isotope of strontium, strontium-87. Now, in this case, the daughter isotope is not a gas but a metal. And, furthermore, strontium-87 occurs naturally in lunar rocks, albeit at low concentrations. So the radioactive decay of rubidium-87 simply enhances the existing ratio of strontium-87 to a non-radiogenic strontium isotope, such as strontium-86.

How, then, can the rubidium–strontium system be used to date the Moon? Well first of all it is important to establish what the primordial lunar ratio of strontium-87 to strontium-86 was. And the simplest approach here is to study rocks which contain appreciable strontium but very little rubidium, in other words the incompatible element deficient anorthosites. Because the lowest strontium-87–strontium-86 ratio to be found in a lunar rock will clearly establish a firm upper limit for this ratio in the primordial Moon as a whole. The radioactive decay of rubidium-87 means that this strontium isotope ratio can only increase with time. But, the lower the rubidium–strontium ratio, the smaller the rate of increase will be.

It turns out, in fact, that the strontium-87–strontium-86 ratios in some lunar anorthosites are so low (0.6990) as to be indistinguishable from the initial ratio calculated for the basaltic achondrite meteorites and, by inference, for the Solar System as a whole. So it seems that the Moon's anorthositic crust must have formed within about 100 million years of the formation of the Solar System. If the chemical differentiation of the Moon had been delayed much longer than this, then its higher overall rubidium content (about 0.4 parts per million) would have ensured that no lunar rocks could have been formed with such a low strontium ratio, let alone survived with it to this day. The very primitive strontium isotope ratios in lunar anorthosites, then, confirm that the Moon must indeed be about as old as the Solar System itself.

### The rubidium—strontium isochron

But anorthosites were not the only rocks to be formed in the early chemical differentiation of the Moon. And their strontium isotope ratios are in fact insufficient to yield a really accurate date for this event. The other differentiation products, however, contained relatively more rubidium, so in these rocks a precise correction must be made for the strontium-87 which has been produced by radioactive decay. The higher the rubidium–strontium ratio and the longer the time since the primordial differentiation event, the higher the strontium-87–strontium-86 ratio will be. So how did the Moon's rubidium and strontium segregate during its early chemical differentiation? And to what extent was this segregation modified by later events, such as the generation of the *mare* basalts? Well much of the strontium ended up in the anorthositic crust. This is because strontium readily replaces calcium in plagioclase feldspar. And, although most of the remaining rubidium was concentrated in the residual KREEP liquids, appreciable quantities must have been incorporated into the dense olivine-rich cumulates which were later remelted to produce the *mare* basalts.

In order to date this early differentiation event it is therefore necessary to find rocks which were derived, directly or indirectly, from these early differentiates, without subsequent rubidium–strontium fractionation. Their rubidium–strontium ratios and strontium isotope compositions should then define a straight line on what is known as the rubidium–strontium isochron diagram.

Consider the three hypothetical rock types shown in Figure 6.1, distinguishable from one another solely on the basis of their rubidium–strontium ratios. Immediately after the chemical differentiation event in which they are produced their strontium isotope ratios will, of course, still be identical. But as time passes their compositions will all evolve along lines with slopes of 45°. In other words, as their rubidium-87–strontium-86 ratios decrease by, say, 0.1, their strontium-87–strontium-86 ratios will increase by exactly the same amount. All of the rubidium-87 which disappears is converted into strontium-87. Now the higher the rubidium–strontium ratio to start with the more rapidly will this composition evolve. And it is therefore not difficult to appreciate that at any given time the compositions of the three samples will still define a straight line, the slope of which must reflect the time since the differentiation event. The longer the time is, the steeper this slope will be. Furthermore, the intersection point on the strontium isotope axis will still correspond to the initial ratio. In effect this point corresponds to the evolution path for a rock which contains no rubidium whatsoever. And the lunar anorthosites are in fact a good approximation to this case.

### Whole rock isochrons

So much then for the theory, what happens in practice? In other words, to what extent do lunar rocks define such a 'whole rock isochron'? Well the simplest approach here is to analyse lunar soils, which are effectively

Figure 6.1. *The rubidium–strontium isochron.* After chemical differentiation, the product rock types ($A$, $B$ and $C$) will all have identical strontium isotope compositions. As rubidium decays, however, compositions will evolve along lines with 45° slopes, but will always define a straight line (or isochron), the slope of which is directly related to the time since the differentiation event. This line will always intersect the strontium isotope axis at the initial strontium composition.

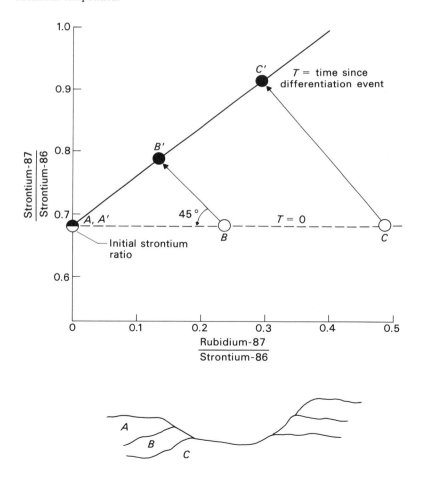

mixtures of rock types from widely different sources. Now when lunar soils are plotted on the rubidium–strontium evolution diagram (Figure 6.2) the points do indeed scatter about the 4600 million year reference line. And this line also defines that very primitive strontium isotope composition.

This then is certainly strong evidence for a very ancient Moon. But it does not mean of course that the soils themselves are necessarily that old. On the contrary, many lunar rocks are in fact much younger than 4000 million years and it is clearly impossible for soils to be older than the rocks from which they were derived. No, what is being dated here is the time when the Moon's rubidium and strontium were first fractionated, not when the products of this primary fractionation finally ended up at the Moon's surface.

Figure 6.2. *Lunar soil 'isochron'*. Compositions of lunar soils scatter around the 4600 million year reference isochron, indicating that the Moon must have differentiated very early in its history. The scatter indicates later differentiation events. Model ages are referred to an initial composition known as BABI, the best initial ratio for basaltic achondrite meteorites. An even lower ratio has been found in the Angra dos Reis meteorite (ADOR). (Courtesy G. J. Wasserburg, California Institute of Technology, Pasadena.)

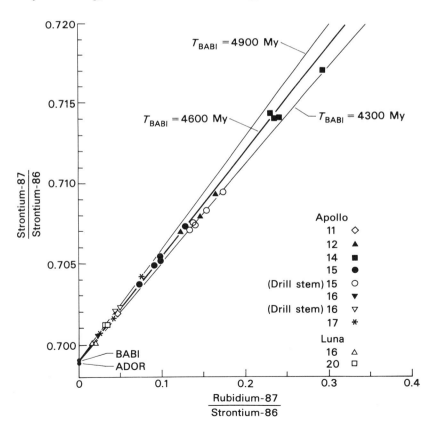

This event could well have been quite long drawn out. Indeed whole rock isochrons of highland rocks (rather than soils) indicate differentiation ages as young as 4300 million years. But as these isochrons are effectively controlled by the rubidium-rich KREEP rocks (which were presumably derived from residual liquids) this date could well just define the termination of this crystallisation sequence.

And this interpretation is certainly supported by initial strontium compositions. One celebrated Apollo 12 sample, known as rock 13, was such a complex breccia that a whole rock isochron could be constructed for different clasts within this single rock. And not only was this age as high as 4450 million years, the initial strontium ratio (0.701) was already highly evolved. In other words the primary differentiation event responsible for the various rock types within this breccia must have occurred even earlier still (Figure 6.3).

The question is: by how much? Well the anorthosites certainly do not contain enough rubidium to be of much use here. And the source zones for the KREEP and granite rocks may not have formed until much later anyway. So what is really required are samples of those early dense cumulates. And as luck would have it such rocks were discovered by the crew of Apollo 17.

As we shall hear in the next section, the Apollo 17 dunite and troctolite rocks finally removed any doubts which may still have existed concerning

Figure 6.3. *Rock 13*. This picture shows an internal surface of the small lunar breccia, 12013. This rock has a complex lithology, containing granite and potassium-rich KREEP. Although only 3900 million years old, the large clasts were not isotopically equilibrated at this time. Instead, they define an age of 4450 million years and an initial strontium ratio that, at 0.701, was already highly developed. (NASA, JSC–LRL.)

the Moon's great antiquity. For these two samples are both as ancient as the Solar System itself, 4550 and 4610 million years respectively. And their initial strontium isotope compositions are just as low as those in the anorthosites.

*Secondary differentiation events*

So much then for the direct sampling of the early dense cumulates. But what about the *mare* basalts which supposedly originated from them by partial melting when the Moon's interior gradually heated up? Well it so happens that most *mare* basalts also plot close to the 4600 million year isochron. And this is in marked contrast to terrestrial experience. It means, in fact, that there was little later fractionation of rubidium from strontium.

There are exceptions to this rule, however. The source magmas for the Apollo 11 basalts, for example, show clear evidence for chemical fractionation during, or just prior to, *mare* eruption. Such secondary events have the affect of scattering the chemical and isotopic compositions along lines of shallower slope (i.e. lower age) on either side of the primary isochron (Figure 6.4). And these secondary lines will intersect the strontium isotope axis at higher (that is more evolved) values. Depending on whether a particular rock type was enriched or depleted in rubidium at this time, it may appear older or younger than 4600 million years. The source magmas for the Apollo 11 rocks, in fact, were enriched in rubidium by a factor of 4 or 5 at this time. So in these cases the rocks have model ages (in other words, single stage evolution ages from the most primitive strontium isotope composition) which are more comparable with the individual rock ages than with the age of the Moon as a whole.

Rocks which have been depleted in rubidium, on the other hand, may have model ages in excess of 5000 million years. In other words, their strontium isotope compositions are much too high to be accounted for by the decay of the rubidium which these rocks (or soils) contain today. Volatilisation of soils can also result in this effect as rubidium is easily vaporised.

*Other radioactivities*

The study of rubidium and strontium in lunar rocks, then, shows us that the Moon suffered a major chemical differentiation event very early in its history, and certainly no later than 4300 million years ago. But rubidium is not the only long-lived radionuclide which has a metallic daughter. Another which has been investigated in a very similar manner is samarium-147, a rare earth isotope which decays to another one, neodymium-143. In this case there was apparently very little chemical fractionation between the two elements during basalt extrusion, even in Mare Tranquillitatis. The two elements were, however, quite clearly fractionated relative to one another during the Moon's early differentiation. And the date for this event can again be placed at about 4400 million years ago.

Finally, of course, there is uranium. And here too there is very strong evidence for a major lunar differentiation event prior to 4400 million years ago. But dating the Moon with uranium is intimately concerned with the dating of individual rocks. And this is the subject to which we must now turn. For having established that the Moon is just as old as the other planets it is now important to map out its subsequent history of bombardment and volcanism.

## 6.2   Dating individual rocks

There is clearly more to lunar dating than simply establishing a lower limit for the age of the Moon as a whole, important as this objective may be. Because only by measuring the ages of individual rocks can a complete lunar chronology possibly be established. Crater frequency plots can certainly reveal subtle age differences between the various lunar formations. But these relative ages must eventually be calibrated by some absolute measurements. And this means dating the rocks which are believed to be representative of those formations.

Figure 6.4. *Secondary differentiation.* If a second differentiation of rubidium and strontium occurs, the products of this second differentiation will define an isochron of lower slope and a more evolved initial strontium isotope composition. The slope of the isochron yields the time since the second differentiation event. Differentiation caused Apollo 11 basalts to have low BABI model ages. The different symbols simply represent the components of the system at different times.

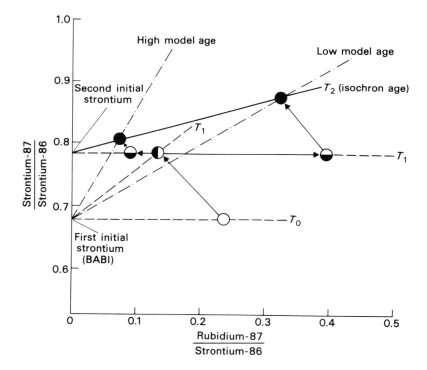

*Rubidium—strontium mineral isochrons*

A number of techniques have been employed for dating individual lunar rocks. Some of these had already been tried and tested before Apollo, while others were completely novel. And perhaps the most rewarding of all of these methods has again been that based on rubidium and strontium.

Now the approach here is very similar to the whole rock isochron technique just described. The only difference is that, instead of complementary lunar rock types, it is the various minerals within a particular rock which are analysed.

To appreciate the full significance of such internal (or mineral) isochrons, consider a *mare* lava just after eruption. Now, because lava is a liquid, its strontium will be isotopically homogenous at this time. In other words, the isotopic composition of the strontium initially incorporated into each mineral will be exactly the same.

But some lunar minerals are richer in rubidium than others. And their strontium isotopic compositions will therefore evolve more rapidly with time. So, by measuring the rubidium and strontium in the constituent minerals of an individual rock, it is possible to construct a mineral isochron for that rock, an isochron which reflects the time at which the parent liquid for that rock crystallised, rather than the time when this liquid first separated out deep inside the Moon.

What, then, do rubidium—strontium mineral isochrons for lunar rocks look like? Well the *mare* basalts and some crystalline highland rocks do indeed yield well-defined mineral isochrons, isochrons which provide us with precise rock ages in which we can have a high degree of confidence.

Figures 6.5 and 6.6 show internal isochrons for an Apollo 11 *mare* basalt and for the Apollo 17 dunite respectively. The dunite presented a particularly difficult experimental challenge because this rock type consists almost entirely of one mineral, olivine.

A comparable challenge was provided by the troctolite, from which it was absolutely necessary to obtain ultrapure mineral separates. In this case it was possible to obtain a good spread in rubidium—strontium ratio only by ensuring that the olivine separates were contaminated with less than 0.1% of plagioclase, the principal carrier of strontium.

But the results from other rock types were sometimes less clear cut. In the case of the lower grade highland breccias, for example, the various minerals sometimes do not define an isochron at all, within the analytical uncertainties. And this must be because strontium isotopes are more easily homogenised in some mineral assemblages than in others. Indeed, the reason why the dunite and troctolite yielded such ancient ages at all may well be just because olivine is such a high temperature phase. Its strontium may not have been affected by those later impact events which reset the rubidium clocks in less temperature resistant minerals. The matrices of the boulders from which these olivine-rich samples were taken as clasts are certainly very much younger than the clasts themselves.

A similar explanation can be put forward to account for the rubidium–strontium results obtained for rock 13. The internal mineral isochrons for individual clasts in this Apollo 12 breccia yielded ages of about 3900 million years, ages typical of rocks thought to be associated with the Imbrium event. And yet the various clasts which comprise this rock yield a whole rock isochron age as high as 4450 million years. Clearly, the impact event which finally welded this rock together was of insufficient intensity to homogenise the strontium isotopes between one clast and another.

### Samarium–neodymium mineral isochron

So much then for rubidium–strontium mineral isochrons. A similar approach with samarium has now borne fruit, the samarium–neodymium method being the most recent of all lunar dating techniques. And, because parent and daughter are both rare earth elements, they behave rather similarly during crystallisation. So their fractionation in lunar minerals was less significant than was that between rubidium and

Figure 6.5. *Mare basalt Rb–Sr mineral isochron.* Mineral separates from this Apollo 11 basalt define a 3710 million year isochron, taken to be the time of basalt eruption. Note that the initial strontium ratio is slightly more evolved than BABI, but that the total rock sample plots on the 4600 million year whole rock isochron. (Courtesy G. J. Wasserburg, California Institute of Technology, Pasadena.)

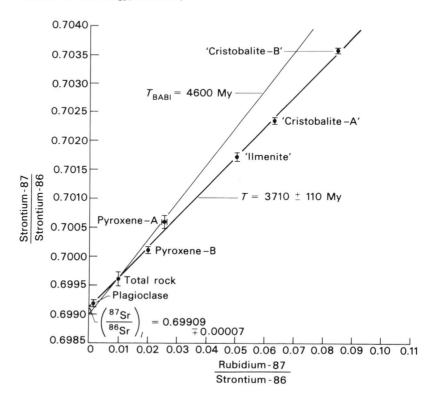

strontium. Nevertheless, the relatively high rare earth contents of lunar rocks means that it has now been possible to obtain some really accurate samarium–neodymium crystallisation ages (Figure 6.7).

Figure 6.6. *Lunar dunite Rb–Sr mineral isochron.* This truly ancient Apollo 17 rock yields a 4550 million year isochron. The inset shows the fractional deviation of each point (and its associated analytical uncertainty) from the isochron. (From Papanastassiou, D. A. & Wasserburg, G. J. (1975). *Proceedings of the Sixth Lunar Science Conference*, pp. 1467–89. New York: Pergamon.)

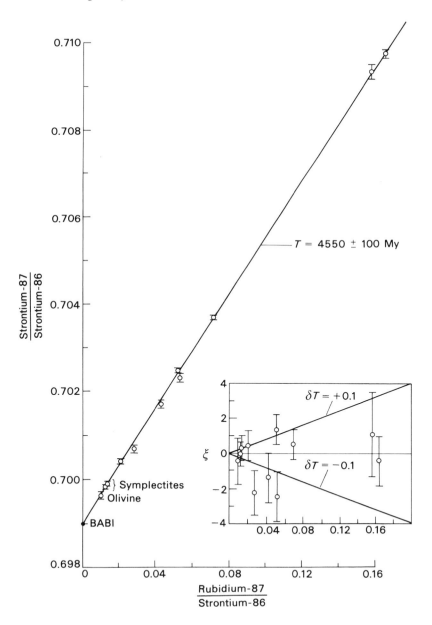

But why should it be necessary to date a rock by more than one method? Surely it is the age itself that matters and not the method by which this age was obtained. Well, to some extent this is certainly true. But the moments at which the various radioactive clocks started ticking away in a particular rock may not have been quite the same. The samarium–neodymium age of the Apollo 17 troctolite, for example, is 4260 million rather than 4610 million years. The potassium–argon and uranium–lead methods also yield 4300 million year ages for this rock.

*Potassium–argon*

The potassium–argon method has the advantage, and the disadvantage, of being based on a gaseous daughter isotope. The advantage is that in general it can be assumed that no correction need be made for initial argon in the rock. In other words, all preexisting argon is assumed to have been lost when the rock was last melted. This must certainly have been true for the *mare* basalts. But it may not have been quite so for the more deep-seated rocks, such as the troctolite and the dunite.

The principal disadvantage of having argon-40 as a daughter isotope, however, is that it could well have diffused out of a lunar rock long after its final crystallisation, particularly if it was subjected to impact shock or thermal cycling under a hot sun. So the main problem inherent in potassium–argon dating is finding some way to detect and account for possible argon loss.

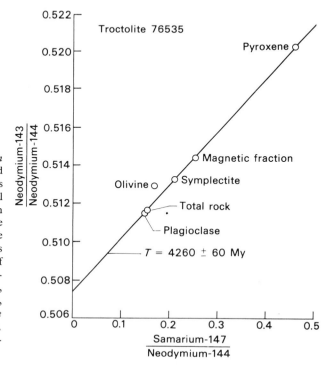

Figure 6.7. *Samarium–Neodymium mineral isochron.* Samarium and neodymium compositions in this lunar troctolite define a mineral isochron age of 4260 ± 60 million years. It is noteworthy that the olivine point does not plot on the isochron, indicating that olivine is resistant to thermal equilibration of rare earth elements as well as strontium. (After Lugmair, G. W., Marti, K., Kurtz, J. P. & Sheinin, N. B. (1976). *Proceedings of the Seventh Lunar Science Conference,* pp. 2009–33. New York: Pergamon.)

Another problem is one of contamination, not only from the Earth's atmosphere, which contains 1% of argon, but also from the Moon's, which is constantly being supplemented by the argon which is even today being vented from the Moon's interior.

That such corrections are indeed required was revealed as soon as the first Apollo 11 rocks were analysed. Their potassium–argon ages were found to be invariably younger than their rubidium–strontium ages. So some of their radiogenic argon had clearly been lost.

One way around this problem of argon loss is to analyse only those minerals in lunar rocks which are known to be argon retentive. Indeed, this is the normal approach in the potassium–argon dating of terrestrial rocks. But apart from requiring laborious mineral separation techniques (during which more argon could easily be lost), there is still no way to be sure that even here some argon loss has not already occurred. And the most successful approach has, in fact, been to analyse the potassium and argon in the most retentive minerals indirectly, by first bombarding the sample with fast neutrons.

### The argon-39–argon-40 method

What happens then is that a small fraction of the potassium in each mineral is converted into a second argon isotope, namely argon-39. In this way an age can then be obtained by making a single measurement, namely that of the ratio of argon-39 to argon-40. The efficiency of the neutron activation process in converting the potassium-39 to argon-39 is monitored by simultaneously irradiating a rock of known age and potassium content.

Now if the argon-39–argon-40 method simply consisted in measuring the total amounts of these two isotopes in an irradiated lunar rock, then it would have no advantage whatever over the normal approach. But argon can only be extracted from a rock by melting it. And, because this argon is being used as a measure of both parent and daughter isotope, the argon is now best extracted in several stages.

Why should this be so? Well by measuring the ratio of the two isotopes in the argon being extracted at steadily increasing temperatures it is possible, in effect, to 'date' gradually more and more retentive lunar minerals. The argon which is released first must come from 'leaky' minerals, in other words those which may have already lost some of their argon-40 while on the Moon.

As the extraction temperature rises, however, the 'age' of the argon being released gradually increases. And if this age eventually reaches a steady maximum then this maximum value must represent the true age of the rock. This is because none of the various minerals responsible for this plateau release can have suffered much natural loss of radiogenic argon (Figure 6.8).

How successfully, then, has this approach been in practice and what has now been learnt about natural argon loss on the Moon? Well the method has certainly been very successful and it was, in fact, the first to reveal that the low potassium Apollo 11 *mare* basalts are some 200–300 million years older than the high potassium variety (Figure 6.9), a conclusion only much later confirmed by the rubidium–strontium laboratories.

In some cases there is indeed strong evidence for argon loss. But, even when this loss amounts to as much as 30% of the total gas present, it is sometimes still possible to obtain precise ages from the high temperature argon release.

Some rocks, however, are much more retentive than others. One Apollo 17 anorthositic gabbro, for example, yielded a plateau age which was identical to the total rock age (Figure 6.8). So, despite its great age (at 4220 million years it was one of the oldest returned) this particular rock had clearly suffered negligible natural losses of radiogenic argon.

Like the rubidium–strontium method, however, the argon-39–argon-40 method is not always as successful as one might hope. In some cases, for example, argon loss has been so extreme that a meaningful age cannot be obtained. And, in the case of the finest-grained KREEP-rich rocks, neutron irradiation effects overcomplicate the age release patterns. Nevertheless, the method has certainly done much to improve our understanding of lunar chronology.

### Uranium–lead systematics

Then of course, there are the two long-lived isotopes of uranium, each of which decays in several stages to a characteristic isotope of lead. Uranium-238 has the longer half-life (Table 6.1), decaying to lead-206, whereas uranium-235, the power source of fission reactors and atomic bombs, ends up as lead-207.

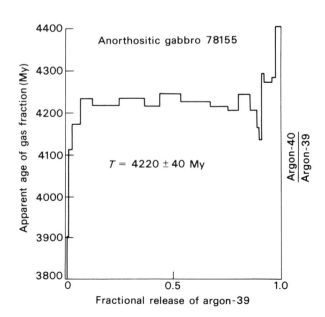

Figure 6.8. *Argon-39–argon-40 dating.* As the argon from this neutron irradiated anorthositic gabbro was extracted, the apparent age was remarkably constant, at 4220 million years, making this rock one of the oldest on the Moon. The plateau ages are identical to one another within the experimental uncertainties of the method.

So what do the uranium and lead in lunar rocks tell us about early lunar history? And how does this picture compare with our view based on other dating techniques? Well, first of all, the uranium–lead method is potentially the most powerful technique for investigating the Moon's first half billion years. This is simply because there are two independent decay sequences.

Having two uranium isotopes means, in fact, that it is even possible to date rocks just by measuring their lead isotopes. We could ignore their uranium contents completely. But this approach can easily lead to an oversimplification of the true picture. There could, for example, be too little uranium in a rock to account for its overall content of lead. And it turns out, in fact, that the Moon's early history must indeed have been rather complex. So the more information that can be extracted about the distribution of its uranium and lead the better, even if some of this information does in the end turn out to be redundant.

*The concordia diagram*

The evolution of uranium and lead on the Moon can perhaps best be appreciated on what is known as the concordia diagram (Figure 6.10). In this plot a curved line, known as concordia, defines the gradual build-up of the two radiogenic lead isotopes in an initially lead-free rock. It can perhaps best be viewed as the path which the origin of the plot has traced out through geological time, as each lead isotope slowly accumulated with the radioactive decay of its respective parent. The point marked 3000, for example, marks where the origin of the plot was 3000 million years ago.

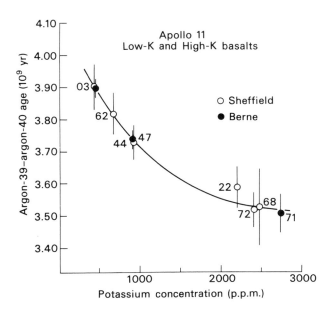

Figure 6.9. *Ages of Apollo 11 basalts.* There is a good correlation among the Apollo 11 *mare* basalts between age and potassium content. Note that the oldest rock, 10003, is even older than the 3880 million year old Apollo 14 breccias thought to date the Imbrium impact. (From Stettler, A., Eberhardt, P., Geiss, J., Grögler, N. & Maurer, P. (1974). *Proceedings of the Fifth Lunar Science Conference,* pp. 1857–77. New York: Pergamon.)

The reason why concordia is a curve rather than a straight line is simply that the two uranium isotopes have very different half-lives (Table 6.1). The amount of uranium-238 in the Moon has only halved since the planet first formed, while its content of uranium-235 has decreased a hundredfold. As a result, the lead being produced 3000 million years ago was very much richer in lead-207 (relative to lead-206) than it is today. Consequently the concordia line was much flatter in those early years. Nowadays most of the radiogenic lead being produced is lead-206.

### Concordant ages

Where then do lunar rocks plot on this concordia diagram, and what do such plots reveal about early lunar history? Well it turns out that a few lunar rocks do indeed plot on, or very close to, the concordia curve. In other words, the time required for their lead-207 to have accumulated from the available uranium-235 is exactly the same as that necessary to account for the production of their lead-206 from their uranium-238.

And where these ages are concordant at about 3900 million years, as is the case for an Apollo 17 anorthositic gabbro, there is concordancy not only with rubidium–strontium and potassium–argon ages, but also with the ages based on a third heavy element decay series, namely thorium–lead. Thorium-232 decays to yet another radiogenic lead isotope, lead-208. As the thorium in a lunar rock might well have been chemically fractionated with respect to uranium, however, thorium–lead dating has been rather neglected.

But most lunar rocks plot far away from concordia, or else yield concordant ages which, at 4420 million years or so, are greatly in excess of their ages based on other dating methods. So how can we possibly account for these high (or often grossly discordant) uranium–lead ages? Well the answer lies, of course, with our basic assumption that lunar rocks were originally lead-free. Just as initial strontium is so important in rubidium–strontium dating, so is initial lead crucial to our understanding of the Moon's uranium and lead budget. Only by considering initial lead can we possibly make sense of these discordant 'ages'.

Figure 6.10. *The concordia curve.* Rocks with identical uranium-235–lead-207 and uranium-238–lead-206 ages should plot on a curve known as concordia. The ages marked are in millions of years, and effectively mark where the origin of concordia was at the time. The changing slope of the curve is due to the different half-lives of the two uranium isotopes. 3000 million years ago, for example, the rate of lead-207 production was much higher than it is today, so the curve was less steep, as shown in the dashed box.

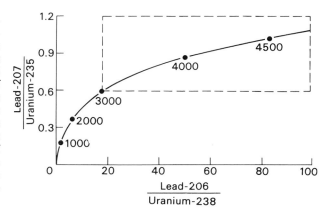

*Initial lead – the theory*

In order to make the concept of initial lead a little clearer it is helpful to redraw the concordia diagram using slightly different axes (Figure 6.11). Rocks having concordant uranium–lead ages will once again fall on the concordia curve at the appropriate points. And it can be seen straight away from this plot that, just as we would expect, the older the rock, the higher its lead-207–lead-206 ratio will be.

Now what happens if such a rock was not initially lead-free? What if it incorporated (or already contained) some radiogenic lead when it first formed? Well it will then contain a straight mixture of (*a*) this initial lead and (*b*) the lead which has accumulated since crystallisation as its own uranium decayed. And any such simple two-component mixture should plot somewhere on a line connecting (*a*) the age of the rock on the concordia curve and (*b*) the initial lead composition on the vertical axis

Figure 6.11. *The highland uranium–lead isochron.* Lunar highland rocks tend not to fall on the concordia curve, because they contain initial lead. The rocks which are most discordant are those with low uranium contents (parts per billion figure in parenthesis). Precise ages cannot be calculated because the isochron is almost tangential to the concordia curve. The lower intersection with the curve is, however, consistent with the majority of highland ages determined using other methods. (Courtesy G. J. Wasserburg, California Institute of Technology, Pasadena.)

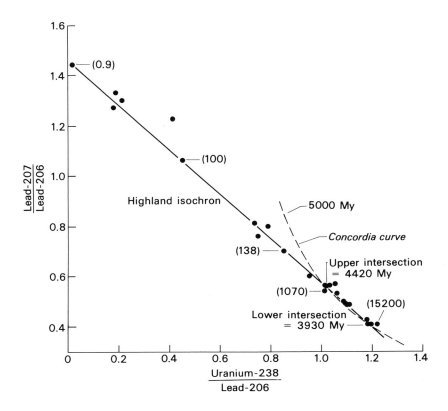

(Figure 6.11). Exactly where it will plot depends on (*a*) its uranium content and (*b*) on how much initial lead is present. A rock in which there is absolutely no uranium can clearly only contain initial lead. A rock with a high uranium content (relative to its initial lead), however, will plot close to, or on, the concordia curve.

Lunar rocks may of course contain primordial lead in addition to that which has been produced inside the Moon by radioactive decay. By primordial lead we mean that incorporated into the Moon when it first formed.

Fortunately, it is quite easy to correct for primordial lead, because iron meteorites contain practically no uranium whatsoever. But, like lunar rocks, they do contain small amounts of the only isotope of lead which is *not* radiogenic, namely lead-204. So the primordial lead correction in a lunar rock can be made on the basis of (*a*) its content of lead-204 and (*b*) the ratio of lead-207 (or lead-206) to lead-204 in iron meteorites. And it is these corrected ratios which are always plotted on concordia diagrams. The initial lead in a rock really means its initial radiogenic lead.

### The highland rock 'isochron'

So what happens in practice? Well, it turns out that the initial radiogenic lead in most highland rocks has much the same isotopic composition (Figure 6.11). Most of these samples (including mineral separates from individual rocks) plot on, or very close to, a single mixing line, a line which defines an initial lead-207–lead-206 ratio of 1.45. Its other end (in other words its lower intersection with concordia) is not well defined in terms of age because this particular mixing line is almost a tangent to the concordia curve. So the 200 million years age spread obtained for the bulk of highland samples using other methods (i.e. 3850–4050 million years) is quite insufficient to displace these samples much from this mixing (or concordia) line. The goodness of the fit, however, certainly confirms that most lunar highland rocks are indeed about 3950 million years old.

The question to be asked now is: what significance, if any, should be attached to the upper intersection with the concordia curve at 4420 million years? Well one possible explanation is that the lunar highlands are really this old and that the impact events which reset the rubidium–strontium and potassium–argon clocks in highland rocks redistributed the lead which had accumulated throughout the intervening period. Partial lead loss, for example, would simply have displaced the rock some way down the discordia line towards its lower intersection point.

But there is a marked anticorrelation between discordancy and uranium content. This suggests that there is no direct relationship between uranium content and initial lead. The lower the uranium content of a highland rock the more discordant it tends to be. And this can best be explained if lead volatilisation and subsequent redeposition was very effective prior to 3950 million years ago. Such a remobilisation process is strongly supported by the fact that the initial lead is sometimes easily leached out of highland rocks by acid and must therefore be largely a surface deposit. Furthermore, rocks which plot to the left of the concordia curve must have suffered a net gain rather than just partial loss of radiogenic lead. And this is not surprising because, although relatively volatile, lead, unlike argon, will clearly never be lost from the Moon completely. It is interesting to note that the lead in a hypothetical piece of pure uranium left to decay for 470 million years 4420 million years ago would have had just the same lead-207 to lead-206 ratio (namely 1.45) as that defined by the intercept on the lead isotope axis. This composition also happens to be the average composition of the lead in the uranium-free Genesis Rock.

This picture of lead remobilisation is clearly an oversimplification of reality. For one thing the initial lead composition in Genesis Rock is quite variable, from 1.42 to 1.63. Similarly, other highland rocks show evidence for much more recent lead remobilisation, possibly associated with the excavation of nearby impact craters, such as Aristillus and Autolycus at the Apollo 15 site and North Ray at the Apollo 16 site. But more detailed studies are still required to clarify these specific problems.

*Implications for early lunar history*
In summary, the highland 'isochron' implies that the lunar crust is certainly as old as 4420 million years and that there must have been extensive remobilisation of lead throughout the next 500 million years. But what about the radiogenic lead which was produced during the first 200 million years of lunar history? The dunite and troctolite rubidium–strontium data suggest that the Moon must already have had a well-developed crust by this time. There are a number of intriguing possibilities here, one of which is that the early radiogenic lead was removed from the crust by dissolution in 'pools' of dense iron sulphide which subsequently sank. What is more likely, however, is that the 4420 million years age is really just an average age for the lunar crust. The formation of this crust could clearly not have been instantaneous and could quite well have continued for several hundred million years. It should be noted, however, that the uranium–lead results do not actually exclude the possibility that the Moon may be significantly younger than the rest of the Solar System.

*The* mare *basalts*
One piece of evidence which does lend strong support to the 4420 million age being that of a well-defined event comes from the analyses of *mare* basalts. As already known from other dating methods, the lunar *maria* formed much later than the highlands. But, like the highland samples, the *mare* basalts also contain initial radiogenic lead, and generally turn out to be discordant (Figure 6.12). Here, however, the initial lead, like the initial strontium, must have been inherited from the *mare* basalt source regions deep down inside the Moon. The excess lead is unlikely to have been added during lava extrusion, although there is the remote possibility that it could have been picked up as the magmas penetrated

through the highland crust. The inheritance of deep-seated initial lead is however supported by the fact that the *mare* basalts never plot to the left of the concordia curve.

What is most remarkable here, however, is that mineral separates from individual *mare* basalts define isochrons which also have an upper intersection point with concordia at 4420 million years (Figure 6.12). So the source zones for the *mare* basalts must clearly have formed at around this time. And their initial lead is just that which accumulated in these source zones prior to lava extrusion. It need hardly be added, then, that the lead-207–lead-206 ratio of this initial lead is lower than in that highland rocks. This is because, on average, it was produced later.

The fact that a couple of Apollo 17 *mare* basalts yield concordant ages at 4420 ages means that these basalts must have been generated without appreciable uranium–lead fractionation, a most remarkable finding, in view of the considerable geochemical differences which exist between these two elements.

Figure 6.12. *Mare basalt uranium–lead internal isochron.* The uranium–lead data for Apollo 15 *mare* basalt 15555 (Great Scott Rock) confirm an age of 3300 million years for the sample. The upper intersection with concordia is identical to that of the highland rock isochron, indicating that *mare* basalt lavas must have inherited initial lead from a source region that became differentiated from the highlands at this time. (Courtesy G. J. Wasserburg, California Institute of Technology, Pasadena.)

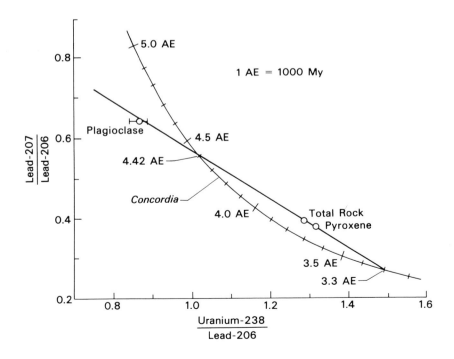

*Fission track dating*

Dating rocks using uranium is not just restricted to analysing lead isotopes. This is because, in addition to its radioactive decay, uranium also undergoes spontaneous fission, albeit at a much slower rate. And it turns out that this fission process can provide some really valuable complementary information about the Moon's early history.

So what is fission, what evidence might we expect to find for it on the Moon and how can this evidence be used as a dating tool? Well, unlike radioactive decay, which consists of the ejection from the nucleus of only small subatomic fragments (in other words, electrons and alpha particles), fission refers to the splitting of a nucleus into two smaller nuclei of comparable size, together with the ejection of neutrons. This process releases abundant nuclear energy. And it is of course this energy which is harnessed in a fission reactor and expended in an uncontrolled fashion in the atomic bomb. In these cases, however, fission is greatly accelerated by the presence of the neutron byproducts. These are capable of inducing uranium-235 to fission at a much greater rate than it otherwise would, a process known as a chain reaction.

When the daughter nuclei are formed during fission they carry some of the energy released away with them. And, as with those heavy cosmic rays described in Chapter 4, these fast moving ions will leave behind them dislocations, which can be revealed by etching as 'tracks' and counted. The older the rocks and the more uranium it contains, the more of these 'fission tracks' we would expect to find. But how can we distinguish them from cosmic ray tracks? And just how long might we expect fission tracks to survive?

The best approach is clearly to examine those minerals which are richest in uranium. Whitlockite, apatite and zircon are generally found to be most suitable in this respect. The cosmic ray track background in these minerals is then simply taken to be that in an adjacent uranium-free mineral.

But this is not the only complication to be allowed for. Because, just as neutrons can induce fission in a nuclear reactor, so can they also on the Moon. Those secondary neutrons produced by cosmic rays can conceivably cause uranium, thorium and even bismuth to fission. But this induced fission contribution is thought to be rather minor, particularly for rocks with low exposure ages.

What then have we learnt from fission track dating and what problems still remain? Well in some cases fission track ages are certainly very similar to those obtained by other methods. The Apollo 17 orange glass, for example, has a fission track age of 3700 million years, quite compatible with its potassium–argon age.

But with older rocks there seems to be an excess of fission tracks. In some cases there are far too many to be accounted for by the uranium-238 present, even if the rock was 4600 million years old. It seems then that there must have been some other fissioning isotope in these rocks. And that can only mean the now long-extinct radionuclide, plutonium-244.

*Plutonium fission*

Plutonium-244, the longest lived of all the plutonium isotopes, has a half-life of only 81.8 million years. And this means that practically all of the Moon's plutonium must have decayed away by the time the first *mare* lavas erupted.

But, in rocks 4000 million years old or older, the fission tracks due to plutonium-244 should be by no means negligible. And the older the rock is, the higher we would expect the proportion of these excess fission tracks to be.

But exactly how many fission tracks we will find in a rock of a given age will depend on two crucial factors. Firstly, what was the ratio of plutonium to uranium in the Moon when it first formed? And secondly, to what extent does plutonium resemble uranium geochemically? In other words, during the crystallisation of a particular rock, did plutonium and uranium become incorporated in exactly the same proportions in all minerals? Not surprisingly, the geochemical behaviour of plutonium is not well known, it being one of the most toxic substances known to man. Excess fission tracks on the Moon therefore provide us with a useful opportunity to investigate this geochemical behaviour.

When rocks as old as 4000 million years are studied, about a third of the fission tracks present must be due to plutonium fission. Track densities in the 3950 million years old Apollo 14 breccia known as Big Bertha, in fact, imply that the Moon's original plutonium–uranium ratio must have been about 0.02, a value quite consistent with meteorite studies.

This assumes, of course, that the Moon's plutonium did not fractionate with respect to uranium. And there is now some evidence to support this. In Big Bertha, for example, the excess factor in zircon is exactly the same as that in whitlockite. So it looks as though plutonium is geochemically more similar to uranium than it is to, say, thorium. For the thorium–uranium ratio in whitlockite is much higher than it is in zircon.

But this cannot be the whole story. Because excess fission tracks are unexpectedly abundant in fragments of crystalline Apollo 15 KREEP basalt, several of which have been dated by other methods at 3900 million years. So it could be that plutonium was greatly enriched in the Moon's residual KREEP liquids, even more so in fact than uranium was.

Another complication is that of track fading. A fission track age clearly refers to the time since the temperature of the rock last fell below the annealing temperature for fission tracks. This is the temperature at which the dislocations in the mineral under study can no longer be made visible by etching techniques. And it turns out that, for some rocks and minerals, fission track ages are markedly lower than, say, their potassium–argon ages. So we would generally expect the fission track clock to have been the last one to start. Consequently fission tracks are potentially useful as geothermometers.

But their most important application obviously concerns early lunar history. Because dating with plutonium, with its 82 million year half-life, is extremely sensitive to small time differences during the first 600 million years. And it is during this period of course that the age uncertainties using other dating methods tend to be largest.

But excess fission tracks are not the only consequences of the former existence of plutonium on the Moon. There are also the chemical remnants from fission. And it is in the rare gas laboratories that further crucial information about plutonium has now been obtained.

*Fission xenon*

Although krypton and xenon are both produced during uranium fission, it is xenon, with the larger number of stable isotopes, which has proved to be more profitable in terms of new information gained. And one of the reasons for this is that, although there are several other potential sources for xenon in lunar rocks (e.g. primordial, solar wind, cosmic ray spallation and neutron capture, as described in Chapter 4), fission xenon has a very distinctive isotopic composition. And, because each xenon component tends to be concentrated in different mineral sites within the rock, the various components can in certain cases be efficiently separated by stepwise heating (Figure 6.13). They can therefore be characterised isotopically. In certain rocks, for example, the fission xenon is released at temperatures of only a few hundred degrees. Spallation xenon, on the other hand, may not be expelled until the rock is near its melting point.

The importance of being able to characterise this fission xenon component is not only concerned with determining its magnitude, and hence the fission gas retention age of the rock. Much more important is to

Figure 6.13. *Evidence for fission xenon.* If the xenon in this breccia was either trapped (solar wind) or spallation, all gas extracts would plot on a single mixing line. The displacement of some points to the right indicates the presence of excess xenon-136, which can only have come from fission. Temperatures are given in °C. (After Drozd, R., Hohenberg, C. & Morgan, C. (1975). *Proceedings of the Sixth Lunar Science Conference*, pp. 1857–77. New York: Pergamon.)

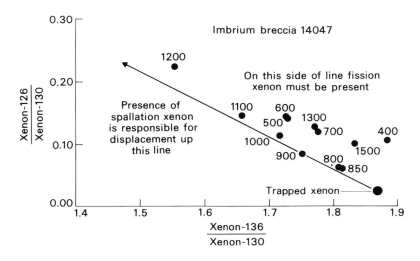

determine the source of the gas. And in the case of the Apollo 14 breccias the fission xenon is in fact intermediate in isotopic composition between that expected from uranium and that determined experimentally from plutonium-244. In one nail-biting experiment some plutonium-244 was isolated from a nuclear reactor and allowed to decay for two years in a sealed vessel. After this period of time there was just enough xenon generated by its spontaneous fission for the isotopic composition of 'plutonogenic' xenon to be determined.

### 'Parentless' xenon

The amounts and isotopic composition of fission xenon in lunar rocks more than 3900 million years old are certainly consistent with the primordial plutonium–uranium ratios postulated on the basis of the fission track densities in these rocks. But the fission xenon story does not end here. Because in some rocks the fission xenon is much more abundant than can possibly be accounted for by *in situ* fission, even after invoking the former existence of plutonium-244. In one poorly consolidated Apollo 14 breccia, for example, there was nearly eighty times more fission xenon than would have been expected just from uranium-238. And although the composition of this excess component was indeed that expected from plutonium-244 and the total quantity could just have accumulated in 4600 million years, other dating methods show this particular specimen to be 700 million years too young. So this extra excess fission xenon must originate from elsewhere. It is usually described as parentless.

So where did this parentless xenon come from? And what, if anything, can it tell us about early lunar history? Well, for one thing, like solar wind xenon, it seems to be concentrated on grain surfaces. The greatest abundances occur in the finest grains. And this certainly suggests an external source, possibly the lunar atmosphere.

Just as lead vapour once vapourised must become readsorbed, so also must xenon. A gas as heavy as xenon can never be lost from the Moon completely.

But the most intriguing aspect of this parentless xenon problem is that the component is not just fission xenon. It also contains xenon-129, the daughter isotope of that 16 million year half-life extinct radioisotope, iodine-129. And what is most remarkable is that the proportion of this component to plutonogenic xenon in different rocks varies. In some rocks there is proportionally more xenon-129 than in others.

And what does this mean? Well, first of all it means that the Moon really must be about 4600 million years old. Because if it was younger than this, any xenon-129 that the Moon succeeded in trapping when it first formed would have been thoroughly mixed up with any other xenon present at that time. The only way in which it can have been kept separate from plutogenic xenon, was in an already solid moon. But exactly where is still a mystery. Somehow the early Moon must have harboured gaseous xenon reservoirs of rather different ages, from which gas was released and somehow reincorporated into the regolith.

*Estimate of age uncertainty*

This then all but concludes this brief discussion of lunar dating methods. The aim throughout has been to show that dating lunar rocks is by no means a straightforward analytical operation, but does instead require a deep insight into physical processes which have operated on the Moon in the remote past.

Only by discussing each dating method in some detail can we be in a good position to compare ages obtained by the various methods. For how else could we possibly account for the fact that some *mare* basalts yield apparently well-defined uranium ages which are some 600 million years older than their rubidium–strontium ages?

But, before moving on to discuss lunar history in general, there is one point that should be clarified. None of the ages quoted so far has been qualified by any analytical (or interpretative) uncertainty. But such uncertainties are very important in dating methods. Because how else can we, for example, decide whether one rock (say the dunite at 4550 million years old) is really just as old as another (such as the troctolite, nominally 60 million years older). In fact the uncertainties in these two particular dates are sufficiently large to make them indistinguishable from one another. An uncertainty of 100 million years in 4500 million years is really not very significant, however large it may seem in absolute terms.

In general, lunar ages can be calculated with an accuracy of better than 50 million years. But this may depend on factors other than just analytical ones. It may be possible to measure argon-40–argon-39 ratios in a rock to an accuracy better than this, for example, but argon losses from the rock may make a really accurate age difficult to obtain.

In some cases, however, precise ages can be obtained using several different methods and comparisons between these ages help us to calculate more precise half-lives for the long-lived radionuclides, some of which are still rather uncertain. This can be seen as one of the very useful spinoffs from lunar research.

## 6.3    The Moon's origin

The preceding sections of this chapter have detailed the methods by which the ages of the Moon and its constituent rocks have been established. It is now appropriate to discuss these results in a somewhat broader context. In particular, where and how did the Moon originate as a planet? And to what extent is its subsequent history now fully understood?

The first question was, of course, a popular one to ask before the days of Apollo. Indeed it was frequently cited as one of Apollo's principal scientific objectives. But being able to formulate a question does not necessarily make it any easier to answer. Indeed, very little is known about the origin of the Earth, despite the fact that geologists have been studying it for hundreds of years.

Table 6.2. *Some of the geochemical and geophysical constraints which must be accommodated into models for the formation of the Moon and its subsequent evolution*

| Constraint | Implication |
|---|---|
| Geophysical constraints | |
|    Moment of inertia | Lack of a large iron core |
|    Density | Consistent with pyroxenite |
|    Magnetic permeability | Maximum free iron = 4.5% |
| | Maximum total iron = 13.5% |
|    Secular acceleration | Moon once closer to Earth |
|    Late bombardment | Late capture? |
|    Extinct radionuclides and 4600 million year rocks | Moon as old as Earth and meteorites |
| Geochemical constraints | |
|    Depletion of early crust in siderophiles (e.g. Ni) | Removal of iron metal prior to accretion |
|    Depletion in volatiles (e.g. K) | Formation at high temperature |
|    Enrichment in refractories (e.g. rare earths) | Condensation of protolunar cloud at high temperature |
|    Constancy of elemental ratios (e.g. K/La) | Homogeneous accretion |
|    Highland rocks and *mare* basalts complementary | Moon melted down to a depth of 200 miles after formation |
|    Essentially solar isotope ratios | Formation within the Solar System |

Some very firm constraints have, however, now been established on the Moon's possible origins (Table 6.2). And the questions now being asked tend to be rather more precise than they used to be. So what are these constraints, and just how well can they now be accommodated by the three principal contending theories: fission; formation in Earth orbit; or formation elsewhere in the Solar System followed by capture by the Earth?

*Geophysical constraints*

The first crucial factor is of course the great antiquity of the Moon as a planet. It is difficult to avoid the conclusion that there has been a chemically differentiated Moon for at least 4400 million years. And it is also highly probable that, by that time, the Moon had already been in existence as a discrete planetary body for some 200 million years. All the evidence, then, points to a Moon that is just about as old as the oldest meteorites.

Another major geophysical constraint is provided by the Moon's mass and moments of inertia. These imply that the Moon approximates to a uniform sphere (with a density equal to that of the Earth's mantle) and preclude the existence of a dense metallic iron core of any appreciable size. So theories for lunar origin must be capable of explaining why the Moon contains relatively less iron than the Earth. The Moon's mass is also very large compared with the Earth. No other planet has such a relatively large satellite.

Finally, of course, there is the question of the Moon's orbit. Unlike some other satellites in the Solar System (such as Neptune's Triton), the Moon revolves about the Earth in the same direction as the Earth revolves around the sun. Triton's retrograde orbit is sometimes taken to indicate that it must have been captured by Neptune. Perhaps significantly, Triton is also large when compared to its parent planet. In this respect it is second only to Earth's moon.

But it is not just the direction of the Moon's orbital motion that is important here. Other elements of its orbit are known to have evolved with time, largely because of tidal forces. And the most important of these effects is the Moon's overall recession, a recession which currently amounts to nearly 2 inches per year. If tidal friction has always been as effective as it is currently, then the Moon and Earth must have been in close proximity at some time during the last 2000 million years. There is clearly no basis for such an assumption, but it nevertheless seems inescapable that the Moon was once very much closer to the Earth than it is today.

### Geochemical constraints

But the Moon's low iron content is not the only well-established geochemical constraint on its origin. The constancy of a number of other elemental ratios in all lunar rocks provides additional information about the Moon's overall chemical composition and, by inference, its origin. This is because the extent to which this global chemistry differs from that of the chondritic meteorites is generally taken to be a direct reflection of the physical and chemical conditions which prevailed during the Moon's formation. Temperature, pressure and the availability of free oxygen, for example, would all have determined the rate at which the various elements accumulated. They would have done this by controlling which mineral assemblages could form at any particular stage of the condensation sequence.

Perhaps the most important constraints here are the Moon's relative deficiencies in siderophile and chalcophile elements (such as platinum and lead) and those which are easily volatilised (such as nitrogen). There is also a slight overall enrichment in refractory elements, including uranium and the rare earths. So once again there are significant differences here between the Moon and the Earth, differences which would seem to preclude a common origin for both planets. It should be pointed out, however, that a number of investigators believe that the primordial Moon was not homogeneous. It is proposed that the Moon's apparent enrichment in refractories could represent just a late-stage addition of refractory material. It seems odd, however, that the refractories should have been accreted last rather than first.

Isotopic studies, however, make it quite clear that the Moon must certainly have originated relatively locally. There is absolutely no question of it having been captured from outside the Solar System.

So what are the various hypotheses for lunar origin and which ones now stand up best against the constraints imposed upon them?

### Fission

Unquestionably the best-known origin theory is that the Moon was somehow pulled out of the Earth's crust, leaving behind a deep scar which is still clearly recognisable as the Pacific Ocean. An extension of this bold idea is that the fission process also involved Mars. In this case the Moon would have been just a 'droplet' left behind as the two planets separated from one another.

The attractiveness of this fission hypothesis has recently been tarnished to some extent by the verification of continental drift. The ceaseless movement of the continents implies that the circular outline of the Pacific Ocean is undoubtedly an ephemeral feature on a geological time scale. Its form can have no longer-term significance.

But quite apart from this particular shortcoming of the fission hypothesis, there are still a number of other serious problems with it. How, for example, did the primitive Earth ever gain sufficient angular momentum for crustal material to be thrown off from its equator? If the Earth had originally rotated sufficiently rapidly for this to occur, then how did this extra matter ever accumulate in the first place?

An intriguing possibility here is that fission was a direct consequence of the formation of the Earth's core. The sinking of dense iron metal would inevitably have led to an increase in the Earth's rotation rate. This is exactly the same principle as that by which a skater can rotate more rapidly by bringing the arms in close to the body.

Another early proposal was that strong solar tides could have induced resonant oscillations which might eventually have resulted in droplets being thrown off. But such an idea totally ignores the question of the inevitable damping of such oscillations by internal friction. So the tidal disruption theory can no longer be considered tenable.

The fission hypothesis as a whole is now clearly the least-popular theory for lunar origin, for geochemical as well as dynamical reasons. Because, while the theory certainly does have a certain aesthetic attraction, and this is admittedly consistent with the Moon having been much closer to the Earth in the distant past, the geochemical differences between the Earth and the Moon are just too great. Core separation prior to fission could perhaps account for its relative deficiency in iron and siderophile elements, but the Moon is chemically very unlike the Earth's crust and is clearly composed of more refractory elements than the Earth as a whole. It must therefore have formed in a rather higher temperature environment.

### Lunar accretion – hot or cold, local or remote?

If the Moon was not plucked out of the Earth, Velikovsky-style, then it must have formed more or less independently. Either it came into being as a result of the steady accumulation of particles already in Earth orbit, or else it formed as a separate planet and was later captured by the Earth.

In the first case one might expect there to be at least some geochemical similarities between the Earth and the Moon. Formation as a separate planet, however, could have occurred in a much hotter environment, as indeed is suggested by the Moon's enrichment in refractories.

Deciding exactly where the Moon formed in the first place is not an easy task. And there are certainly no compelling reasons for preferring one particular environment to another. What does seem quite certain, however, is that the final accretion process was completed quite rapidly once it had begun. All the evidence, in fact, points to a hot origin rather than to a cold one.

At one time it was thought possible that the Moon might have formed slowly, perhaps enshrouded in a large gas cloud (like that surrounding Jupiter) which was subsequently dispersed. But the geochemical evidence implies that the final stages of accretion were so rapid that the kinetic energy released during this period was sufficient to melt the Moon down to a depth of perhaps 200 miles.

Some of this early heat might possibly have come from the radioactive decay of that short-lived radionuclide aluminium-26. But this isotope would presumably have been distributed throughout the Moon. The release of gravitational kinetic energy, on the other hand, would only have been significant during the final stages of planet formation, by which time the accumulated mass of the Moon was sufficient to exert a strong gravitational pull on the remaining infalling material. The inferred preferential melting of the Moon's outer layers, then, suggests that lunar accretion was the primary heat source, and that it may have been essentially complete 4500 million years ago. So the absence of volatile materials on the Moon (such as those found in carbonaceous chondrite meteorites) puts the final nail in the coffin for the Cold Moon hypothesis.

But the fact that large quantities of heat were generated during the final stages of accretion does not help us to establish where this accretion actually took place. In other words, where did the primordial mineral grains (which were later to come together to form the Moon) condense out of the gas cloud surrounding the Sun, from which all planets were ultimately derived? Were they part of the same protoplanetary cloud from which the material of the Earth first condensed, or was there a completely separate cloud for the Moon?

It is at this point that the Moon's slight enrichment in refractory elements can help. Because, although the Moon's deficiency in volatiles could simply reflect their loss during the final stages of planetary accretion, the same cannot be said for the Moon's enrichment in refractories. This can only be accounted for if the mineral grains that were later to come together to form the Moon were already enriched in refractories. And this can only mean that the Moon's protoplanetary cloud was at a somewhat higher temperature than that from which the Earth formed. So, if the Moon as a whole is indeed enriched in refractory elements compared to the Earth (and this is still a contentious point), the Moon must have formed elsewhere in the Solar System. And this means that it must subsequently have been captured by the Earth.

*The dynamics of capture*

This is clearly not the place for a mathematical treatment of the dynamics of lunar capture, the subject of numerous more or less controversial papers in lunar science. Suffice it to say, however, that there are still a number of difficult problems. First of all the Moon must have originally occupied an orbit which crossed that of the Earth. Now this is by no means impossible. The orbit of Pluto, for example, passes within that of Neptune. And Pluto may, in fact, originally have been one of Neptune's moons. But, as far as the inner planets are concerned, the only bodies which are 'Earth-crossers' today are the so-called Apollo objects. And these are all rather insignificant asteroids. This does not mean, however, that there might not have been a number of much larger Earth-crossing minor planets in the past.

But there is clearly a lot more to lunar capture than simply bringing two planets close together in space. Unless the Moon was slowed down somehow, it would immediately have moved away again, at the same rate at which it approached. Either there was a transfer of energy from one planet to another, or else a third body must have taken away some of the excess energy.

Most capture theories rely on tidal energy transfer to slow the Moon down sufficiently for it to be captured, if not immediately, then on some subsequent close approach. And there are two possibilities here. Either the Moon was captured in an orbit that was retrograde (like that of Triton) and the orbital plane subsequently flipped over the Earth's pole, or else the Moon has always revolved about the Earth in the same sense as it does now. As both scenarios are mathematically quite feasible, it is difficult at the moment to choose between them.

*Disruptive capture*

But perhaps the most attractive tidal capture theory is the most extreme one. Because, in order for tidal friction to have been really efficient, the Moon must have passed very close to the Earth indeed. And if it approached within four Earth radii (a distance known as the Roche limit), then the gravitational pull exerted on the near side of the Moon would have been so much greater than that on the far side that the Moon would have been torn apart. Its own self-gravitation would have been insufficient to hold it together under these conditions. Each separate chunk of Moon would then have begun to move along its own particular trajectory.

Once these chunks had passed by the Earth, of course, they would undoubtedly have come back together again to reform the Moon. But meanwhile there could have been a gravitational separation of dense core materials from the lighter mantle rocks. It could just be, in fact, that the Moon did not completely reform. And this could certainly account for why the Moon is deficient in iron compared with the Earth. Its original iron could well have been added to the Earth, or alternatively might never have been reaccumulated, as the Moon reformed in space. So could we have the answer here to the origin of the iron meteorites? It is certainly a possibility worth considering.

### Collisional capture

Finally, there is the hypothesis that the Moon was captured when it collided with one or more preexisting satellites, during one of its close approaches to Earth, producing the lunar basins.

The author has even gone so far as to propose that the formation of the Gargantuan Basin could well have been the critical event in this capture scenario. Either the Gargantuan collision was what pushed the Moon into an Earth-crossing orbit in the first place or, alternatively, the Gargantuan projectile may have been the original satellite of Earth which, by colliding with the passing Moon, resulted in its capture. The age for this huge crater could well be 4300 million years, the most likely time when KREEP lavas were first erupted on to the lunar surface. So this event might correspond to the time of lunar capture.

### Weighing the evidence

It cannot be pretended that any one theory for lunar origin has much more support than any other. And, even if it did, this would still not automatically mean that it was the correct one. For the correctness of any theory cannot be judged solely from its numerical support. What is needed instead is a collection of hard facts which that theory can account for better than any other.

The problem with the question of the Moon's origin is that few of the known facts are very hard and all of the theories are based on suppositions that may or may not be strictly valid. It all happened such a long time ago that none of the assumptions is now directly testable. How can we ever discover, for example, whether or not the Earth once held a family of small satellites?

One must try, however, to come to some firm conclusion. And, in the author's opinion, the evidence seems to point to capture as being the most likely means by which the Moon came to be in orbit around the Earth. Its large size, and the fact that the Earth is alone among the terrestrial planets in possessing a major satellite, goes some way towards outweighing the dynamical improbability of the capture hypothesis. As the Earth's moon is unique in many ways in the Solar System, a dynamically improbable origin becomes somewhat less improbable.

The capture hypothesis also allows the Moon to have formed elsewhere in the Solar System. And this is desirable in order to account for the global chemical differences that exist between the Moon and the Earth. The 'formation in Earth orbit' hypothesis has to be stretched to its credibility limits in order to accommodate these chemical differences. Furthermore, disruptive capture has the added advantage that it provides a mechanism for the removal of large quantities of metallic iron from an already differentiated Moon. It also implies that the Moon must have been captured close to the Earth, from which it has been receding ever since.

But exactly where it formed in the first place is still a complete mystery. And even the time of capture is not tightly constrained. All that we can be certain of is that the Earth has had its moon for at least 4000 million years. If capture had occurred any later than this, then we would have expected to observe the effects of such a violent event on the Moon's surface. Craters would have been distorted and the crust would doubtless have been remelted. That such effects are not observed pushes lunar capture right back into the first dark era in lunar history.

This then completes this brief summary of lunar origin theories. We can now move further forward in time to discuss the Moon's subsequent evolution. And we shall see that after the first few hundred million years our picture begins to clear slightly. By the time the *mare* basalts began to be erupted, the Moon's history is no longer deeply shrouded in mystery; its recent history is now becoming increasingly well defined.

## 6.4  A summary of lunar history

The final synthesis must now begin. In other words, the absolute ages of lunar rocks must now be combined with all the other geophysical and geochemical data in an attempt to construct a cohesive picture of lunar history. How, for example, has the rate at which the Moon has been bombarded with meteorites changed over the past four and a half billion years? And what does this cratering history tell us about the evolution of the Solar System as a whole? What can the Moon's history of volcanism tell us about the thermal and geochemical evolution of the deep interior? And just how young are those fresh-looking volcanic provinces in the western hemisphere? These are just some of the crucial questions to which satisfactory answers must now be found.

The oldest of all lunar rocks (the dunite and troctolite) may indeed be almost as old as the Moon itself. But these two rocks really are exceptional, having survived relatively intact from the very earliest phase of crust formation, a period when most of the olivine which crystallised then promptly sank to the base of the postulated magma ocean.

*The highland rock age histogram*
By far the majority of highland rocks dated so far are younger than 4100 million years old. And very few of them indeed are more than 200 million years older than this. The general picture, then, is one of a crust composed largely of anorthositic gabbro which for a long period was being thoroughly reworked by the impacts of meteorites, large and small. Only when this really intense bombardment started to tail off did the radiometric clocks in highland rocks have much chance of not being reset again.

So how well does this picture agree with highland ages as a whole? Well, it turns out that there is a 'peak' in the highland rock age histogram at about 3950 million years (Figure 6.14). In fact this 'peak' spreads out more than 100 million years on either side of the mean value.

But much of this spread is simply due to those inevitable analytical errors. It is also accentuated by the systematic differences which arise between the different dating laboratories. The ages obtained for a particular suite of highland rocks by one laboratory using rubidium–strontium, for example, may be consistently higher (say by 50 million years) than those obtained by another group using potassium–argon.

### Basin chronology

There are, however, some geologically significant age differences within this peak. The Apollo 17 boulders, for example, all tend to be nearly 4000 million years old, whereas rocks associated with the formation of the Imbrium and Crisium Basins (from Apollo 14 and Luna 20 respectively) are nearer the 3900 million year mark.

This correlation is particularly significant because lunar photogeologists can distinguish the stratigraphic relationships which exist between all of the younger nearside basins. The Nectaris, Crisium and Humorum Basins, for example, are clearly stratigraphically older than Imbrium and the still younger Orientale. But all are younger than the less well defined Serenitatis Basin(s).

Now this is very interesting indeed. Because what it means is that these six basins (and presumably a number of others on the far side as well) must all have been excavated in the space of less than 200 million years or so. Bearing in mind the typical age uncertainty of 50 million years, dates of 3980, 3950, 3910, 3880 and 3850 million years can be ascribed to the Serenitatis, Nectaris, Crisium, Imbrium and Orientale events respectively.

Figure 6.14. *Age histogram for highland rocks.* This compilation of argon-39–argon-40 ages shows that there is a peak in age at about 3950 million years, which must correspond to the time of bombardment of the Moon by numerous large projectiles, including Imbrium and Serenitatis. The secondary maximum at about 4200 million years may mean that there was a late burst of bombardment, rather than just a monotonically decreasing bombardment.

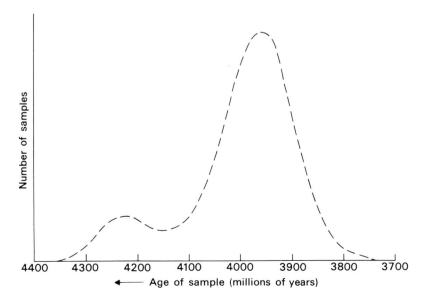

Figure 6.15. *A summary of lunar history.* This timetable summarises dated events in lunar history, from the formation of the Moon 4600 million years ago, to the excavation of the youngest crater at the Apollo 16 site, South Ray. Note how the Moon's internal activity was confined to the first 35% of its history.

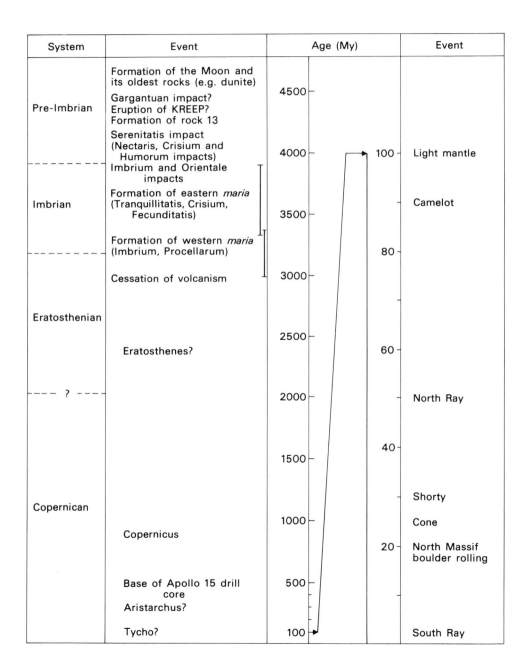

| System | Event | Age (My) | | Event |
|---|---|---|---|---|
| Pre-Imbrian | Formation of the Moon and its oldest rocks (e.g. dunite) | 4500 | | |
| | Gargantuan impact?<br>Eruption of KREEP?<br>Formation of rock 13 | | | |
| | Serenitatis impact (Nectaris, Crisium and Humorum impacts) | 4000 | 100 | Light mantle |
| Imbrian | Imbrium and Orientale impacts | | | |
| | Formation of eastern *maria* (Tranquillitatis, Crisium, Fecunditatis) | 3500 | | Camelot |
| | Formation of western *maria* (Imbrium, Procellarum) | | 80 | |
| | Cessation of volcanism | 3000 | | |
| Eratosthenian | | 2500 | | |
| | Eratosthenes? | | 60 | |
| ? | | 2000 | | North Ray |
| | | 1500 | 40 | |
| Copernican | | | | Shorty |
| | | 1000 | | Cone |
| | Copernicus | | 20 | North Massif boulder rolling |
| | Base of Apollo 15 drill core | 500 | | |
| | Aristarchus? | | | |
| | Tycho? | 100 | | South Ray |

Although no unequivocal Nectaris ejecta was recovered by Apollo 16, 3950 million year ages did predominate there, and the age of Nectaris is in any case very tightly constrained by the dates for Serenitatis and Imbrium. There is certainly no firm evidence to support its greater antiquity.

The lower age limit for Orientale is also just as tightly constrained as its upper one, in this case the Imbrium event. The reason for this is that Orientale must be older than the oldest known *mare* surface on the Moon which, as we shall see, may be the 3840 million year old basalt fill in the Taurus–Littrow valley. If Orientale had been excavated more recently than this then it could be argued that the Apollo 17 site should be covered by a thin layer of young highland rock fragments. And no such layer was found.

### The lunar cataclysm

What, then, are the implications of this proposed basin chronology? How can we possibly account for the intense bombardment still being suffered by the Moon some 600 million years after its formation? There are two possible scenarios. Either the formation of the Moon was a long drawn out process, or else the most recent basins are the outward expression of a late burst of activity, a cataclysmic bombardment several hundred million years after the end of lunar accretion.

In the first case there is the problem of explaining why the gradual accretion process ended so abruptly with Orientale. With the lunar cataclysm hypothesis, however, there has to be some place in the Solar System where the late basin-forming projectiles could have been 'stored' for several hundred million years before they were all collected in rapid succession.

The most popular idea is a hybrid one, in which lunar accretion was indeed a long drawn out process, but one which finally ended with a flourish, usually referred to as a 'terminal lunar cataclysm'. But this compromise still does not solve the problem of projectile storage.

In the author's, view the most attractive solution here is that these bodies were terrestrial satellites, swept up by the Moon as it rapidly receded following capture. Not only does this idea explain how the basin projectiles could have survived collision with the Moon for so long (in other words for all the time prior to lunar capture), it can also account for why the cataclysm ended so abruptly. It was simply that the Orientale projectile was the last (that is the most distant) of the Earth's originally numerous moons.

### Other early impacts

The Moon was not, of course, just bombarded with large projectiles during its early history. There must have been innumerable smaller impacts as well. This is certainly attested to by the saturation of the highlands with craters several miles in diameter. And there were, of course, some very major impacts prior to Serenitatis.

But characterising this early bombardment by simply dating lunar rocks is not at all straightforward. Because there is no obvious relationship between rocks and individual craters. And in the case of the smaller cratering events the radiometric clocks in the affected rocks may not have been reset. In fact even the association of Apollo 17 boulder ages with the formation of the Serenitatis Basin is still somewhat contentious. Some workers believe this basin to be at least 4200 million years old. But it is then very difficult to explain the preponderance of 4000 million year ages at the site. If anything could be associated with the formation of the Serenitatis Basin then it must surely be the massifs themselves, rather than a handful of small fragments lying around on their lower slopes.

One could argue that the highland age histogram is itself sufficient to define the decreasing cratering rate. And up to a point it can. The rarity of highland rocks much younger than 3900 million years shows quite clearly that the terminal lunar cataclysm must have extended to smaller impact events. And this is of course consistent with the photogeological observations that the number of large craters formed since Orientale, compared with those contemporaneous with this and other basins, is really very small.

But there is a serious sampling problem here. Because the Apollo sites were all carefully selected in order to investigate particular basins; Imbrium by Apollo 14, for example, and Serenitatis by Apollo 17. And even the collection of rocks at these sites was by no means random. A disproportionate number of Apollo 17 rocks, for example, were chipped from the large Massif boulders. The highland rock histogram could therefore be dominated by rocks whose radioactive clocks were reset by a small number of major impact events.

### Older highland rocks

This uncertainty about rock origins makes it all the more difficult to interpret that little cluster of highland ages which is clearly apparent between 4200 and 4300 million years (Figure 6.14). If our sampling of the highlands had been truly random then this secondary peak could not possibly be accounted for in terms of a monotonically decreasing cratering rate. It could only imply that the Moon's cratering rate decreased to a minimum some 4100 million years ago and then increased again. As it is, our limited sampling of the Moon is totally inadequate to enable us to decide if there really was such a lunar cataclysm.

It is clearly still desirable to obtain rocks from areas which are far removed from the younger basins. There are areas towards the south pole, for example, which geologists believe to be much more ancient than the sites studied so far. So perhaps it is from here that some of those enigmatic 4200 million year old rocks are ultimately derived. But this last proposal cannot be put to the test, of course, until man is in a position to continue his direct exploration of the Moon.

It could just be a sampling problem, then, that is all that prevents us from understanding fully the earliest period of lunar history. If it is difficult to associate the ages of large boulders with specific cratering events then it is hardly surprising that the ages of small rock fragments can only be treated statistically. And, interestingly enough, it is among the pebbles that most of the oldest samples turn up. Maybe this is because they tend to be the very rocks that have travelled the greatest distances.

So only when genuine outcrops of crustal rock can be studied (by climbing mountains and drilling deep cores) is this uncertainty ever likely to be removed. Only then will we be really confident that what we are dating is indeed a recognisable lunar formation.

### Early lunar volcanism

So much, then, for the first few hundred million years of lunar history. But what happened after that? Well, it has already been pointed out that the oldest *mare* surface dated, the Taurus–Littrow valley, is almost as old as the Orientale Basin. Is this a coincidence or might there be still older *maria* on the Moon? Well a number of Apollo 11 low potassium basalts may well be older than Orientale, and some 3950 million year old clasts of aluminous basalt were collected at the Apollo 14 site, embedded in breccias which were some 70 million years younger. So we can be certain that there was *mare* volcanism prior to Imbrium, if only on a limited scale.

But it is difficult to be sure about exactly when this early volcanism began. Because any really ancient *maria* will have long since been obliterated by the terminal lunar cataclysm and blanketed by its associated ejecta. Extensive early lunar volcanism is therefore now only potentially discernible through the chemical signature which it could have written across wide areas of the Moon's surface. And it might be possible to interpret such a signature in terms of the rock types which are represented as hand specimens, pebbles and tiny fragments returned by Apollo. Only by this combination of remote sensing and experimental geochemistry, then, can we ever hope to characterise fully and map the Moon's earliest phase of volcanic activity. Photogeology is clearly limited when it comes to discovering such ancient, and therefore much disturbed, rock units.

### KREEP lavas

Fortunately, one of the earliest volcanic rock types, namely KREEP basalt, is chemically very distinct from the others. And its relatively high content of natural radioactivity means that its presence is clearly recognisable from lunar orbit. Indeed, the thorium maps generated by the gamma ray spectrometer experiments effectively reveal the areal distribution of KREEP at the lunar surface (Figure 5.22). And what is clearly apparent from these maps is that there is a marked localisation of this rock type in the Imbrium–Procellarum area.

Now it should be pointed out here that there is a school of thought which interprets the present day surface distribution of KREEP solely in terms of the Imbrium event. The idea here is that solid pockets of KREEP were melted and excavated from beneath the highland crust by this, the largest of the more recent lunar basins, and then spread over a large area.

But KREEP is by no means symmetrically distributed around the Imbrium Basin as we would expect from simple cratering mechanics. And there are a few crystalline fragments of KREEP which are almost certainly volcanic in origin. These particular fragments contain none of those extralunar element excesses which are so characteristic of impact melts and highland breccias. Their chemistry is instead just what we would expect for a basalt magma produced at shallow depths by partial melting.

And why should one have to resort to impact melting when early volcanism is such a perfectly acceptable alternative? There was certainly no shortage of heat sources in those early years, so why should part of the Moon's near side not have been extensively flooded by a sequence of KREEP lavas, some richer in potassium than others?

The author postulated the existence of the Gargantuan Basin in order to account for the observed KREEP distribution. It was proposed that KREEP lavas filled the vast Gargantuan depression in much the same way that *mare* lavas later filled the more recent, and now better defined, nearside basins.

But just as there is no direct temporal relationship between the *maria* and the impact basins which they filled, so is there no necessity to associate the time of KREEP extrusion with the age of the Gargantuan event, although the basin must of course be older than the *mare*.

It seems that one reason why this Gargantuan Basin hypothesis has not so far gained wider acceptance is that there are now no KREEP lavas to be seen on the Moon. But this is hardly surprising. The surface of this early KREEP *mare* would, after eruption, have constituted a relatively low stratigraphic horizon, a horizon which would have been the first surface to be covered up by subsequent volcanic eruptions.

In certain parts of Mare Imbrium and Oceanus Procellarum, however, the normal lava stratigraphy has been severely upset. Underlying layers have been excavated by cratering events, the largest of which was, of course, the Imbrium impact itself. There were also several smaller impacts within the confines of the Gargantuan Basin, notably Copernicus, Aristarchus and Archimedes. And sure enough, associated with these craters are unusually high excesses of natural radioactivity. Much of the Copernican ray material transported to the Apollo 12 site consists of KREEP glass melted and blasted away by the Copernicus impact event. And Aristarchus, Archimedes and the Fra Mauro formation all show up as impressive radioactive hotspots in the gamma ray maps.

But just how old are the KREEP basalts? This is a question to which there is as yet no really satisfactory answer. Most KREEP rocks have ages quite consistent with the Imbrium event. But these are all either breccia clasts or small pebbles, all of which could easily have had their radioactive clocks reset by the Imbrium event. KREEP could in fact be as old as 4300 million years or so if the KREEP whole rock isochrons correspond to the time of lava extrusion as well as magma separation.

Of all lunar rocks, KREEP is perhaps the most sensitive as far as their internal radioactive clocks are concerned. What we really require here are samples of KREEP from beneath western Oceanus Procellarum. But how can such samples ever be obtained? Certainly not by impact excavation. Because the KREEP glass at the Apollo 12 site, presumed to have originated as impact ejecta from the Copernicus Crater, is only some 850 million years old. And if Copernicus was capable of resetting the radioactive clocks in KREEP basalts at this time than it should come as no surprise to find that no really ancient KREEP rocks turned up at the Apollo 14 or 15 sites. Imbrium was much larger than Copernicus. So the search for crystalline KREEP basalts will no doubt remain a major objective for some time to come.

As for other pre-Imbrium basalt flows, the situation is still more uncertain and likely to remain so. We know that ancient aluminous *maria* must have existed, but finding them could be a major problem. One tool which could be of some use here is the spectral reflectance technique discussed in Chapter 5. This approach has certainly shown that rock types other than those returned to date do indeed exist on the Moon. So it is now up to future lunar missions to search for and return the corresponding hand specimens.

*The eastern* maria

Turning now to lavas extruded since the formation of the Orientale Basin, the fog begins to clear a little. The oldest *mare* rocks returned from the Moon so far are the low potassium basalts from Mare Tranquillitatis (some of which are more than 3900 million years old) together with the equally titanium-rich 3800 million year old lavas from the Taurus–Littrow valley, on the borders of Mare Serenitatis.

In the case of the Taurus–Littrow basalts the spread in ages is not really very significant. Taking into account the experimental errors in age measurement, the lavas in this valley could quite easily have been extruded one after another about 3780 million years ago. Alternatively, and this now seems more likely, their extrusion may have begun as early as 3850 million years ago and continued for a further 150 million years (Figure 6.16). The trouble here is that the analytical uncertainties are simply not sufficient to enable us to resolve such relatively minor time intervals. 150 million years may seem like a long period in present day terms (being in fact twice the time which has elapsed since the extinction of the dinosaurs), but it is really quite insignificant when compared with the great antiquity of the lunar *maria*.

In the case of Mare Tranquillitatis there is no such ambiguity. This is because the spread in ages here is that much greater, the Apollo 11 basalts ranging in age from less than 3600 to more than 3900 million years. And there is also a marked anticorrelation here between age and potassium content (Figure 6.9), the younger rocks being very much richer in potassium than the oldest ones. So the iron-rich lavas in Mare Tranquillitatis were clearly extruded over a period of at least 300 million years, becoming increasingly rich in potassium towards the end of this period, a geochemically reasonable sequence.

It was during this extrusion of the Tranquillitatis lavas that three other eastern *maria*, namely Nectaris, Fecunditatis and Crisium were also formed. But, for these *maria*, sampling was much less comprehensive and was totally restricted to small, and possibly unrepresentative, fragments from the regolith. One such *mare* basalt particle found at the Apollo 16 highland site, for example, may well have been transported there all the way from Mare Nectaris to the southeast. So we can infer from its age that *mare* lava flows were filling this basin some 3790 million years ago. If this particle can be shown to have originated from Mare Tranquillitatis to the northeast, however, we must accept that the age of Mare Nectaris is still an unknown quantity.

In the case of the other two eastern *maria* the sampling was rather more direct. The Mare Fecunditatis regolith was probed by Luna 16, whereas Mare Crisium was the landing site for the most recent lunar landing mission, namely Luna 24. The low titanium basalt fragments from Crisium are from 3300 to 3650 million years old, whereas the more aluminous rocks brought back by Luna 16 are about 3500 million years old.

So we can conclude from all these ages that, although the Moon's eastern *maria* began to be formed immediately after, or even before, the excavation of the Orientale Basin, this widespread volcanism in the east certainly did not cease until about 3300 million years ago (Figure 6.16).

But how does this absolute chronology compare with the relative ages of the *maria* based on studies of crater densities? And what about those unsampled *mare* units? Well, Mare Tranquillitatis certainly does exhibit a high density of craters for a *mare*, particularly at larger crater densities. And, taking into account the subduing effects of the dark mantle, the Taurus–Littrow valley does indeed look just as old, if not older.

We would of course expect the Apollo 17 site to appear slightly older, because the uppermost surface of Mare Tranquillitatis must consist of

Figure 6.16. *Mare basalt age histogram.* This is a compilation of Argon-39–Argon-40 ages for *mare* basalts from all missions. The numbers represent the Apollo mission, except for those prefixed by 'L', which are Luna samples. Note the wide spread in age for the Apollo 11 suite. The Apollo 14 sample was a *mare* basalt clast in a Fra Mauro breccia, whereas the Apollo 16 sample was an exotic fragment, possibly from Mare Nectaris.

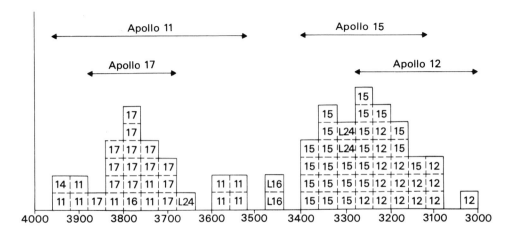

the most recent lava flows in the area, represented among the Apollo 11 rocks by the 3500 to 3600 million year old high potassium basalts.

But the Apollo 17 rocks clearly cannot be taken as representative of Mare Serenitatis as a whole. For it is quite apparent from orbital photographs that there are some very major differences in age and chemistry among the various lava flows in this area (Figure 1.32). Until astronauts can visit the apparently young centre of Mare Serenitatis, however, we will be unable to establish just how recently the last dribbles of lava trickled out there. Similar, but less extreme age differences are observed in Mare Tranquillitatis and this may reflect the wide range of absolute ages obtained among the Apollo 11 basalts.

Perhaps the most important point to be stressed here is that the period of *mare* filling must have been at least as extensive as the age range based on rocks collected at the surface. Because only really large meteorite impacts are capable of excavating *mare* basalts from below depths of several miles. And such impacts are sufficiently large to have reset the very radioactive clocks that we wish to read.

So perhaps only when we have learnt more about the mechanics of mascons, and why they sink, will we be in a position to place further constraints on the period of *mare* volcanism in the Moon's eastern *maria*. Mare Crisium and Mare Smythii (the latter as yet unsampled) are topographically the lowest *maria* on the Moon and could therefore well be the oldest, despite the evidence to the contrary provided by the returned rocks.

### *The western* maria

The centre of Mare Serenitatis may not have been sampled by Apollo, but it is of course not the only young lava unit on the Moon. Some of the *maria* in the Moon's western hemisphere appear to be just as young, if not younger, if we are to believe crater studies. And Mare Imbrium (or at least Palus Putredinis) and Oceanus Procellarum were sampled directly by Apollo 15 and Apollo 12 respectively.

In the case of Hadley Rille, the spread in *mare* ages is again very small, practically all of the rocks having ages within 100 million years of 3300 million years. Furthermore, there is no obvious correlation within this narrow period between age and the stratigraphic sequence for the site as inferred from the distribution of rock types.

Much the same is true for the Apollo 12 site, except that the age range is even narrower here, and the mean age is younger still, less than 3200 million years. So this is the age that is generally taken to represent the approximate time of lava extrusion in Oceanus Procellarum.

But there are parts of Mare Imbrium and Oceanus Procellarum which appear to be younger than both areas. Indeed, some believe that *mare* volcanism on the Moon may have continued until as recently as 2500 million years ago. And although these 'recent' lava flows were not sampled directly by Apollo a distinct possibility was that rock fragments from these areas had been transported over great distances. If so, then such rocks might well be present as small basalt fragments in the regolith at both of the western *mare* sites.

It was to this end that exotic *mare* fragments were searched for in the Apollo 12 soils. But while some significantly younger rocks were found, the very youngest were still some 3000 million years old. So there seems to be no way to escape from the conclusion that *mare* volcanism on the Moon effectively ceased at around this time. Why should the western *maria* be so much younger? Well it could quite well have been due to the thinning of the crust by the Gargantuan Basin. This would certainly have made it that much easier for basalt magmas to reach the surface from great depths.

*Recent lunar history*

So what has happened on the Moon during the last 3000 million years? Well once *mare* volcanism ceased there was no real possibility that it might start up again. The minimum depth at which the Moon was at least partly melted must have moved steadily inwards towards the core, from less than 200 miles, 3000 million years ago, to about 600 miles today. There have been occasional venting of gases (this apparently continues to this day in conjunction with the isostatic readjustment of mascons) but no widespread eruption of magma.

All that has really happened during the last 3000 million years is that the Moon has been subjected to a continuous bombardment by meteorites, large and small. This bombardment may not have been anywhere near as intense as that suffered during the first half billion years or so, but it was quite sufficient to saturate the surfaces of the *maria* with small craters and to generate a thick regolith over the entire Moon. It is generally believed that this bombardment has steadily decreased in intensity throughout this period, but there is some evidence to support a relatively recent peak in intensity.

What is certain, however, is that what appear to be very fresh craters are still extremely old by terrestrial standards. As already mentioned, Copernicus is thought to be 850 million years old and even Tycho (one of the two most recent large craters on the Moon) is at least 100 million years old, if our interpretation of the exposure ages of rocks associated with the Tycho ray which crosses the Apollo 17 site is correct. The Copernican era on the Moon, then, far from being the lunar equivalent of the Pleistocene, covers a time period that is longer than the entire Phanerozoic.

The evolution of the regolith (and hence the rate of small scale cratering) has been just as gradual. Samples from a depth of 5 feet in the Apollo 15 core tube, for example, have lain essentially undisturbed for 500 million years. So these cores contain a continuous half billion year record of the history of solar and micrometeorite activity just waiting to be interpreted.

Moving closer to the present time, we have been able to establish the dates of a number of still more recent events (Figure 6.15). These dates are inferred from the times of exposure of the lunar rocks associated with these events to cosmic rays and micrometeorites.

The White Rocks at the Fra Mauro site, for example, reveal that Cone Crater must be 25 million years old (Figure 2.18). At 50 million years old, North Ray Crater at the Apollo 16 site is twice as old as Cone Crater, whereas the Light Mantle and Camelot Craters at Taurus–Littrow are both as old as Tycho.

The Tycho impact was not, however, the youngest event to influence the Apollo 17 site. The impact that produced Shorty, the source of the orange soil, penetrated right through the light mantle to the *mare* and this is apparent from its younger age, 30 million years. A still more recent event was the one which made the Station 6 boulder (Figure 2.34) roll down the side of the North Massif 10 million years later. But the title of youngest accurately dated crater on the Moon must surely go to South Ray at the Apollo 16 site. Based on the concordant exposure ages of a number of rocks strewn across the southern traverse, South Ray must be only 2 million years old. In other words this particular crater was actually formed during the time that man has existed on Earth.

### Current lunar activity
South Ray is certainly the youngest crater on the Moon that has been dated through rock exposure ages. But the Moon is obviously being bombarded to this day with smaller bodies, just as meteorites are today still falling to Earth.

And one such impact may in fact have been recorded by man in the pre-telescopic era. There are some medieval records which describe fiery sparks on the northeastern limb of the Moon, which may well correspond to the excavation of that obviously extremely young farside crater Giordano Bruno. If this is so then this must go down as an incredible coincidence. Craters of that size are only formed every few tens of millions of years. So the odds against such a crater being formed, let alone noticed, during the last two thousand years are thousands to one.

Present day lunar bombardment activity is, of course, not just limited to direct observations from Earth. During the space era our sister planet has been struck by a number of man-made projectiles, and these impacts, together with many more natural ones, were recorded by the Apollo seismic network. The largest such natural impact must have produced a crater several hundred feet across. One of these may have been responsible for that flash of light in the Orientale Basin which was observed from lunar orbit.

And these impacts become increasingly abundant with decreasing size. Indeed, the influx of micrometeorites is so great that several microcraters were recorded in the glass of the Apollo windows. Although by no means a rain, in the terrestrial sense, it is these microcraters which are today responsible for most of the erosive activity on the Moon.

This then concludes our brief summary of lunar history. But what of the future? What steps are being taken to explore further?

## 6.5   The future of lunar exploration

How can the vast expense of a manned (or, for that matter, an unmanned) lunar exploration program possibly be justified? In other words, how could we have spent all those billions of dollars on Apollo when there was such widespread poverty back here on Earth? And how can we now reconcile this 'investment' with our current concern over the world's limited resources of energy and minerals? Several attempts are still made to justify the Apollo program in these and other terms, but some of these efforts are clearly more valid than others.

One approach has been to belittle the huge cost of manned spaceflight altogether, by comparing it to the much larger US defence budget, or by equating it to the money spent each year on, say, cosmetics or cigarettes. Another popular argument is the purely commercial one. In other words the benefits to the economy which have arisen directly from space exploration (through communication, navigation, meteorological and resource mapping satellites, for example) far outweigh the total cost of Apollo. Whether or not this second argument is sound, it is certainly strengthened by the economic advantages of some less obvious technological spinoffs. Because, although we might still be managing quite well without non-stick frying pans and ballpoint pens that write upside down, where would we be today without our pocket calculators and microprocessors?

It is noteworthy that, in the face of the economic recession which has prevailed since the premature cutback of the Apollo program, it is the computer (rather than the aerospace) industry that is still flourishing today. The Apollo program certainly provided widespread employment during its lifespan. And introducing the concept of zero defect technology, it also did much to improve the quality and reliability of hard and soft engineering products.

It might, of course, be argued that we could quite easily have invented the non-stick frying pan and the pocket calculator without going to the trouble of sending men to the Moon. And this may well be true. But, just like a war effort, the Apollo program provided that ideal environment in which initially vague ideas could quickly become realities, frequently as intermediate stages in the development of more obvious products. Under these highly favourable conditions, precise objectives and constraints can be defined and almost unlimited finance can be made available.

### The politics behind Apollo

But the Apollo program was not, of course, instigated as a purely commercial enterprise. It could well be argued, in fact, that as much, if not more, in the way of technological spinoff could have been achieved if a concerted effort had instead been made to colonise the ocean depths (sometimes referred to as inner space), to solve the problem of the world's chronic shortage of energy, or to conquer cancer and heart disease.

Landing a man on the Moon and returning him safely to Earth, however, was such an easily understood and technically feasible project that it became a natural political tool, particularly as the Russians were apparently intent on achieving the same aim.

The Apollo program, then, was a prestige project, a show of strength designed to restore national morale (particularly low after the Bay of Pigs fiasco), and to show the rest of the world what America really could achieve if it tried. In other words, the first manned lunar landing was to be the ultimate weapon in that peaceful war known as the space race, a race that was never really lost, because the Russians never explicitly referred to their involvement in it.

As it turned out, of course, the rest of the world was indeed very much impressed by the American achievements on the Moon. But Apollo 11 was difficult to follow and there was that inevitable backlash against all things technological. It could be argued in fact that the conservation and ecology movements were a direct consequence of Apollo. The explosion of awareness of such matters certainly followed very close on the heels of the first Moon landing and was much heightened by those awe-inspiring views of our precious blue-green planet from outer space.

But without that intense political pressure behind it, it is likely that the Apollo program as such would never have succeeded, and it may never even have been embarked upon in the first place. The Moon may still have been conquered eventually, of course, but only as part of a more leisurely and coherent plan for the exploration of the Solar System, not as an isolated exercise. Only now, in fact, with the advent of the Space Shuttle, can the manned exploration of the Solar System begin to develop again in a systematic way.

*The science behind Apollo*

Having made the Apollo program into a national goal, however, it suddenly became necessary to justify it in terms which were not just political. And this, of course, is where science came in. Indeed, the very continuation of the program after Apollo 11 was almost impossible to justify in any terms other than scientific ones. Why go to the Moon twice?

A number of more or less scientific questions were therefore asked, and the answers to these questions were considered by many to constitute the ultimate objectives of Apollo. Needless to say, very little attention was actually paid to these objectives by the media, and attention was focussed throughout on the technology of spaceflight and on the personalities of the astronauts. Indeed science came a very poor second to technology in Apollo.

It was this low priority from the outset which meant that only after five successful landing missions did a scientist-astronaut finally go to the Moon. And by that time the program was practically over. The pace was such that a mission had to be fully planned long before the previous one had borne scientific fruit. It is certainly true that more science would have come out of Apollo if the program had been spread out over say six rather than three, years. The limited time available for astronaut activity could then have been spent more constructively and in a less emotional atmosphere. That spent planting flags, stamping letters and playing games could have been expended more productively in the pursuit of scientific understanding.

To what extent, then, can the Apollo program now be justified in terms of its scientific return? In other words what did we really expect (or hope) to discover and how successful was Apollo when viewed in these terms?

### The unanswered questions

The questions to be answered by Apollo fell into two broad categories: (1) fundamental questions relating to the Moon as a planet, and to its place in the Solar System; and (2) practical questions having a direct bearing on the future colonisation of the Moon and the possible exploitation of its natural resources.

As far as these purely scientific questions are concerned, we now know a great deal about the nature of the Moon's interior and about the different sorts of rocks that comprise its surface. We have also learnt a lot about the Moon's bombardment history and also about its interactions with charged particles, not only from the Sun but from beyond the Solar System as well.

Our sampling of the Moon by Apollo (and Luna) was, however, necessarily rather limited in areal extent. And, similarly, it was only possible to set up those crucial surface experiments (such as magnetometers and heatflow probes) at a very small number of spots on the lunar surface. It is certainly true that remote sensing experiments have allowed us to extrapolate some of these direct chemical and physical measurements to other areas. But, even with this extended orbital coverage, the total area studied is still not very large, because by far the major fraction of the Moon's surface was never even overflown by Apollo.

So how important is this gap in our current knowledge and how can it now best be filled? Well the problem may not be quite as serious as it may seem. Because the surface of the Moon is certainly less varied than that of the Earth. There may well be marked differences between the lunar *maria* and the highlands, but these differences are really quite minor when compared with those which exist on Earth between, say, a desert and a coral atoll, the Arctic icecap and a river delta, or a volcanic island and an ancient craton. So, in order to obtain a representative sample of Earth, extraterrestrials would certainly require more than just six landing missions to our planet. In the case of us and the Moon, on the other hand, we may not be missing very much.

### Water on the Moon

But there is at least one lunar environment (namely the polar regions) which may be very different from those sampled so far, at least from a chemical viewpoint. This is because the lunar poles, unlike the Earth's are almost perpendicularly aligned to the ecliptic. And this means that they cannot benefit so much from the warming effects of the midsummer Sun.

Indeed, temperatures at the lunar poles may well never exceed that of liquid air. There must be innumerable cracks and crevices there, in fact, which never experience direct sunlight at all from one year to the next. And these frigid conditions could well have prevailed for several billion years, only occasionally punctuated by local impact heating.

This means, of course, that the Moon's poles could have acted as 'cold fingers', capable of trapping out any volatiles that happened to escape from the Moon's deep interior during this long period.

We now know that radon and argon are among the gases still being vented from inside the Moon today. The big question is: could there also be some water vapour, assuming of course that the Moon did indeed contain some to start with?

The lunar rocks returned so far were all collected from equatorial regions and are essentially devoid of water. But it is just possible that there may be surface (or perhaps subsurface) deposits of permafrost at the Moon's poles. Indeed some investigators think that this is quite likely. And it is certainly a question to resolve one way or another before we make our first attempt to establish a permanent lunar colony. And the way to do this is to study the polar regolith, either indirectly from a satellite or directly by means of an unmanned lander or a remote controlled roving vehicle such as the Russian Lunokhod. At one point, in fact, there was a suggestion to fully automate the Apollo 17 Rover, but this did not receive adequate support.

*A polar orbiting observatory*

The concept of a Lunar Polar Orbiter (or LPO) has existed for many years. Its main tasks would be to tie up all those loose ends left behind after Apollo, including the water question (Figure 6.17).

Although quite easy to justify on a purely scientific basis (in that it would be much more cost effective than Apollo), there is a feeling among those who control the NASA budget that quite enough has already been spent on going to the Moon. In other words, lunar scientists should have made better use of the facilities provided by Apollo and cannot expect still more money to be channelled into lunar research. So, while other more prestigious projects (such as Voyager and Viking) have gone ahead, the LPO has remained firmly on the drawing board and is likely to remain there for the forseeable future. It is just possible, however, that its objectives could be taken over by a joint European satellite, launched by the US Space Shuttle or the Ariane rocket.

So what are these objectives and how can an LPO attain them? Well the proposed experiments are, in fact, just extensions of those flown on Lunar Orbiter and Apollo. The orbital plane of the satellite will be inclined by 95° to the equator and will precess at a rate of 11° longitude per revolution. This will ensure total global coverage (from an altitude of 60 miles), with any given area being viewed under the same illumination angle every 14 days. Similar probes are now seen as powerful tools for the investigation of other planets and satellites that lack an appreciable atmosphere, such as Mercury and Jupiter's moons. The LPO, then, is just one of a family of sophisticated planetary missions.

As far as the experiments themselves are concerned, it is envisaged that alpha particle, infrared, visible, X-ray and gamma ray spectrometers will enable us to generate maps of surface natural radioactivity (e.g. thorium), chemistry (including water) and heat flow. Similarly, we would expect to use magnetometers to probe the Moon's magnetism further, stereophotography and laser altimetry to study its surface topography and radiotracking data to investigate the mascons. Perhaps most important of all, we would then be in a better position to correlate these properties one with another. How does heatflow vary with natural radioactivity, for example, and are there any more obvious relationships between surface chemistry, topography and mascons?

In the case of the Moon's farside topography and gravity field it is, of course, necessary to be able to track the LPO while it is behind the Moon. And the idea here is to have a second satellite (in the same orbital plane as the LPO but much farther out) which will act as a radio relay. If we can then succeed in obtaining this crucial farside data, we will be in a position to establish some much firmer constraints on our models of the Moon's interior structure. Why exactly is the Moon's centre of mass displaced from its centre of figure, and by how much?

Figure 6.17. *The proposed Lunar Polar Orbiter.* The object of this mission (involving the primary spacecraft (*a*) and a relay satellite (*b*)) would be to extend remote sensing to the entire lunar surface. To achieve this, the orbiter would be in a low orbit, whereas the relay satellite would be in a much higher one (*c*). (NASA.)

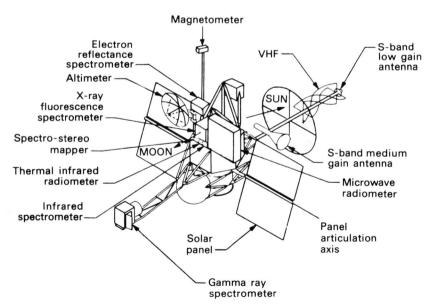

(*a*)

*Beyond the LPO*

Plans for the exploration of the Moon beyond the LPO are naturally still rather vague. There is certainly no support for a continuation of Apollo-type manned landings in the forseeable future, at least in the US. As far as the Russians are concerned there may well be more softlanding Lunas. But this particular spacecraft configuration is such that only the Moon's eastern limb can be easily sampled.

Moving more into the realms of science fiction, the emphasis has recently been moving towards the possible economic uses of the Moon. For if there is ever going to be a semipermanent base on the Moon (or out in space for that matter) then it must be more or less self-supporting as far as energy and raw materials are concerned.

The Moon has a clear advantage over the Earth as a source of raw materials in space, in that it has low surface gravity and a negligible atmosphere. This means that, if only ways can be found to mine and

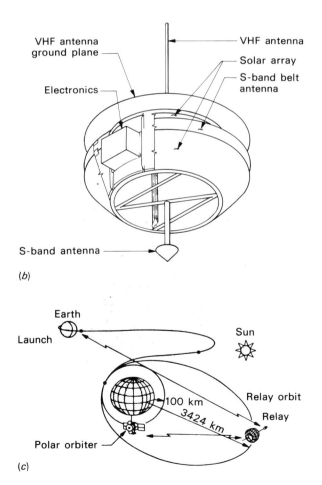

(b)

(c)

possibly refine lunar materials, it would be much simpler and less energy intensive to send them into space from the Moon than it would be to launch them by rocket from Earth. The plan here would be to accelerate magnetically levitated buckets of lunar rock from a launch site on the far side and to catch them in lunar orbit, where they could be further processed. As much as a million tons could be launched in this way in a single year at a cost of a few cents per kilogram.

But there is clearly much more to mining on the Moon than performing a comparable operation here on Earth. For one thing it would be necessary to establish permanent living quarters for the miners, because such a major operation could not possibly be carried out by remote control. And this means importing (or making) such critical raw materials as air and water. Without water it would not be possible to grow food on the Moon, and having to import food and water from Earth would certainly make lunar mining unfeasible.

So what prospects are there for obtaining the necessary raw materials on the Moon and would a lunar colony based on them really be feasible?

### The Moon's resources

As far as lunar water is concerned, it may of course be totally nonexistent, even at the poles. But all may not be lost because as a last resort we could make it. All rocks contain oxygen and there is abundant solar wind hydrogen in lunar soil. Indeed, the reduction of lunar iron oxides would yield not only water, but also metallic iron, another valuable raw material.

Figure 6.18. *Lunar bulldozer*. The fine-grained nature of the lunar surface means that crushing of feedstock for mineral processing would not be necessary. The lunar regolith could simply be collected using bulldozers. (NASA.)

As far as mining iron oxide is concerned there really is no problem. Unlike terrestrial mining there is of course no need to excavate and crush lunar rock prior to smelting it. All we would need to do is collect the ready pulverised regolith using special lunar bulldozers (Figure 6.18). Some pretreatment might, however, be desirable. We could, for example, enrich the feedstock in glassy agglutinates (known to be rich in hydrogen and metallic iron) by using strong magnets (Figure 6.19).

The techniques for mineral processing on the Moon would of course be quite novel. For one thing there would be no water available for washing waste products away. Indeed, waste removal on the Moon would be a problem. The surface of the Moon is a very dirty place and keeping a lunar base free of lunar dust would be a major challenge in itself. Anyone who has worked with it in the laboratory will know just how 'sticky' it is.

The problem of water shortage could mean that metals are purified in the vapour phase. Here, the high vacuum of the lunar atmosphere would come in useful. But vapourising lunar rock is highly energy intensive and could probably not be carried out near the poles, where the first lunar base could well be, in order to have immediate access to any water that

Figure 6.19. *Enrichment of feedstock.* This flow diagram shows a possible scheme for extracting hydrogen and iron from an enriched lunar soil. These two elements are most abundant in the low density magnetic fraction. The recovery of the water used in the extraction process would clearly have to be extremely efficient. (NASA.)

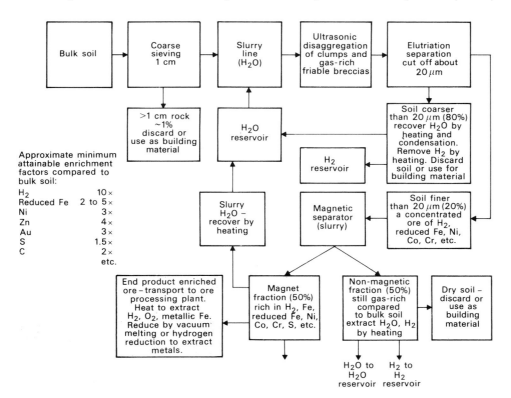

may exist there. Closer to the equator, however, we can envisage the conversion of small craters into parabolic solar heat collectors.

The absence of water on the Moon is of course responsible for the absence of high grade ores on the Moon. Whereas valuable metal-bearing minerals on Earth are precipitated in seas and rivers, and concentrated inside the crust by hydrothermal activity on Earth, the Moon's minerals tend to be finely disseminated throughout the crust. Indeed, nowhere do we find high concentrations of any particularly useful lunar mineral. So there seems to be no way to escape the fact that mineral processing on the Moon will require a great deal of energy. But there is still no reason why it cannot be done.

### Future trends

As far as the long term future of man and the Moon is concerned there are many possibilities. It could well be, for example, that we might wish to use the far side to build a huge radiotelescope array to search for extraterrestrial civilisation, well away from radio interference from Earth. The Moon's far side is certainly ideal for many types of astronomical observation.

Another possibility is that the Moon will become a staging post for long distance space travel. One exotic suggestion is that the Moon could be used to supply nuclear fuels for fission propulsion systems. The idea here is that atomic bombs would be exploded above the regolith, where fissionable materials such as plutonium would be generated and could subsequently be extracted. Although the idea of contaminating our sister planet in this way may seem abhorrent to some, there is clearly much to be said for keeping large quantities of nuclear fuel away from our precious Earth.

In the event of not being able to find any indigenous water on the Moon, it has been pointed out that this could be supplied by capturing a comet or two. An ample supply of water, carbon and nitrogen in lunar orbit would greatly help with the processing of lunar materials. It is here, perhaps, that we may find ourselves building the first starships. Powered by ion rockets, or riding the shock waves from little fusion bombs, these vehicles could well be our ultimate means for escaping from the Solar System. If this comes about then Neil Armstrong's small step for a man will indeed have turned out to be one giant leap for mankind.

# Glossary

| | |
|---|---|
| accretion | formation of a planet from smaller objects |
| agglutinate | glass-bonded soil particle |
| aggregate | *see* agglutinate |
| albedo | efficiency with which a body reflects light |
| alpha particle | energetic helium nucleus |
| ALSEP | Apollo Lunar Science Experiment Package |
| ALSE | Apollo Lunar Sounder Experiment |
| ALSRC | Apollo Lunar Sample Return Container |
| annual equation (annual inequality) | lunar inequality modulation due to the eccentricity of the Earth's orbit |
| anomalistic month | time for successive lunar passages through perigee |
| anorthosite | rock composed largely of calcium feldspar |
| anorthositic gabbro | rock with 65–77.5% calcium feldspar |
| apogee | nearest point on orbit to Earth |
| Apollo | American manned lunar program |
| apparent libration | *see* optical libration |
| armalcolite | a uniquely lunar titanium oxide mineral |
| active seismometry | seismometry using artificial sources |
| augmentation | the difference at a particular time between the Moon's topocentric and geocentric semi-diameters |
| basalt | fine-grained volcanic rock, containing plagioclase and pyroxene as main minerals |
| basin | large crater with multiple rings |
| blue moon | colouration of Moon due to terrestrial dust |
| breccia | rock with large angular grains cemented together by a finer grained matrix |
| CAPCOM | capsule communicator |
| cataclastic | friable texture produced by shock |
| CDR | mission commander |
| central peak | mountain at centre of a crater |
| chain craters | series of craters aligned in a chain |
| cleft | *see* rille |
| clinopyroxene | a monoclinic iron/magnesium/calcium silicate |
| cold cathode gauge | instrument for studying lunar atmosphere |
| colour difference | technique for revealing *mare* units |
| command module | primary Apollo spacecraft |
| Copernican system | most recent era in lunar history |
| core | section of regolith in drive tube |
| cosmic ray | high energy ion in space |
| crater | depression caused by impact (or volcanism) |
| CRD | Cosmic Ray Detector experiment |
| crater chain | *see* chain craters |
| crater cluster | group of secondary craters |
| crater counting | technique for relative dating |

| | |
|---|---|
| CSM | Command and Service Module |
| cuff checklist | reminder of tasks to be accomplished |
| cumulate | rock formed by crystal settling |
| cusps | extremities of a crescent phase |
| DAC | Data Acquisition Camera |
| dark halo crater | lunar crater surrounded by dark material |
| devitrification | recrystallisation of glass |
| Descartes formation | bright rock unit at Apollo 16 site |
| diaplectic glass | *see* maskelynite |
| diurnal libration | daily libration of the Moon |
| diurnal inequality | differences in height between successive tides |
| dome | low hill with upward convex profile |
| Doppler effect | change in radiation frequency caused by motion |
| DPS | Descent Propulsion System |
| draconic month | time for successive lunar passages through node |
| drill stem | deep regolith coring device |
| dune | transverse feature of ejecta blanket |
| dunite | rock composed largely of olivine |
| earthshine | illumination of Moon by the Earth |
| ecliptic | plane of Earth's orbit around the Sun |
| ejecta | rocks thrown out of a crater |
| ejecta blanket | layer of ejecta |
| emersion | reappearance of a body after occultation |
| equation of the centre | description of lunar motion in orbit |
| equator of illumination | plane of symmetry of a phase |
| Eratosthenian system | intermediate era in lunar history |
| EVA | Extra-Vehicular Activity |
| evection | lunar inequality |
| Explorer | series of Earth/lunar satellites |
| exposure age | length of time spent at lunar surface |
| far side | the lunar hemisphere turned away from Earth |
| far UV spectrometer | ultraviolet spectrometer on CSM |
| fault | displacement of crust |
| feldspar | aluminosilicate mineral |
| fillet | bank of soil next to rock |
| first quarter | half illuminated Moon following new moon |
| Fra Mauro formation | Imbrium ejecta sampled by Apollo 14 |
| fines | lunar soil finer than 1 millimetre |
| fluorescence | reemission of energy at lower wavelength |
| flux | density of particles per unit of time |
| Fra Mauro basalt | *see* KREEP basalt |
| front | contact between two rock units |
| full moon | totally illuminated Moon |
| furrows | valleys in ejecta blanket |
| gabbroic anorthosite | rock with 77.5–90% calcium feldspar |
| galactic cosmic rays | cosmic rays coming from beyond the Solar System |
| gamma ray | very short wave electromagnetic radiation |
| gamma ray spectrometry | technique for radioactivity mapping |
| gardening | turnover of regolith by meteorites |
| geometrical libration | *see* optical libration |
| Gemini | two-man Earth orbit space program |

| | |
|---|---|
| geophone | microphone used for seismic studies |
| GET | Ground Elapse Time |
| ghost crater | crater mantled by lava or ejecta |
| gibbous phase | phase between quarter and full |
| granite | rock rich in silica and K-feldspar |
| grazing occultation | occultation by lunar mountains only |
| groove | *see* rille |
| harvest moon | full moon nearest to autumnal equinox |
| high alumina basalt | basalt with 45–60% calcium feldspar |
| high tide | when ocean tide reaches highest level |
| highlands | pale coloured regions on the Moon |
| highland basalt | aluminous non-*mare* basalt |
| highland light plains | pale, level areas in the lunar highlands |
| horns | *see* cusps |
| hunter's moon | full moon following harvest moon |
| ilmenite | iron/titanium oxide mineral |
| Infrared Scanning Radiometer | instrument for measuring temperature of the lunar surface |
| immersion | entry of body into an occultation |
| impact melt | rock melted by meteorite impact |
| Imbrian system | intermediate era in lunar history |
| irregular *maria* | non-circular *maria* without mascons |
| JSC | Johnson Space Centre |
| KREEP | rock type rich in potassium, rare earth elements and phosphorus |
| LACE | atmospheric mass spectrometer |
| laser altimeter | instrument for measuring altitude |
| last quarter | half moon following full moon |
| LEAM | Lunar Ejecta and Meteorites experiment |
| LEM | *see* LM |
| LESC | Lunar Environment Sample Container |
| libration | means by which 59% of Moon is made visible |
| light mantle | highland avalanche at the Apollo 17 site |
| limb | edge of Moon as seen from Earth |
| line of cusps | plane passing through cusps |
| lithology | nature of rock |
| LOI | Lunar Orbit Insertion |
| low tide | when ocean tide is at lowest level |
| Luna (Lunik) | Soviet unmanned lunar space program |
| LM | Lunar Module |
| LMP | Lunar Module Pilot |
| lunar grid | hypothetical arrangement of lunar lineaments |
| Lunar Neutron Probe | experiment for measuring neutrons in regolith |
| Lunar Orbiter | US lunar photoreconnaissance program |
| LPM | Lunar Portable Magnetometer |
| Lunokhod | Russian roving vehicle |
| LRL | Lunar Receiving Laboratory |
| LRRR (LR$^3$) | Lunar Laser Ranging Retroreflector |
| LSAPT | Lunar Sample Analysis Planning Team |
| LRV | Lunar Roving Vehicle |
| LSI (now LPI) | Lunar Science Institute |
| LSPE | Lunar Seismic Profiling Experiment |

| | |
|---|---|
| LSPET | Lunar Sample Preliminary Examination Team |
| lunation | the complete cycle of lunar phases |
| LUNI | the lowest strontium isotope ratio on the Moon |
| luny rock | Apollo 11 KREEP fragment |
| magcon | lunar magnetic anomaly |
| magnetometry | measurements of magnetic fields |
| mantle | zone between crust and core |
| | *or* superficial layer (e.g. dark mantle) |
| mapping camera | *see* metric camera |
| *mare, maria* | dark areas of iron-rich basalt |
| *mare* basalt | basalt from *maria*, rich in iron and titanium |
| mascon | positive gravitational anomaly |
| mass spectrometer | instrument for measuring isotopes |
| maskelynite | plagioclase converted to glass by shock |
| megaregolith | fragmented upper lunar crust |
| Mercury | innermost planet *and* first manned space program |
| MESA | Modularised Equipment Stowage Assembly |
| mesostasis | glassy residuum in lunar basalts |
| MET | Modular Equipment Transporter |
| meteorite | solid object in space |
| meteorite component | component of lunar soil originating from meteorites |
| metric camera | CSM camera for lunar mapping purposes |
| microcrater | crater produced by micrometeorite |
| micrometeorite | very small meteorite |
| mineralogy | the study of minerals |
| mission control | the centre for directing Apollo missions |
| monomict breccia | a breccia formed from a single rock type |
| moonquake | natural seismic disturbance within the Moon |
| mortar | device for generating artificial seismic sources |
| mugshot | photograph of lunar rock at LRL |
| nadir | point immediately below spacecraft |
| NASA | National Aeronautics and Space Administration |
| neap tide | exceptionally low tide |
| near side | lunar hemisphere turned towards Earth |
| neutron | uncharged subatomic particle |
| new moon | phase when Moon is not illuminated |
| non-*mare* basalt | iron-poor basalt from the highlands |
| nucleus | the centre of an atom |
| norite | highland rock of basic composition |
| occultation | the passage of one body behind another |
| olivine | ferromagnesian silicate mineral |
| optical libration | libration due solely to observer's position |
| orthopyroxene | ferromagnesian silicate mineral |
| padded bag | bag used to protect a lunar rock from abrasion |
| panoramic camera | CSM camera used for high resolution studies |
| peak ring | small basin with a ring of central peaks |
| parallactic inequality | lunar inequality due to solar gravity |
| PSE | Passive Seismic Experiment |
| penumbra | shadow of partially eclipsed Sun |
| perigee | Moon's nearest point to Earth |

| | |
|---|---|
| perihelion | planet's nearest point to Sun |
| permeability | magnetic property of a material |
| phase | extent of lunar illumination by Sun |
| physical libration | libration due to Moon's forced vibrations |
| petrography | study of rock texture |
| PLSS | Portable Life-Support System |
| Pioneer | early US deep space program |
| poikilitic | texture in which one mineral encloses another |
| polymict breccia | breccia with multiple source rocks |
| plagioclase | calcium feldspar mineral |
| pre-Imbrian system | the earliest era in lunar history |
| primordial | pertaining to the earliest Moon |
| principal investigator | scientist involved in lunar analysis |
| proton | positively charged subatomic particle |
| pyroxene | calcium/iron/magnesium silicate mineral |
| pyroxferroite | uniquely lunar silicate mineral |
| quintessence | incompatible element-rich rock residuum |
| rake | device used to collect lunar pebbles |
| RTG | Radioactive Thermoelectric Generator |
| Ranger | early US lunar program |
| ray | streak of ejecta from crater |
| regolith | fine-grained lunar surface layer |
| regolith breccia | breccia formed by sintering of soil |
| retardation | Moon's rising time difference on successive nights |
| ridge | elevated linear feature on the Moon |
| rill, rille, rima | depressed linear feature on the Moon |
| ring plains | *see* walled plains |
| ring mountains | mountain arcs defining lunar basins |
| Rover | *see* LRV |
| Saturn 5 | Apollo launch vehicle |
| scattering zone | upper crust, with its unusual seismic behaviour |
| scarp | *see* ridge |
| scoop | device used for lunar sample collection |
| Scientific Instrument Module | bay in CSM for orbital studies |
| seas | *see maria* |
| secondary crater | crater produced by primary ejecta |
| secular acceleration | acceleration of Moon due to tidal friction |
| selenography | study of lunar features |
| selenology | lunar geology |
| seismology | study of shock-wave transmission |
| SIVB | Saturn 5 third stage |
| sinuous rille | collapsed lava tube |
| SM | Service Module |
| sidereal month | time taken by the Moon to return to the same celestial longitude |
| silicate | mineral with a lattice of silicon and oxygen |
| suprathermal ion | ion more energetic than the solar wind |
| soil | fine component of the regolith |
| soil breccia | *see* regolith breccia |
| soil maturity | exposure of soil to extralunar influences |
| solar cosmic rays | energetic ions from the Sun |

| | |
|---|---|
| solar flare ions | *see* solar cosmic rays |
| solar wind | low energy ions from the Sun |
| SWC | Solar Wind Composition experiment |
| spinel | an oxide mineral |
| spring tide | exceptionally high tide |
| stellar camera | camera used for orientation purposes |
| subsatellite | small satellite used for measuring particles and fields near the Moon |
| Surveyor | lunar softlander program |
| sun tan age | time of totally unshielded exposure |
| terminator | boundary of illuminated hemisphere |
| terra | *see* highlands |
| tongs | tool for collecting lunar samples |
| terraced wall | feature indicative of crater slumping |
| thassaloid | unfilled basin |
| tide | distortion of a planet by another |
| tidal lag | difference in time between lunar meridian transit and high tide |
| TLP | Transient Lunar Phenomenon |
| tranquillityite | unique lunar silicate mineral |
| troctolite | rock containing plagioclase and olivine |
| troilite | iron sulphide mineral |
| talus | collection of debris in rille or crater |
| umbra | total shadow |
| valley | *see* rille |
| variation | a lunar inequality |
| VHA basalt | highland basalt with very high alumina content |
| walled plain | large crater with a flat floor |
| vitrification | the glass-making process |
| X-ray | short wave electromagnetic radiation |
| zap pit | microcrater |
| zodiacal light | light scattered by interplanetary dust |
| Zond | Russian deep space probe program |

# Selected Bibliography

## Lunar geography

*The Moon* by P. Moore and H. P. Wilkins, Macmillan N.Y. (1958)
*The Moon Old and New* by V. A. Firsoff, Sidgwick and Jackson (1964)
*Photographic Atlas of the Moon* by Z. Kopal, T. W. Rackham and J. Klepesta, Academic Press (1965)
*The Craters of the Moon* by P. Moore and P. Cattermole, Norton (1967)
*Lunar Atlas* by D. Alter (ed.), Dover (1968)
*The Moon in Focus* by T. W. Rackham, Pergamon (1968)
*Guide to the Moon* by P. Moore, Lutterworth Press (1976)
*Moon, Mars and Venus – a Concise Guide in Colour* by A. Rükl, Hamlyn (1976)

## Lunar geology and geophysics

*The Planets* by H. C. Urey, Yale University Press (1952)
*The Measure of the Moon* by R. B. Baldwin, University of Chicago Press (1963)
*Lunar Geology* by G. Fielder, Lutterworth Press (1965)
*The Moon* by Z. Kopal, Academic Press (1966)
*Introduction to the Study of the Moon* by Z. Kopal, D. Reidel (1969)
*Geology of the Moon – a Stratigraphic View* by T. A. Mutch, Princeton University Press (1970)
*The Lunar Rocks* by B. Mason and W. G. Melson, J. Wiley (1970)
*Moon Rocks and Minerals* by A. A. Levinson and S. R. Taylor, Pergamon (1971)
*Lunar Mineralogy* by J. Frondel, J. Wiley (1975)
*The Moon in the Post-Apollo Era* by Z. Kopal, D. Reidel (1974)
*Lunar Science – a Post-Apollo View* by S. R. Taylor, Pergamon (1975)
*Planetary Geology* by N. M. Short, Prentice–Hall (1975)

## NASA special publications

The following are a selection of publications relating to lunar research, all of which are available from the US Government Printing Office, Washington, DC

*Unmanned missions*
*Ranger IX Photographs of the Moon*, NASA SP–112 (1966)
*Surveyor Program Results*, NASA SP–184 (1969)
*The Moon as Viewed by Lunar Orbiter*, by L. J. Kosovsky and F. El Baz, NASA SP–200 (1970)

*Manned missions*
Apollo 11 Preliminary Science Report NASA SP–214 (1969)
Apollo 12 Preliminary Science Report NASA SP–235 (1970)
Apollo 14 Preliminary Science Report NASA SP–272 (1971)
Apollo 15 Preliminary Science Report NASA SP–289 (1972)
Apollo 16 Preliminary Science Report NASA SP–315 (1972)
Apollo 17 Preliminary Science Report NASA SP–330 (1973)

*Future missions*

The Colonisation of Space – a Design for a Human Community in Space, NASA/Ames final report (1976)

Mission Summary for Lunar Polar Orbiter, JPL document 660–41, NASA–JSC Houston Texas (1976)

## Conference proceedings

The first scientific results from Apollo 11 were published together in a special issue of the journal *Science*, **167** (1970). A more comprehensive account can be found in the Proceedings of the Apollo 11 Lunar Science Conference edited by A. A. Levinson, 3 volumes, Pergamon (1970)

A Lunar Science Conference (now a Lunar and Planetary Science Conference) has been held in Houston each year since 1970, and the results have appeared in the appropriate set of *Proceedings*. The *Proceedings of the Second and Third Lunar Science Conferences* were published by MIT Press.

Abstracts for these, and other, conferences relating to lunar science are available from the Lunar and Planetary Institute.

A special set of abstracts, *The Apollo 15 Lunar Samples*, includes many results not presented at conferences.

Summaries of work involving analyses of Russian samples appear in:

*Proceedings of the Soviet–American Conference on the Cosmochemistry of the Moon and Planets* (Moscow), The Lunar Science Institute (1974)

Mare Crisium – the View from Luna 24, Pergamon (1978)

Philosophical Transactions of the Royal Society of London, **284A** 131–77 (1977)

The Proceedings of the conference 'The moon – a new appraisal from space missions and laboratory analyses' appeared in:

Philosophical Transactions of the Royal Society, **285A** 1–606 (1975)

## Information sources

The principal centre for information about lunar science is The Lunar and Planetary Institute, 3303, NASA Road One, Houston, Texas, 77058 USA

*A Lunar and Planetary Information Bulletin* is published quarterly by LPI (editor: Frances B. Waranius), and includes information about recent publications, forthcoming conferences and recent events.

Details of lunar mapping projects may be obtained from the LPI or from: Lunar Programs Office, NASA Headquarters, Washington, DC USA

Individual maps, and photographs (if identified by number) can be obtained from: National Space Science Data Center (inside USA) or World Data Center A for Rockets and Satellites (if outside USA) Code 601, Goddard Space Flight Center, Greenbelt, Maryland 20771, USA

A wide selection of NASA photographs are available from: NASA, Room 6035, 400, Maryland Avenue SW, Washington DC, USA

## Periodicals and Journals

Popular accounts of lunar science have appeared in:
*New Scientist*
*Scientific American*
*National Geographic Magazine*
*Life Magazine*
*Sky and Telescope*

Specialist reviews appear in:

  *The Moon* (now called *The Moon and the Planets*)
  Space Science Reviews
  Reviews of Geophysics and Space Physics
  Icarus

Detailed technical reports and letters appear in:

  *Earth and Planetary Science Letters*
  *Journal of Geophysical Research*
  *Nature*
  *Geochimica et Cosmochimica Acta*
  *Science*
  *Meteoritics*
  *Planetary and Space Science*

# Index

albedo, 274
alpha particle spectrometry, 81, 282
Alphonsus Crater, 22, 32, 55, 108, 113, 122, 277
ALSEP, 99, 113
annual equation, 261
anomalistic month, 8
anorthosite, 102, 127, 129, 149, 161, 162, 235, 281
anorthositic gabbro, 102, 115, 130, 165, 200, 342
Apennines, 40, 51, 98, 166, 201
Apollo missions, 28, 87, 89, 91, 92, 95, 97, 107, 113, 296, 306
Aristarchus, 10
Aristarchus Crater, 22, 60, 121, 122, 273, 277, 280, 283, 362, 366
Armalcolite, 179
asteroids, 64, 245, 251, 274, 306, 325, 358

Big Bertha, 128, 173, 350
breccia, 94, 96, 103, 113, 128, 147, 163, 172, 337, 352

Cassini, 17
cataclastic rocks, 171, 175
Cayley–Descartes, 56, 72, 107, 174, 301
cinder cones, 55
colour difference photography, 76, 274
comets, 243
Cone Crater, 95, 112. 114, 124, 362, 371
contingency sample, 91, 124
Copernicus Crater, 17, 22, 42, 45, 46, 63, 77, 93, 162, 199, 271, 273, 362, 366
cores, 124, 137, 138, 140, 197, 221, 244
cosmic rays, 203, 214, 222
Crisium, 22, 134, 140, 151, 259, 269, 271, 361, 367
cumulates, 187

dark halo craters, 56, 113, 115, 120
domes, 55
draconic month, 6

Earth–Moon distance, 8, 12
earthshine, 80, 121
ejecta, 42, 58, 66
eclipses, 12, 135, 266, 273, 285, 310
electrical conductivity, 316, 323
Eratosthenes, 9
erosion, 194, 227
evection, 260
Explorer 35, 32, 307, 310, 316, 317
exposure age, 219, 221, 228, 234, 236

Fecunditatis, 22, 137, 151, 362, 367
fillets, 197
fission tracks, 342
Fra Mauro, 22, 46, 70, 94, 280, 366

Galileo, 15, 17, 20, 39, 251, 264
gamma ray spectrometry, 279, 280, 321
Gargantuan Basin, 43, 49, 272, 281, 362, 366, 370
Genesis Rock, 102, 130, 172, 347
ghost crater, 66, 74
glassy aggregate, 198, 209, 211, 243
gnomon, 101, 127, 130
grain sizes, 199
granite, 82, 146, 149, 163, 280, 334
Great Scott, 128, 348
green glass, 103, 116, 153, 166, 183, 191, 199, 210, 325
Grimaldi Crater, 22, 122, 271, 277

Hadley Rille, 49, 51, 53, 97, 104, 114, 300, 369
heat flow, 317
Hevelius, 20
high frequency teleseisms, 300, 304, 306
high lava marks, 48, 271
Hipparchus, 11, 260
hydrogen stripping, 213·

Imbrium, 22, 40, 41, 46, 49, 73, 75, 77, 95, 269, 271, 280, 322, 361
impact melt, 64, 163, 174, 185, 366
infrared scanning, 272, 317

KREEP, 148, 162, 174, 178, 189, 192, 197, 199, 200, 210, 243, 272, 321, 331, 334, 342, 350, 365

landing sites, 28, 90, 93, 94, 98, 106, 114
laser altimetry, 77, 121, 254, 256
laser ranging, 91, 140, 256, 267
lava flows, 41, 49, 56, 66, 74, 77, 101, 104, 123, 153, 156, 256, 367, 369
lava tube, 51, 105, 271
librations, 19, 252, 267
light mantle, 114, 362, 371
lineamants, 38, 54
line of apsides, 8
LPO, 269, 273, 280, 314, 375
LRL, 129, 131
luminescence, 277
Luna, 25, 28, 34, 79, 134, 135, 167, 219, 307, 361
lunar atmosphere, 283, 285
lunar cataclysm, 363